The Snoring Bird

also by BERND HEINRICH

The Geese of Beaver Bog

Winter World

Why We Run (previously titled *Racing the Antelope*)

Mind of the Raven

Bumblebee Economics

One Man's Owl

Ravens in Winter

A Year in the Maine Woods

The Trees in My Forest

The Hot-Blooded Insects

In a Patch of Fireweed

The Thermal Warriors

MY FAMILY'S JOURNEY
THROUGH A CENTURY
OF BIOLOGY

ecco

An Imprint of HarperCollinsPublishers

The Snoring Bird

BERND HEINRICH

HarperCollins books may be purchased for educational, business, or
sales promotional use. For information, please write: Special Markets
Department, HarperCollins Publishers, 10 East 53rd Street, New York,
NY 10022.

FIRST EDITION

Book design by Shubhani Sarkar.
Maps by Christine Fellenz, Springer Cartographics LLC.

Library of Congress Cataloging-in-Publication Data

Heinrich, Bernd.
    The snoring bird : my family's journey through a century of biology /
Bernd Heinrich.—1st ed.
        p.   cm.
ISBN: 978-0-06-074215-7
ISBN-10: 0-06-074215-1
    1. Heinrich, Bernd, 1940– .   2. Biologists—United States—
Biography.   I. Title.
QH31.H356 A3   2006
570.92B22—dc22   2006050808

1 2 3 4 5 6 7 8 9 10 DIX/RRD 11 10 09 08 07

To Ulla

$\mathscr{A}$cknowledgments

I THANK MARY LECROY FOR SHOWING ME THE SNORING BIRD SPECI-men, *Aramidopsis plateni*, in the American Museum of Natural History in New York and for giving me a list of Papa's publications on birds. Erich Diller, Stefan Schmidt, and Johannes Schuberth I thank for sending electronic images and information about Papa's ichneumons maintained in the Bavarian State Museum, Munich. Nuria Selva, Adam Wajrak, and Wieslaw Bogdanowicz and the Institute of Zoology in Warsaw, Poland, offered generous help in making available to me pictures of Papa's collection in Warsaw and the map that allowed the recovery of the buried collection, as well as help in translations from the Polish. I thank Randolf and Mechtild Menzel for their hospitality in Berlin. I am deeply indebted to the Handschriftenabteilung of the Stiftung Preussischer Kulturbesitz of the State Library of Berlin for its gift of my father's correspondence with Erwin Stresemann, and specifically to the late Eva Ziesche and subsequently Iris Lorenz of that same organization. They all helped in deciphering the handwriting of correspondence. Copies of this correspondence, as well as the personal letters, will be archived in the Special Collections of the Bailey/Howe Library at the University of Vermont in Burlington. I thank Violetta Tomaszewska and Malgorzata Adam-czewska of the Zoological Museum of the Polish Academy of Sciences of Warsaw, Poland, for their help in procuring images of specimens from the part of my father's ichneumon collection that is still in Poland now. My mother, Christiane Buchinger-Marks, Christel Lehmann, and Rolf Grantsau provided family photographs and memorabilia, and I thank also Hermen Steinert-Starke for permission to quote from the personal family memoirs of her late sister Beate Richter-Starke. Ralph du Roi Droege of Trittau, Germany, provided information and German translations. Dieter Radke told me his experiences and gave me kind permission to read his private memoir. Thank you to Alice Calaprice for taped interviews with my mother. Elva Paulson gave me letters by my father and information regarding how American ornithologists came to the aid of their German colleagues after the war. Sylke Frahnert of the Museum für Naturkunde, Berlin, and Jared Diamond and K. David Bishop and Paul Sweet provided information and images of Papa's birds from Sulawesi. David Willard of the Field Museum in Chicago provided pictures of *Cossypha heinrichi* and other Angolan birds. Gary Volker provided information on *Heinrichia*. I thank Kristof Zyskowski at Yale University for help on Papa's birds from Tanzania and Sulawesi, and John Strazanak and David B. Wahl for providing unpublished information of ichneumons and J. Curt Stager for information on African paleoclimates.

I also here take the opportunity—possibly the only one I'll ever have—of expressing my appreciation for the unsurpassed kindness and generosity of the Mainers my family met when we first arrived as immigrants in America. They helped us to build a new life and inspired me in almost everything I have to say in this book. Particular thanks go to the families of Floyd and Leona Adams, Phil and Myrtle Potter, and the Curriers, Gilmores, Ellriches, Brooks, and the Fretz family from Indiana—neighbors and friends all. The Good Will Home Association of Hinckley, Maine, where I lived and studied for six years, helped me to grow up—and gave me enough stories for a lifetime. I owe it and the many dedicated teachers and role models there a great debt of gratitude. I thank my college teachers and academic colleagues, especially my master of science adviser and inspiring mentor at the University of Maine, the late James R. Cook, and his family. I also thank George Bartholomew and Franz Engelmann, my PhD advisers at UCLA, and Robert Colby and Coach Edmund Styrna.

My agent, Sandra Dijkstra, helped launch this project and Dan Halpern at Ecco had the faith to make it real. Gheña Glijansky, Lisa Chase, and Susan Gamer gave many suggestions, and improved it with their thoughtful and insightful editing, and Emily Takoudes' and David Koral's expertise and patience, which saw it through the final trying process to production. Last but not least, I thank my wife, Rachel Smolker, for her enthusiastic encouragement, and indeed her urging me to write this story, her years of patience while I was tussling with it, and her many valuable suggestions that guided it.

Contents

$\mathcal{P}$reface

WHEN WE CAME TO THIS GREAT COUNTRY IN 1951, MY FATHER, GERD, was already a mature man and an accomplished naturalist. Much of his life had passed. But that past was ever-present. Papa and our mother, Hildegarde (*Mamusha*, as we called her), talked often about their lives in Poland and Germany in the years leading up to and during the Second World War. They had many stories to tell.

During our first winter in Maine, there was no television, no radio, no electricity. To save on kerosene, we would blow out our lamp early on most nights. After it was dark, my sister Marianne and I would snuggle together in bed, warmed by a couple of bricks heated on the kitchen stove and placed under the covers at our feet, as Papa told us stories. We were old enough to have vivid memories of our very recent fairy-tale existence in a northern German forest, where our family had been refugees, but we were also young enough to be excited by the prospect of new beginnings in the land of hummingbirds, rattlesnakes, Indians, and skyscrapers.

Still, we were rapt as Papa talked about the Old World; about Poland, where we had lived; and of far-off lands in the tropics (the scene of his own youthful adventures) inhabited by entrancing birds and butterflies. My father was a naturalist, and no story was as gripping as that of the Snoring Bird, a species of ground-living jungle rail from Indonesia that had been thought to be extinct. The title of this book is a direct heist from Papa's *Der Vogel Schnarch* (*The Snoring Bird*), published in 1932, in which he recounts his experiences searching the jungle, at the behest of Dr. Leonard Sanford of the American Museum of Natural History in New York, for the Snoring Bird. I am sad to say that Papa's book has drifted into obscurity; it became unavailable shortly after it was published, and the publishing house (Dieter Reimer, Berlin) cannot issue a reprint, because it no longer exists—the publishing house was presumably bombed into oblivion during the war. But the title is still excellent, and so is his inscription in the copy of this book that he gave to me. Translated into English, it says: "To Bernd Heinrich, 'zugeeignet' [adapted] for the memory of the pioneer time of the discovery of the world's animal variety, and for his father, who took part in it, 24 XII, 1972." I here take him up on his thought. Papa spent two very strenuous years in his quest for the secretive snoring bird in Celebes (now Sulawesi) while also hunting for his beloved ichneumon wasps—strange and exotic creatures that became something like family icons for us, as they had, in many ways, determined the trajectory of our lives as he pursued them at great cost over several continents.

My Snoring Bird is the story of that path, of our family's surviving through two world wars, fleeing from Poland and Germany, and eventually coming to America. That I later became a scientist and a writer had something to do with Papa, and our family's history was influenced not only by the wars, but by the experiences and views of people who passionately desired to understand the workings of nature during a time when biology as a science was undergoing huge transformation.

The events that I describe are real, and I feel compelled to say so up front, because some of what you are about to read may seem like fiction. But this is not a novel. It is an account that is as factually accurate as I could make it. I originally wrote this book for my children, who knew nothing about their grandparents and little about my own personal story. I started writing decades ago, to preserve what I increasingly saw as an interesting tale. Bits and pieces of it have been published in my writings before, but here, I have put the puzzle together and shown the connections that were apparent to me only much later, after I became interested in the past and its amazing and sometimes startling effect on the future. So this book has literally been a work in progress for decades.

Initially I was inspired by the stories I had heard, first from Papa; then from my mother; and also from Papa's former wife, Anneliese, and her daughter Ursula (Ulla), my half sister. I want to clarify that the quoted exchanges between people in this memoir were not recorded. They are my memory of what was said, though I did often take notes after a conversation, and much of what I heard that was of significance could also be confirmed in our family correspondence. The Heinrichs have always corresponded regularly by means of handwritten letters. Even now, in the age of e-mail, we still do. Fortunately for me, these letters were saved and serve as a rich source of information for this book. I also quote from my father's unpublished memoir—written while he was in his mid-eighties, shortly before he died—and from the works of my mother, and from family friends. I reviewed thousands of pages of correspondence. Most of this passed between Papa and the late Professor Erwin Stresemann—his lifelong friend and ornithological mentor from the Museum für Naturkunde (nature study) in Berlin. Because Papa and I were apart more than we were together, there was also much correspondence between him and me.

I have tried as best as I could to decipher the script and to translate the German into English. I hope I did justice to my father's, Stresemann's, and many others' words in German. My translations are not entirely literal, but I was true to the facts and intended meanings. Aside from Papa's *Der Vogel Schnarch*, I have also drawn from two of my father's other books, *Von den Fronten des Krieges und der Wissenschaft* (*From the Fronts of War and Science*; D. Reimer, Berlin, 1937), and *In Burmas Bergwäldern* (*In Burma's Mountain Forests*; D. Reimer, Berlin, 1940).

part ONE

The Old World

one   A Visit Home

When you start your journey to Ithaca
then pray that the road is long,
full of adventure, full of knowledge.
Do not fear the Lestrygonians
and the Cyclops and the angry Poseidon.

—"ITHACA," CONSTANTINE P. CAVAFY

MAMUSHA IS JUST SETTLING DOWN ON HER BED
to watch the evening news when I arrive. Two cans of Coors, which she
has opened with the point of a pair of scissors, are on the table next to
her, along with a box of German chocolates. She used to make her own
beer, but now, in her mid-eighties, she likes Coors from a can; and because her gnarled hands are too weak, she cannot pull off the tabs. Duke,
the huge shepherd-hound that she rescued from the pound, is at her feet,
and a one-legged chicken lies cradled in her lap. She is mildly irritated at
me for arriving unannounced (I have a tendency either to just show up or
to come an hour later than I've promised, which annoys her also), but
soon I have placated her and she offers me a beer.

Mamusha is the Polish word for mama or mommy. Mamusha was
born and raised in what is now Poland, and despite her willingness to
consume Coors, she remains, in her memories and her ways, a product
of the Old World. My visits to her are usually spent listening to stories
about the past. We are sitting in the low-ceilinged brick room that Papa

3

built decades earlier for the purpose of protecting his precious wasp collections from fire; the rest of the house might be consumed, but his ichneumons would be safe. When we moved to this house near Wilton, Maine, in 1951, it was a simple saltbox-style farmhouse with six rooms. Since Papa's death, Mamusha has added on haphazardly, so that the house is now a collage of thirteen rooms. The walls are decorated with pictures of flowers that she has purchased, although one wall sports a portrait of George and Laura Bush that she received free in the mail. This small brick room is her main habitation, which she shares with her dog and house chickens. I was met in the entryway by three hens, perched on the dresser. In the living room, one drawer sits partway open to accommodate a setting hen that has made her nest there. "My chickens outsmarted me again," Mamusha says. "Yesterday I found one upstairs in a corner of the bedroom. I have too many. Next time bring your shotgun and at least help me to get rid of a few roosters. They are all so pretty—brown, black, speckled, some with feathers on their toes, some without. I can't decide which ones to get rid of."

Mamusha keeps chickens in her barn and chicken house, but her house chickens are often the ill ones. Once in a while, for some reason, some of the newly hatched chicks have trouble with their legs. They splay out to the side, and the chicks can barely stand, much less walk. Mamusha has discovered that if you cradle the chick in your arms for a week or two, and sleep with it cuddled up next to you in bed, it will eventually improve and become a fully functional house pet. By the warmth of the woodstove, with feed scattered across the floor, the chickens are quite comfortable. This is Mamusha's twist on the concept of "survival of the fittest." The most afflicted chickens receive the best care, and with the warm fire nearby they breed year-round, producing more and more afflicted youngsters for Mamusha to look after. Their eggs do have the deepest-yellow yolks, which Mamusha brags about.

I WAS SOMEWHAT AGHAST AT THE CHICKEN BUSIness until my wife, Rachel, pointed out that taking care of animals is good for one's mental health. OK, at least my mother is not taking antidepressants or tranquilizers. Mamusha never learned how to drive and so rarely leaves home anyway. And I'm not the one tied down every day to a clucking chicken! All I have to do is periodically bring my gun when I visit.

Papa would never have lived this way. He was a neat freak. He saved every scrap of paper, saved every receipt, kept a ledger book with every single penny accounted for. After his death in 1984 Mamusha claimed for years to be in the process of sorting through his stuff, but I never saw any signs of progress. It was of course none of my business what she saved or disposed of; she has reminded me for decades that, after all, I disagreed with Papa far more than a son should. This was taken in the family to be a sign of my disrespect. I know better than to ask Mamusha for any of Papa's scraps of paper, or to snoop around the household junk piles.

The barn, where the rest of the chickens live, along with the accumulated artifacts of decades of human activity, used to house pigeons when I was little. It began with a single bird, one that had escaped from Murray Foss. Murray was the son of Dot Foss, who owned Dot's Place, a filling station and tiny store selling cigarettes and candy at the intersection a mile up the road. Murray ran away to Los Angeles, and before he left, he gathered up each of his beloved racing pigeons and one by one broke their necks. One, however, escaped, and fled to the top of our barn. Mamusha put out some grain and it eventually fed. A while later another pigeon appeared, obviously of the opposite sex, and they began to reproduce like crazy. The grain available for the pigeons also attracted and fed the mice, chipmunks, house sparrows, and rats in the barn, all of which prospered. I remember catching twenty rats in a grain barrel in one night. Next the weasels cashed in, and a pair of goshawks discovered the riches as well. They sat and waited hidden in a nearby tree; and when a pigeon came out of the barn it stood a good chance of becoming food for the hawk's young, huddling in their nest of sticks high in a pine tree on Picker Hill. Mamusha wanted to shoot the hawk to protect her pigeons, but Papa was adamantly opposed. In a letter to me he had written: "The hawk is ruining the pigeon flock; he takes one every day, but we rarely see him. We have given up. Let him take them all."

After Mamusha and I had finished our beer, I wandered out to the barn and climbed up the steep stairs to the old hayloft. The hay had long since been removed, but battered remains of my old beehives from at least forty years ago lay strewn about. A porcupine (perhaps the one I had raised as a pet?) had chewed on some of them and on the edge of the steps to the loft. The windowpanes on both ends were broken when we'd arrived here in 1951, and so they remained. I recalled the barn swallows, phoebes, and robins that used to fly in and out of the windows, nesting on the barn beams. There was dust and a thick layer of guano, four decades' worth, all over the place, as well as some boxes filled with Papa's things, papers and personal effects. It looked as though chickens had been scratching among the papers, as there were bits of eggshells and some long-rotten eggs in the boxes, some of which had spilled their contents. Here, I could certainly poke around without Mamusha's recriminations. Peeking out of one box was Papa's old suit. I can only dimly remember him ever wearing it. Mostly I remember him in white shorts and a faded green baseball cap.

Inside the box I found his old insect traps. I also found several of the rattraps he had carried with us during our flight from the Red Army, the ones he had with him on his expeditions to Celebes (now Sulawesi), Persia (now Iran), Burma (now Myanmar), Borneo, and other exotic places. I had watched him set these traps in the Hahnheide forest in the late 1940s in Germany when I was a young boy. Another box contained copies of reprints of his scientific papers: page after page, now coated in dust and yellowed. These were the culmination of his lifework—the work that had stood at the center of his entire existence—prepared with such painstaking devotion and sacrifice. I felt saddened to see them now discarded and decaying in the barn attic. Probably fewer than a dozen people had ever read any of these papers. Most people have never even heard of ichneumon wasps, and many of those who have would be likely to disparage them as "flies."

Ichneumons ("ick-new-mons") are mysterious and exotic wasps that are parasitic connoisseurs. Each specializes in one specific host (or a very few hosts). Ichneumons are delicately sculptured and patterned with almost infinite variety and subtlety. There are tens of thousands of species, and they are difficult to differentiate; males and females within a species often come in different shapes and sizes, dimensions of appendages, and colors (from yellow and brown to brilliant shiny metallic-blue with contrasting black and white patterns). The female possesses an ovipositor that projects out of her rear end. Depending on the species, the ovipositor can be so short as to be barely visible, or as long or longer than the wasp's body itself. The ovipositor doubles as a stinger, but primarily it is a tool for injecting eggs into the body of her host.

Papa described more than 1,000 new species (many from museum specimens) during his lifetime, and all of us in the family who took up the net have been immortalized with a species of ichneumon wasp bearing our name that we had captured in the field. Papa gave four species the generic name *berndi*: one that I caught in New Guinea, one in Africa, one from Los Angeles, and another from near our farm in Maine. To my knowledge, nobody else has found or recognized any of these since, but each species is undoubtedly going about its business over large areas of its respective continent. And so it is with many other species still to be discovered all over the planet. By virtue of their diversity and unobtrusive manner, these gorgeous, delicate, wondrous insects remain largely unknown and unseen by humans—except Papa. He saw the world in terms of ichneumon habitats. He could name the hundreds of minute details that distinguish one species from another. He could recall precisely when and where he acquired each of his specimens. He could predict, like no human before or since, the likely places to find an ichneumon.

Truth be told, as a kid I could discern little of interest about ichneumon wasps. Furthermore, I sometimes objected to being dragged along on outings with my father to chop open rotting logs in search of hibernating ichneumons in the winter or straining my eyes in search of ichneumons as they alighted on the foliage in the woods. I would have much preferred to be out tracking a weasel or hunting for grouse or rabbits, activities my father shunned, perhaps because they took up a lot of time and often brought few tangible returns. Worst of all were the summertime outings, when he would take me on a long car ride to some godforsaken swamp where we would be eaten alive by black flies and mosquitoes as we searched for one of his special wasps. They were his passion, not mine. Yet I must admit that when I did catch a "cracker"—one that was needed for his collection and, if found, earned me a tasty treat—I felt a certain sense of accomplishment and exhilaration despite my best efforts to be nonchalant.

After the hunt, we would return home and Papa would retire to his study to mount his specimens. He would painstakingly arrange each of the wings and the dozen-odd appendages (yes, dozen) just so, using tiny pins to hold them in place. He treasured his little beauties; nothing was too good for them. Pinned in neat little rows and labeled in his miniature script, they were placed in glass-covered boxes of a special airtight design that he ordered specially from Germany, and then stored in the fireproof brick room that had cost him his last penny to construct. He had not a single dollar to spare to send any of his five offspring (although I mention only four in this account, since one we never knew) to college, but he had to have that brick room for his wasps. Watching him bent over this arduous and demanding task, I often wondered, Why? What was the point of it all? Why go to so much trouble?

As he worked in the evenings with lamp and microscope, I would sometimes sit across from him, and he would try to get me interested in family history by pulling out photographs and pointing to strange faces. I was polite, but no more. It was his history; I wanted to be accepted into the world of the future. My heroes were the neighboring farm boys who played baseball and ran track, who went fishing in the spring, deer hunting in the fall, coon hunting at night, ice fishing and trapping in winter. Papa, in comparison, was a hopeless bug watcher of whom at times I felt ashamed. He had several war medals in a small case hanging on the wall, and was not above pointing them out to neighbors, along with various animal curiosities from Africa and Mexico. Invariably, he always showed off his bug collections to our visitors, whether they were students, farmers, workers at the local woolen mill, or teachers at the college in nearby Farmington. Some of these people actually behaved as though they were interested. I cringed. Didn't he understand that they didn't give a damn for his flies? Yet his flies were the one essential ingredient of our lives that brought us together—to the extent that we were together. I suspect that were it not for my keen eyesight and my acquiescence in helping him with his hunts, I would have had very little contact with Papa. He made no secret of the fact that he hoped I would one day take over his collection, but I had no interest in or intention of fulfilling that hope. As I grew up, I distanced myself from him and from his wasps; I had other things to do. Later I did go on to study biology, though not his sort of biology. I think he was deeply disappointed, and I was too young and immersed in my own life to notice.

Looking at the boxes full of old, unreadable reprints in Mamusha's hayloft, I now felt a tinge of regret. Poking around, I noticed a book with a green cover. It was water-stained from snow that had come in through the open windows for decades. It was also garnished with gobs of dried bird manure, but underneath in gold lettering I read: "Burmesische Ichneumoninae, Part 1, by Gerd Heinrich, Dryden [referring to our one-room post office next to the town of Wilton], Maine, U.S.A." A second volume was labeled "Part 2."

To the undiscerning eye, these two books would have no more meaning than they did to the chickens that had scratched among their pages. Yet I knew what they had meant to Papa. He had been a combatant in the two world wars. He understood all too well the fleeting nature of existence. Somehow, against all odds, he not only had survived, escaped, and started, with nothing, all over again, but had gone on to produce this work that would long outlive him. And here it was, covered in chicken shit, scattered in the dust and unread. I thought, How "immortal" is a lifework if the subject matter is so obscure that hardly a soul takes an interest in it? His passion for these wasps had been the single thread of continuity as everything else—his home, his family, his loves—was heaved around by world events beyond his control. The wasps had been his anchor in the storms of his life. Was this why he so loved the order and control and steady progress that the wasps demanded of him?

I REMEMBER MY FATHER MOSTLY AS AN OLD MAN. When we immigrated to the United States in 1951 he was already fifty-five years old. He spent hours hunched over his collections, his eyesight failing. Soon his mental facilities would follow. It was not until I was at least as old as he that I came to appreciate the long journey that had brought him to this place. He had not begun life as a rather eccentric, bitter old man. What I should have seen was the remarkable strength and perseverance and passion that had imbued his life, one of great losses, with adventure and accomplishment.

By the 1980s Papa was still working on his primary manuscript—the one on the Oriental Ichneumoninae. It was the prospect of publishing this manuscript that had induced him to come to America after innumerable and seemingly insurmountable difficulties in Europe—difficulties that continued even here. As macular degeneration took his vision, he was racing against time to complete the manuscript, and other manuscripts, on the African, North American, and Russian ichneumons. Rather than trying to publish the entire survey of Oriental ichneumons, he decided to pull out the new discoveries from Burma and publish those, to rescue at least something. By this time, however, taxonomy had gone out of fashion in America; ecology, physiology, and molecular biology were "in"; and the scholarly presses had lost interest in publishing taxonomic works. Papa finally turned back to Europe, and so the manuscript was published there. A Swedish journal published the first seven installments (in German), and when its editors decided they could do no more, Papa turned to a Polish journal to publish the remaining four parts, for a total of 1,258 pages.

Among the boxes in the hayloft were Papa's expense ledgers. He had encouraged me to keep similar accounts of my spending, but I could never bring myself to record anything to do with money. I did keep a ledger of sorts, a notebook, but I filled it with records of where and when I found different birds' nests and what they contained, and the types of bees I found visiting different kinds of flowers. In his latter years, Papa had very little money. Every penny counted. When we came to the United States, Papa had hopes of finding some academic position where he could receive a salary for doing the sort of work he enjoyed most and was most able to do: studying his wasps and collecting specimens for museums. But lacking any certification in the form of an academic degree, he was usually met with skepticism and less than enthusiastic support. He and Mamusha were forced to do menial work in order to make those few pennies he kept such careful records of, before he was being recognized as "a collector" and hiring out to museums and earning most of his money collecting small birds and mammals for their inventories.

Poking around in the barn, I also found some old insect nets and boxes of unused shotgun shells that were left over from the expedition to Africa where I joined Papa and Mamusha for two summers and the intervening year during my undergraduate studies. (I was unsalaried, of course.) There were two sturdy reinforced wooden crates that had been used on the expedition to Tanzania. Papa had them specially constructed for transporting equipment, and they were painted dark green with the words "Yale University," with Papa's initials, G. H., written in bold white lettering.

Here lay the artifacts of my family's history, scattered and decaying. I suddenly felt that I must act and act now to understand that history and to pull it all together for myself and for my children's sake, before it was lost. Twenty years earlier I had sketchily written about my family's escape from Poland in In a Patch of Fireweed, but Mamusha, on reading it, accused me of "lies."

Her verdict intrigued me. "What lies?" I asked.

There was a long silence, then: "It was not you who picked the forsythia flower before going into the bomb shelter—it was Marianne." It was true that my memory was a bit vague about which of us picked the forsythia branch. I do not remember which of us found it and what we did with it. I do remember taking the flower with us and also writing that I developed an aversion to this plant. And though Mamusha had never planted a sprig of forsythia before, the next spring she had planted a big bush of it next to the house. That seemed like a direct rebuke to me, although she said she just "felt like" planting it. When I now told her that I wanted to write the Heinrich family story, she said flatly, "No, you can't."

I was stunned, but not totally surprised. I thought she would be glad to have her son express such interest in the past that she and Papa had always spoken of, and had struggled to engage me in.

"Why not?" I asked, but I had already anticipated the answer.

"Because it would not be the way I want it." And that's God's honest truth. She would not approve me as the author.

"You can tell me what you know and help me make it better," I persisted, realizing I had simplified, left a lot unsaid, and had relied on hearsay. I added: "Or better yet, write it down and send it to me." I was hoping she might let me see the guest book that she had taken from our Polish estate, Borowke, when my family fled before the advancing Russian army. The book had served as something of a diary during the three-month exodus.

"If I write it down I might as well write the book myself," she replied, and added, "I'm so busy, I don't have time." If she doesn't have time now, in her mid-eighties, I wondered, when would she?

"Anyway," she added, "I don't like your writing."

BACK IN THE BARN, MY EYE WAS CAUGHT BY A badly stained and tattered blue folder tied shut with a piece of string. On

Liebet papa wilst du die Haselmause mit nemen und Le dän big mit hir her pringen und be halten und ich hab aach eine Fledermaus und si ist auch le bändig in Raupenkasten und ich fütte und ich habe auch ein Teich für dich gefangen so einz Wi ich in forjh jar gefangen hap es ist schon apgezogen und ich grüse Papa und Mamuscha.

A letter I wrote Papa from the Hahnheide
while he was away collecting.

the outside was a label written in Papa's handwriting. It said "*Erinnerungs-Correspondenz* [remembrance letters], 1945–1950." Reverently, I picked up the frayed and weathered packet. Here was communication, direct from him. I opened the folder and the first letter I saw was one I had written to him when I was about seven years old. The letter, in German, was sloppily composed on uneven lines that I had drawn on a piece of blank paper. About half of the words were misspelled. It said: "Dear Papa, will you bring with you the hazel mouse and bring it with you alive and keep it and I also have a bat and it is also alive in a caterpillar cage and I feed it and I have also caught a moorhen for you like the one I caught before it is skinned already and I send greetings Papa and also Mamusha." I recalled even now catching that gallinule. It had tried to hide under an overhanging bank by a pond. Seeing this letter that I had written to Papa while he and Mamusha were collecting in southern Germany in the Alps—and while his wife Anneliese (whom Mamusha referred to as "the wicked witch") was left to care for us in a hut in the forest in northern Germany—brought back a flood of memories and gave me an odd feeling. Had Papa somehow intended for me to find these long after his death?

Possibly my first drawing, a stag beetle.

Later, I leafed through the many letters in the folder. Although I had spoken little German for more than half a century, I still knew enough to understand much of what was before me. My parents had conversed mostly in German, except when discussing something they did not want

Berlin N 4, den ___ Sept. 1957
Invalidenstraße 43
Fernruf: 42 71 52

Tgb. N.

A letter from Erwin Stresemann, addressed
to Papa: "Dear Mousecatcher and Ratking."

Marianne and me to understand, at which time they would switch to Polish. One letter was from a man named Ernst Schäfer. The name was familiar. I think Papa had spoken to me about him, but I don't remember
why. He was a member not only of the Nazi Party but also of the SS,
Hitler's most notoriously brutal thugs. Why had Papa corresponded with

him? There were also handwritten letters from Papa's good friend and colleague, Erwin Stresemann. Stresemann had been an ornithologist at the Berlin Museum für Naturkunde (nature study) and also a professor at Humboldt University. He had been one of the world's most influential biologists, and had initiated the modern study of birds. I recognized the letterhead, from the museum, but most of the rest was indecipherable to me, and I was frustrated.

Several days later I received a letter from a woman named Eva

Ziesche. She lived in Berlin; had recently read my book *Ravens in Winter*; and, like so many other people, had written to tell me about her own experiences with a pet raven that she had owned. When she mentioned that she worked for the *Handschriftenabteilung* (handwriting division) of the Prussian State Library, I grew very excited. I had never heard of such a department in a library or anywhere else, but if the ancient handwritten letters I had just obtained were any indication, I could understand why one might be needed. I contacted her immediately and asked if she could help me to decipher a few letters. Her answer came a week or so later:

> I have a surprise for you. The library here owns a fairly complete bequest of Erwin Stresemann. There are two thick files of correspondences from Gerd Heinrich to Erwin Stresemann, dated from February 1926 through November 1963! I think that you have the counter letters in your file. I have examined the Stresemann handwriting and I do not think it would be a problem to decipher.

We immediately arranged an exchange of copies. "I am dumbfounded how you, through my raven, have found such a highly interesting correspondence. Simply fabulous!" Eva wrote. Fabulous, and fantastical, I thought.

I copied Stresemann's letters and sent the originals to be archived in Berlin, where they belonged. Soon, I received a heavy package with some 1,500 photocopied pages of Papa's letters. This correspondence would provide me with a firsthand account of some of Papa's experiences and feelings as expressed to his most trusted colleague, friend, and fellow German Luftwaffe officer through World War II. In them I found vivid descriptions of his travels to remote lands, his beloved Borowke estate, and his struggle to remain not only alive but engaged in his studies through two wars and life as a refugee.

The improbability of having this rich source of insight fall into my hands seemed to me an omen that could not be ignored. So many of the stories Papa had told to me about his life had seemed too full of coincidence and strokes of fabulous luck to be believable. Now, my own incredible stroke of luck would permit me to discover the truth about his life. And so I began putting the pieces together.

two   Borowke

WHEN PAPA REFERRED TO BOROWKE, THE AGRI-
cultural estate in Poland on which he grew up and spent much of his life,
he called it the "home of my soul." It was, he wrote, "the deeply loved
homeland to which I cling with all the tendons of my heart." He had been
driven from Borowke by war, an act he equated with murder. My only
childhood memory of Borowke is of the snowy day when we fled from the
Russians. I was four years old at the time. But I have heard about life there
for as long as I can remember; even as an old man on our farm in Maine,
Papa described scenes from his life at Borowke with reverence and nostalgia. From what I have heard from him and others who had visited there,
it was as close to utopia as anything this earth had to offer. Everyone who
visited there spoke of it as if it were his or her very own Shangri-la.

Some of what I know about Borowke comes from an artistic render-
ing of a map of the estate belonging to my half sister Ursula, or Ulla. It
depicts forests and fields, moors and hills, and is decorated with sketches
of foxes and storks. The map also shows the barns and outbuildings,
the main house, and the quarters for the many workers who lived on the
estate. There are people sprinkled across the landscape in appropriate
places—cutting grain in the fields with scythes, raking it into bundles,
and loading them onto horse-drawn wagons. Eleven horses in all are pictured.

"Horses," Ulla said, "were the love of my life. Papa often said that
only a moron could be so interested in horses. Since there was no doubt
about my inclinations, his conclusion was obvious. A German poet once

said, *Und Glück dieser Erde, liegt auf den Rücken der Pferde* ('Luck on this earth lies on the backs of horses'). The moments of ultimate happiness that I remember were racing through the open fields on horseback—or trying to outrun a thunderstorm and getting caught in it. I love all animals, but the horse is creation's pearl." In one corner of Ulla's map there is a tiny picture of two people side by side on horseback. "They are Ruth and me," Ulla told me. Her friend Ruth Michelly (now Gilbert), who had visited her at Borowke, had made the sketch. I begin this story with Borowke itself because Borowke is the source of much that is to follow.

The estate came into our family because my grandmother (we call her Oma) was born female, not male. Her ancestors originally came to Poland from Germany, when, in 1697, August the Strong (he could reputedly bend an iron horseshoe with his hands and was reputed to have sired 365 to 382 children) was made king of Poland. The new king arrived from Saxony with his entourage of functionaries, including a finance minister named von Tepper. The king rewarded von Tepper with the use of a large tract of land, called Trzebon, in northwestern Poland. The land still belonged to the king, but it was administered, with no strings attached, by the family to whom it was given. It was passed down through the generations, but only through male heirs.

One of the von Tepper heirs, lacking a son, adopted a Scottish immigrant named Ferguson, who was then called von Tepper-Ferguson. He himself begot a son, Adolf Trzebon von Tepper-Ferguson. Adolf married a woman named Valeria Schlesinger, who was Jewish, and they had a daughter. This daughter, born April 9, 1868, was Papa's mother, my Oma, Margarethe von Tepper-Ferguson. Because she could not inherit land, her aristocrat parents purchased Borowke for her, an area of 1,600 morgen (3,360 acres). Margarethe loved Borowke deeply, and although she attended art school in Paris, she returned to Borowke afterward and never left again. In 1889, she married the thirty-year-old *Stabartz* (military doctor), Hermann Heinrich, my grandfather, Opa.

Opa was a stiff and portly man, and a very successful physician. He had a practice in Berlin and treated patients from all over Europe. His father was a Lutheran pastor, and in addition to Hermann there were four other sons to feed. Hermann's path to success lay with the military, which financed his medical training. Papa's second wife, Anneliese, described him in a letter as "an original, and not a very pleasant one. An 'über' Prussian with high-standing moralistic views. Because of the generation gap, we found him old fashioned. He was *entsetzlich* [dreadfully] stubborn and *rechthaberich* [dogmatic]. May he rest in peace."

Margarethe and Hermann spent much time together at Borowke, where first a daughter, Charlotte (known to us as Lotte), was born (November 3, 1890); and then my father, Gerd (November 7, 1896). The children had private tutors until about the age of ten, when they were sent to school in Berlin. There Lotte fell in love when she was nineteen years old, with Ernst Redlich, a law student from the Latvian city of Riga. Ernst introduced himself to Margarethe and Hermann and was accepted, and the couple were engaged to be married when Ernst fell ill with typhus. Hermann, knowing typhus to be highly contagious, forbade Lotte to visit her fiancé. Margarethe, however, encouraged Lotte to follow her heart, and she did, visiting Ernst daily in his sickroom. Lotte contracted typhus and died, while Ernst survived. Lotte was buried at Borowke in a grave that her mother decorated with a replica of the statue of Nike of Samothrace, the Winged Victory. When I was young, Papa frequently mentioned this grave statuary to me, and though it seemed irrelevant at the time, it would reemerge in important and unexpected ways when I grew up.

Both parents were overcome with grief, and Hermann was furious at Margarethe for encouraging Lotte to visit her sick fiancé. A rift developed that could never heal, and Margarethe and Hermann separated. Margarethe reverted to her maiden name and lived out her years in a small cottage and art studio at Borowke that stood at the edge of a pond at the end of a row of old chestnut trees. She spent most of her time painting and sometimes did not even come up to the main house for meals, but rather had a maid bring them to her. Hermann spent most of his time at his apartment and office in Berlin.

Life at Borowke continued, even thrived. Each day at dawn, Franka, known as the "dwarf," came downstairs to start the fires in the big brick stove, decorated with tiles, in the kitchen. Helen, a full-time cook, then made breakfast, while Erna, the chambermaid, set the tables and served the morning meal of bread and butter and sausages. These three each had a room in the main house. They were an extension of the family and of the estate, which assumed responsibility for them. Twelve families of field hands lived in a "village" of six duplex cottages. Each family had some land of their own where they raised their own milking cows, pigs, chickens, and geese, and tended their own gardens. A carpenter and a blacksmith were employed full-time for cash befitting skilled labor. The field hands got only modest pay, but they did not have to worry about house payments, rent, grocery bills, taxes, or medical care. If someone fell sick, "Debu," as Dr. Debushewitz was called, came from the village. He'd see the patient, play a few games of Ping-Pong, and stay for the next meal. He was paid with a side of ham or a bag of potatoes.

Each day a gong sounded the wake-up call, lunch break, and supper time. Mamusha, who'd spent some years at Borowke, said the routine was "the same and ageless. There was no reason to change it because always the same work had to be done." The fields produced hay, potatoes, rye, sugar beets, and hemp. There were fifty to sixty cows and 200 sheep, as well as pigs, chickens, and geese. Forty horses were maintained for work; two of Ulla's beloved Arabian riding horses were also stabled on the estate, as were two carriage horses with a coachman to tend them. All the work was done with horses and hands. Beets and potatoes were planted one by one, and the hay and rye were scythed by hand. The hay was pitched into ricks with a fork and the rye loaded into bundles, and then both were loaded onto carts to be carried to the barn. The rye was threshed using a horse-powered machine. There was no need of petrochemicals to run machinery, or of pesticide for the fields, since Borowke had practiced careful crop rotation for decades, maybe even centuries. Beets and potatoes were dug up, then loaded into sacks and onto the wagons to be carried home and to market. After the fall harvest, geese, herded by two boys, were let loose in the fields to fatten up on the leftover grain. The geese were later slaughtered, and their down was plucked to be used to make blankets and pillows. Under Anneliese's direction, the women all worked together to convert the pigs into sausages and hams, using the emptied intestines for casings. All the meat was smoked and stored upstairs in the house.

There were luxuries as well. Ice cream, made using ice taken from the pond in winter and stored under thick layers of insulating sawdust, was a common dessert, even in the summer. Water was brought in from the shallow well using a bucket dangling by a rope from a long pole. The well was centrally located for access to both the house and the barn. From a nearby swamp, where cranes nested each year, peat was dug and dried into briquettes for heating. Meals were offered around a huge round table, which Oma had decorated by painting delicate pink and purple flowers around the perimeter. Mamusha says, "There were always guests. Often they were artists, naturalists, and zoologists from abroad, and there was dancing in the evenings. Neighbors stayed for half the day and others from out of the country came to stay for a week or a month at a time. They had their own rooms and were on their own. They met at meals, coffee, dances, hunts . . . unlike now, when guests sit in front of you the entire time and must be attended to constantly." Borowke was tranquil and charming and beautiful. It was a comfortable life. It was the "old" way that had been practiced throughout Europe for centuries, but I knew little of its flavor.

Shortly after I began the process of writing this book, I received a phone call from a woman who identified herself as Christiane Buchinger. She had, while sitting in a dentist's office in Maryland, read an account in the Smithsonian magazine of my research on ravens. Seeing the name "Heinrich," she wondered if there was a connection to the Heinrichs her grandmother, Erna Starke, had told her about. "My grandmother had a close friendship with Margarethe von Tepper-Ferguson, who had been married to a Heinrich at Borowke," Christiane told me, and she gave me a copy of an account of life at Borowke that her grandmother had written. Erna describes in detail the rooms of the house, painted in unusual colors by Margarethe. She describes how there were sheepskins and velvet-upholstered chairs arranged around the crackling, aromatic fireplace where people gathered in the evenings, and how everything in the house was made from the products of the surrounding woods and fields. She described the huge dining room table, the middle part made of raw planed wood, polished occasionally with wax, while the edge was smooth and painted light gray. The border between the wood and the gray edge was painted, by Margarethe, with delicate roses. "These were not sweet little decorations, but in deep almost dark colors, of flowering roses with buds and with leaves. It is difficult to compare with the ordinary what one saw at Borowke." Nothing is left of that table, but one of the roses survives! Margarethe had given her friend Erna a painting of a rose that matches the description of those on the table. Before Erna died in 1947, she passed this present on to her sister Ingrid, who immigrated to the United States before the war. She came along with her husband and their two daughters, one of whom was Christiane, who inherited the picture and who gave it to me. I have since passed it on to my daughter, Erica Heinrich, now (at least as of the writing of this book) on a fellowship in hematopathology at the Ann Arbor, Michigan, medical center.

Erna recalls Margarethe (who at the time of their acquaintance must have been in her sixties) as

> a small courageous woman with a narrow face under dense, curly gray hair. She wore wooden clogs and a wool scarf like all the village women, but never once would she be addressed in the fields as anything other than *gnädige Frau* [gracious lady]. Always in her hand she held a knotty walking stick, and under the wool scarf hanging down over her hand twinkled a large ring (with a black onyx inset). Her voice was soft and tender, but when she gave an order, one followed it. She left the house early in the morning, to start the daily routine. The first walk was into the 'park,' which was something like a cultivated wilderness. The park lost itself into the landscape where the moorland began and there was a chain of wooded hills. The first of the little hills was a cemetery, surrounded by trees. This place was sunny and there was only one grave (Lotte, her daughter's). That was the destination of her walk.

Erna Starke also mentions Papa in her account; he would have been a young man at the time of their acquaintance:

> In his study there were many collections, all very much in order, of insects, butterflies, and birds' eggs. They were, naturally, blown out. Pauline [the cook then] made scrambled eggs from the innards. And so we had eaten eggs of jay, snipe, and even owls; never mind those of the lapwing, which are famous. Gerd was an ornithologist autodidactically. But soon he got involved with research expeditions. Once he had to go for an American collector to hunt for a primeval forest rail that for a hundred years had been thought extinct. He spent two years in the wilderness of Celebes, but that was much later after we were already grown up.
>
> We little kids called him Uncle Gerd. But as we became young girls, who one then called *Backfisch* [teenage girl], he didn't want to be called uncle anymore, since he was still a young man. Our sister,

Hermen, who was then not yet *Backfisch*, but already getting witty, said, "he has de-uncled himself."

Later we came to admire the dancing skill of the de-uncled Gerd. He needed no partner, because the rapid accentuated movements of the Cossack dance are enough in themselves, and fill the whole room. Quickly one after another follows the angles of the practiced limbs required of the moving image. This, so I would say today, has a "cubic" quality.

Unforgettable is Gerd's costume that he often wore in the house and which was especially decorative in the dance. It was a white linen shirt, very classy, without tie, naturally, but instead richly decorated with red ornaments. The pants, shorts in the English style, were only unusual in that they were strung together at the side with a red cord. This looked nice with the white linen. Then the boots! They were riding boots of the kind that the Polish cobblers produce, custom-made.

Despite his occasional attacks of exuberance—it also happened, and I have seen it, that Gerd made a handstand on the fully set breakfast table, as the entering guests wished him a good morning—Gerd was not a loud person. To the contrary, he was usually quiet—a son of the forest, a born hunter. Naturally he looked handsome. Very dark hair, very light, intense eyes—the eyes of a sure shot. His hunting passion was applied more and more primarily to insects. "Subtle hunt" as this is called. But he was also at home in the so-called humanities. I recall he once visited us in the city and I was in the dark working on a Latin text that I had to translate. Gerd looked over my shoulder and so fluidly translated ex tempore the whole page that I could hardly keep up in writing it down. So my school assignment was finished quickly. Gerd would one day have become master of Borowke, if not for the horrors of the Second World War and the Hitler time, which had destroyed everything.

My father must have been a very different person back then, at Borowke. It is difficult for me to imagine him doing handstands on the table, or performing a Cossack dance. Neither can I picture Oma skinny-dipping at midnight with the Spanish painter Castelucio (which according to rumor she did), whose paintings of children decorated the walls.

But all during Papa's long life, he felt that his appreciation for the natural world came from his mother, whom he called "a true nature artist."

As he reached maturity, Papa assumed more and more responsibility for the management of Borowke. He was heir to the estate. Yet he still had time for excursions into the surrounding countryside. He later described to me the beauty of the swamp, which assumed a special significance in my mind, because it was where a pair of cranes nested each spring, and as I'll explain much later, where Papa hid to escape enemies. He talked of the "masuries"—the open swampy meadows full of buttercups, primulas, and other wild flowers, where birds sang and many varieties of insects swarmed. He went out into the field in style, riding in a carriage, often with a female companion, taking along fresh raspberry ice cream topped off with light cream to drink, homemade bread and butter, salami, and rich cakes for dessert. He, like everyone else around him, was sure that life would go on like this forever.

Yet when Papa was a young man, the system of the landed gentry was already facing transition. At this time, Borowke was in what was then called West Prussia, a province of Germany. (After World War I it was deeded to Poland by the Versailles Treaty; and World War II started when Hitler invaded West Prussia under the guise of reclaiming it for Germany. At the end of the war it reverted back to Poland.) The Bolsheviks viewed this system as a form of exploitation of the lower classes. Their solution was the "collective," a presumed utopia in which nobody owned anything and everyone worked for a common benefit. But as their massive and tragic experiment showed, the grandest ideals often do not mesh with reality. The European peasants were "freed," only to become even poorer. Twenty million Russian peasants did not survive, and ultimately, the Soviet Union, where this system was forcibly tried on a large scale, collapsed.

I cannot judge whether the workers at Borowke felt exploited, but from all accounts I have read, they along with everyone else who spent time there felt a strong sense of community and of pride in being able to fulfill all their needs directly from the land. They joined in the working of the estate with enthusiasm, holding spirited scything competitions and celebrating their accomplishments alongside the landowners, even though they did not have potential to become landowners themselves. They held great harvest dances, with the fiddlers playing on the dance floor in the barn loft. I suspect that they, like Papa, exulted in hearing the cranes trumpeting in the swamp, and the larks singing from the sky above scented meadows. As they toiled in the fields to sow the rye, the workers may have taken pleasure in hearing the skylarks singing far overhead. As the Yokut Indian prayer for good fortune says, "My life is tied as one with the great mountains, with the great rocks, with the great trees in one with my body. And my heart."

We humans have experimented with various social systems; some have endured and others not. I believe, however, that our well-being is tied not so much to the structure of our society and the politics that determine it, as to our ability to maintain contact with nature, to feel that we are part of the natural order and that we are capable of making a living within it.

three Boy Warrior

"Patriotism ruins history."

—ATTRIBUTED TO JOHANN WOLFGANG
VON GOETHE (1749–1832)

PAPA WAS BORN IN BERLIN, BUT HE GREW UP AT
Borowke, where he led a rather sheltered existence. He entertained himself by wandering about the land and observing the profusion of nature
that surrounded him. He made collections of insects, especially the flashy
butterflies and beetles. On his birthdays and at Christmas, he wished
for (and received from his parents, who bought them for ten to twenty
goldmarks each, equivalent to about $150 to $300 today), exotic tropical
beetles, particularly the great "antlered" stag beetles and brilliant multicolored cetoniid flower beetles. A Goliath beetle that Papa especially cov-
eted for his collection cost more. Papa had been taking violin lessons and
told me that when he was ten years old, he hit on the idea of putting on a
concert, with posted notices and paid admission. He didn't tell me who
attended or what he earned, but the notice (which I saw recently) mentions the date (September 30, 1905) and the price of admission (three
marks first class, one mark second class, and twenty-five pfennig standing). As far as I know, this was his only musical exposition.

As a boy living in the forest right after the war I also collected beetles;
though it was not possible to purchase exotic tropical beetles, I collected

the local native ones. I recall especially the great variety of carabid beetles. I have in my possession now a drawing of a stag beetle I made for Papa when I was about seven or eight years old. Like Papa's tropical beetles, this was a "dream" beetle, because I had never seen one. But I wished for it, possibly after seeing an illustration of it in the book of beetles that Papa had given me as a present.

In his youth, Papa also started an herbarium and collected study cases full of stuffed skins from small birds and mammals. He even collected flies and then tapeworms, which fascinated him because of their complex biology. (His father, Hermann, the doctor, helped him to obtain some rare specimens.) Wondering where all this collecting might lead him, Papa at age fifteen knocked on the door of the curator of insects, Professor Heymons, at the Berlin Museum für Naturkunde. Papa asked Heymons which group of insects was least known, and Heymons suggested the Hymenoptera (bees, wasps, and ants). The ichneumoninae, a group of wasps that were then (and still are) very little known, captivated his interest, and "from that day to this," Papa wrote in his unpublished memoir, "the ichneumoninae have disciplined my time and energy, excepting the interruptions of wars, a revolution, invasion, expulsion, starvation, and emigration." I would make one correction to his statement, and that would be to change the word "excepting" to "throughout." He sold off his entire collection of birds and mice, and his parents purchased for him a collection cabinet with seventy-five drawers (still in use, full of his ichneumons, at the Zoological Museum of the Polish Academy of Sciences in Warsaw). Papa's first real experience in the "outside" world was when he left Borowke in 1906 to attend the Askanasische Gymnasium (high school) in Berlin. He already possessed talent and expertise in biology, but found that few of his classmates shared this passion. In his memoir Papa wrote that "none of their faces seemed to reveal an eagerness or ambition to learn." The curriculum did not focus much upon natural history studies; rather, the emphasis was on reading Latin and Greek as well as French, English, Russian, and German literature. The classroom was highly competitive, with seats assigned and numbered according to achievement. As a newcomer, Papa began in the back, at seat eighty-nine. By the end of his second semester, he had advanced to seat three; and by the third semester, he became the class *primus* (number one).

I suspect that Papa may have felt he was a bit odd among his class-mates, with his obsession for pursuing natural history. He may have viewed himself as one-dimensional or overly intellectual. He may have been *primus*, but in athletics, which mattered at that school, he was ranked last, "the worst gymnast in the class." He was fat and out of shape, and that had to change. With what he admitted was "true fanaticism," he worked to become the best gymnast and a star runner as well. Papa later told me that his transformation was one of his accomplishments in which he took the most pride, and that without it, he might well not have sur-vived what came after.

He wrote in his memoir that save for one, none of the professors, scholars, and researchers whom he met at school were role models for his notion of an ideal "man"—at least not the man he wanted to become. He told me about an incident that occurred in high school, which became a defining moment for him. A student had dusted the teacher's chair with chalk, so that a white rump was revealed when the teacher turned to write on the blackboard. The class roared with laughter. The teacher, finding this less than amusing, demanded that the perpetrator own up. Nobody spoke. The teacher turned to Papa, as *primus*, and demanded: "Do you know who did it?"

Papa unfortunately did know, so the answer could be none other than: "Yes, I do."

"Then who was it?"

"The morale of comradeship forbids me to denounce a classmate," Papa replied. "The culprit should answer for himself."

The teacher had a different opinion, and Papa was ordered to appear at the principal's office the following day. Opa, having heard what hap-pened, accompanied his son and defended him. The principal sided with Papa as well. With that incident, the actual perpetrator, Martin Ernst Karl Dammholz, otherwise known as Mek, became Papa's lifelong friend.

The one teacher who made a moral impact on Papa was Professor Dahms; Papa decribed him as a small hunchback with snow-white hair who, at about eighty years old, could barely see over the desk that he stood behind. When Dahms retired, the gymnasium's 500 or so students and teachers assembled in the auditorium to hear his farewell address, which closed with these words: "Let me give you today one last piece of advice; it is of minor importance what in your later life you may choose as your profession. Nothing is really bad, if you do it thoroughly and right. What really matters is only that you do everything the very best that you can."

Papa confessed that the picture of the small old man, the sound of the shaking voice, and the importance of the message remained etched in his memory. He recalled those words often during his subsequently challenging life, beginning in 1914, even before he finished high school.

Papa recalled the events of 1914 in detail. He was riding on the underground train to the Dahlem district of Berlin for hockey practice when he met a teammate called von Wickede, who, while reading a newspaper, looked up at him and said: "War!"

"Why?" asked Papa.

"Because the crown prince Ferdinand of Austria has been murdered."

Papa did not yet see the connection that von Wickede had astutely made. He left the following day for Borowke, where he spent his time collecting insects and participating in the scything competitions, which were held each year during the rye harvest. In spite of his athletic training, he still found scything day after day from morning to evening to be "the most demanding sport on earth."

One day later that summer, in August, as he and the field hands were coming in from the fields tired and hungry, they found that a red poster had been nailed to the side of the barn. It informed them that war with Russia had been declared, and that all able-bodied German men were to report for duty immediately.

At the age of seventeen, Papa was not sure whether the term "able-bodied man" applied to him yet, and he remained behind as the others left to report for duty. But the following day it became clear that his childhood had ended. The mailman, who walked the five kilometers from Przepalkowo to Borowke, arrived with a telegram for Papa. An emergency graduation exam would be held for students in his class who intended to enlist. Papa knew that the night train would arrive in Berlin at exactly the time the exam was scheduled to begin. He might just be able to make it. He returned his horses to the barn, changed his clothes, and left for the station.

Papa passed the written part of the exam cum laude and so was allowed to skip the oral part and shorten the exams by one day. This would give him time to visit his father in Berlin and return briefly to Borowke to say good-bye to his mother before enlisting, he hoped, in the cavalry, the uhlans, an elite troop of mounted soldiers wielding long lances and riding like a wall of medieval knights in formation. Duty to country was the unquestioned ideal, and his entire class had volunteered, and with so many men eager to enlist, there was some competition. Fortunately, Hermann, the former military doctor, had a patient who was a high-ranking officer and who helped secure Papa's acceptance into the select group of sixty men in the Fourth Uhlan regiment in Thorn (Toruń in Polish).

The next day, Papa caught a train to Thorn and entered the barracks of the regiment, commencing the life of a soldier. He was issued a horse, a uniform, a *czapka* (black helmet), a saddle and headgear for his horse, a saddlebag for carrying a day's fodder, cooking utensils, reserve rations for emergencies, a heavy blanket for the horse, a sword, a long lance, and a carbine with ammunition.

Training started immediately. The recruits, most of whom, like Papa, were from the country and were used to horses, were required to ride bareback down the steeplechase track at a "stretched gallop" and jump over successive obstacles of hedges, ditches, and stone walls. A sergeant, stationed at one of the obstacles, barked out a question to the rider at the moment he was airborne and an answer was expected before he landed. Most of the farm boys passed this test easily, but one, a boy named Bayer, had never before ridden a horse. No special consideration was granted, and he was thrown from his horse and had to be hospitalized. He did not rejoin the regiment until months later, when they were engaged in combat with the Russian infantry on a deserted farm. There, Papa said, "I saw Bayer running across the courtyard toward our trenches and the moment I recognized him, a grenade exploded behind him. Little Bayer was killed on the spot."

At the start of the war, the German general in charge of defending the eastern border was a man named Max von Prittwitz, called *der Dicke* ("the fat one"). He faced the Russian army led by General Pavel von Rennenkampf (an ethnic German who had joined the Russian army at age eighteen, later declined to join the Red Army, and was executed by the Bolsheviks). Von Prittwitz was soon overwhelmed and retreated. The German Kaiser, Wilhelm II, immediately got rid of von Prittwitz and replaced him with General Paul von Hindenburg. Hindenburg's army then forcefully attacked the Russians, fighting courageously because they were convinced they were defending their families, their homes, and their very existence. This was the battle of Tannenberg. It was raging (August 26–30, 1914) when Papa was sent to the front after only brief training.

One day near the end of August, a lieutenant accompanied by two uhlans rode into the Fourth Uhlan training camp seeking replacements because the regiment had suffered heavy losses. Papa wrote: "I gazed with awe at the lance of one of these uhlans, which had been nicked by two bullets." Sergeant Major Mantheu, "a big man with rosy cheeks, a round belly, and a carefully trimmed mustache," was then overseeing the recruits' training. He ordered them all to stand before him in three ranks and declared: "The regiment needs replacements at the front. Six volunteers may go into the field the day after tomorrow. Step up to me if you care." There was a second or two of silence, then a sudden rush as all took their place in front of the sergeant major. Papa remembers that they all volunteered out of bravery. (Could they also have been afraid of being perceived as chickenhearted?)

"Well, so I must make the choice," Sergeant Major Mantheu said, and he called out the names: "Udo Roth, Otto Roth, Hans Wiebe, Georg Feld, Szczyglowki, Gerd Heinrich."

The next day Papa, along with the five other men, rode with ramrod-stiff backs, as befitted soldiers, to the train station. Girls threw flowers and accompanied the heroes-to-be, who boarded the train. The whistle sounded, the big black engine belched smoke, and the train pulled out. Papa remembered the sun shining from a blue sky dotted with sparse white cumulus clouds. He was, as I have said, seventeen.

When the train stopped and Papa and his companions got out, they heard from the east "an uninterrupted, sinister sound, like remote, continuous thunder." It was the cannons roaring in the ongoing battle of Tannenberg. The men mounted their horses and started riding toward the distant rumbling. They passed town after deserted town, all in ruins. "Not a house was standing. There were remains of dead cattle and smashed furniture in the streets, signs of the chaotic and destructive mayhem of an angry retreating mob," Papa wrote. "In the ditch on the right side of the road I saw my first corpse, a German soldier. His beard was red, mouth open, and his eyes and upper skull had been raked off by a grenade splinter [artillery shell]."

By evening they reached a deserted farm and the men bedded down in the house and the barn. But Papa decided instead to overnight by burrowing into a haystack outside, after tying his horse to a nearby fence. Exhausted, he fell into a deep sleep.

When he finally awoke and crawled out of his haystack it was already full daylight. There was no one in sight. His horse was gone. His "iron reserve" rations had been stolen. His gun was gone as well, and a piece of junk had been left in its stead. Famished, he wandered around until he eventually met up with uhlans from another unit who brought him to the front, where he was reunited with the others (who, unable to locate him that morning, had thought him a deserter), and his horse.

Shortly after that, still famished, he saw an officer take a bite from a large salami and then hide the remainder. Papa, who had not eaten anything for forty hours, said: "In those few days it became clear to me that I would perish if I did not learn to defend and care for myself in the same way as others did, and an officer should not deserve to eat if his men didn't." He recalled Goethe's words: *Alle Gewalten zum trotz sich erhalten,* "Maintain yourself despite all." Armed with that rationalization (and with a clear conscience) he stole the rest of the salami and devoured it. Papa had imagined participating in a heroic fighting unit imbued with loyal comradeship. "How differently it actually developed," he wrote. In the meantime, Rennenkampf and his army had been beaten at Tannenberg and were retreating back into Russia. The uhlans drove the Russians back, riding eighty kilometers per day, day after day, with little opposition. Whenever any resistance seemed imminent, the uhlans closed ranks and the order was given: "To attack—lower lances; go!" and off they would charge at full gallop. All resistance vanished. The German army was unstoppable. Nearly 140,000 Russian soldiers were killed or wounded. The assault had turned into a rout. Hindenburg became a folk hero; and the Russian general, Samsonow, committed suicide after his defeat. But the legacy of this massacre would incubate horrible anger on both sides to produce devastating consequences for Germany decades later.

Fall came and went, then the winter of 1915, and the uhlans were still on the march, going ever deeper into Russia. They were even welcomed into Russian huts and enjoyed the company of the peasants, for whom the war was peripheral. But one day while crossing a river, they met something far more powerful than their own mounted cavalry armed with lances. Volleys of Russian bullets, delivered by the newly invented machine gun, erupted from willow thickets on the shore, and Papa watched as half of a unit of men and horses were mowed down.

The uhlans were no longer invincible, and by the autumn of 1916 the charging cavalry with its lowered lances ceased to exist; it was no longer a strategic force on the battlefield. The war, which the uhlans thought would be won after a battle or two, had already extended into its second year. Papa, who was often being sent to reconnoiter as a lone scout, had been promoted to the rank of corporal, and was decorated with the Iron Cross for bravery. At the same time, his tunic had acquired another distinction: a horizontal rent across the chest piece, from a bullet.

He now had eight men under his command and was assigned to hold a sentry post on the Baltic coast near a Latvian fishing village, Kilezeem, on the Gulf of Riga. Here they constructed a warm, comfortable underground shelter in a sand dune with a small window facing out to sea. The walls of their shelter were made of pine logs dragged by hand from the forest bordering nearby Lake Angern. Soon, falling snow covered them, and ice formed along the shoreline. One man at a time remained on duty, while the others stayed warm inside.

Life under the sand dune proved to be snug, but almost unbearably boring, especially to Papa, who never in his life played cards. With nothing to look at but the ice floes, Papa joked that the men ought to go swimming. He bet one of them, Sergeant Spikermann, five marks that he could run naked into the sea and sit among the ice shoals for a full minute, in water up to his neck. That Spikermann wanted to see. Papa disrobed, ran down, and jumped in. Spikermann grinned. Papa managed through sheer determination to stay in. Spikermann produced the five marks.

A few days later some seabirds arrived, and Spikermann, the best sharpshooter among them, managed to shoot one, which fell into the sea. Papa, who wanted it badly, started to swim after it. He was astonished at how quickly his body seemed to thicken from the cold, and he turned back just barely in time. He had almost drowned, and his friends had no easy time reviving him.

One day the sentinel on duty reported seeing a strange form far out on a floating piece of ice. With the aid of binoculars it was identified: a seal. A feast! Sergeant Spikermann was again called to duty, and the animal slipped from the ice to float lifeless among the churning ice blocks. More than for its meat, the seal held considerable interest for Papa as one who had collected mammals; he wanted its skin.

"Tell me, Spikermann, if I get the seal, will it be mine?"

"Certainly!" was the immediate reply.

Papa had a plan. He got a wooden washtub at the nearby village, and nailed two long pine poles onto the sides as outriggers. This time, he kept his clothes on—and good thing, too, because he did not go far before swells dumped water into his craft and it would have become unmanageable if he had stayed in it. Swimming while fully clothed slowed his body's cooling, and he was able to grasp the seal's feet and make it back. He wrote: "The last vestige of energy left me as I stumbled ashore and onto the horse blankets that my men readied. They carried me into the bunker, and rubbed my limbs till I came back to life. It was one of my narrower escapes, but I managed to send home my prized sealskin, a reminder of my watch on the Gulf of Riga."

On the day before Christmas, two riders appeared at the shelter. They had orders for Papa to report to headquarters at Belgallen, about thirty kilometers away, and he was to proceed there pronto. No explanation was given. The sun had already set, but the moon brightened the snow-covered landscape and Papa decided to take a shortcut across Lake Angern. Its frozen surface was dark and glistening. Horse and rider took a few steps on the ice, and all seemed well. But as they ventured farther out onto the "black, endless, and spooky" lake, the ice cracked like rifle shots as breaches spread underfoot. Papa aimed his horse at a light on the opposite shore, and they made it across. There he found a lone hut, and peering through the window saw an old couple drinking beer. He knocked on the door and inquired about directions to Belgallen. The old couple offered him a beer (which Papa told me he did not accept, because he would never drink), as well as directions.

Papa reached headquarters near dawn, fed and stabled his horse, and reported to the adjutant. His orders: "Go first to the grain mill's upper floor. There is a room ready for you. Have some sleep after your long ride."

As ordered, Papa entered the room—and there before him, unbelievably, stood Margarethe, his mother! The very last thing he expected. A miracle.

Slowly she divulged how she had pulled off this surprise. Papa had mentioned to his mother the name of a post office through which to send Christmas greetings, and had said that "Sergeant Knoll" would pick them up there. But Margarethe had, on a crazy impulse, decided to deliver them to him personally. She met Sergeant Knoll, who was on an assignment to dispatch a wagon filled with bales of hay for the horses from the German border. Knoll had made a hiding place within the hay to carry her with him to the front, and he provisioned the hiding place among the hay bales with food for several days. Margarethe was successfully stowed away among the hay for two days and nights, but then at about the halfway point, Knoll made a stop and she left her hiding place and took a walk through a town. She was promptly picked up by a military patrol, who asked to see her civilian pass. She of course had none and was escorted into a guardhouse. Knoll found her there at night, whispered to her to get up, and sneaked her out while the guards were either sleeping or pretending to sleep. Off they went, riding another two days and nights before arriving at the squadron headquarters at Belgallen. The story of her trip had greatly impressed the soldiers at the front, and they had offered her bread and salami.

By the time she was to return home to Borowke, even headquarters had gotten in on the act, ordering Papa to accompany her back to the German border. Papa's official order was to "purchase a riding whip." The trip, about 300 kilometers back to the border, took seven days in an open horse-drawn wagon. Papa, Margarethe, and Knoll spent their nights in peasant huts, huddled, as was customary, atop the huge stone oven in the middle of the house.

On reaching the border, Knoll departed for the Baltic coast, while Papa was to accompany his mother home on the train. But at the train station there was a cordon and everyone was checked for papers. Margarethe had none, and for the first time her confidence faltered and she broke down and cried. Papa took the official who was checking passes aside:

"Comrade, this is my mother. She visited me at the front, obviously with a pass issued by the general command. But enemy artillery hit and burned our quarters, destroying everything. Now I bring her to Tilsit— please let her pass!"

"But if I do she'll only be subject to even sharper scrutiny at the next stop, at the Memelbridge—and then they'll surely take her off the train."

"Kommt Zeit, kommt Rat," replied Papa. ("Comes time, comes counsel.") "Let us board!"

They were permitted to board the train, and when it stopped during the night at the Memelbridge, an official came from one train compartment to the next to check passes. When he entered the compartment where Papa and Margarethe were seated, the lights suddenly went out and he missed checking them.

When I was growing up, I remember Papa often saying: *Der Mensch muss Glück haben* ("One must have luck"), as though one could conjure it up. Of course that's not possible, but then if one is never in a position to need it, one is assured of never having it. And those who were not lucky are not around to tell their tales.

After his return to the Gulf of Riga, Papa stayed until March 1916 with his regiment. He was then sent to join the Twenty-eighth Reserve Infantry Regiment, which was holding a fortified line through the Tirul swamps south of Riga. The Russians had established positions in the woods and fortified them with barbed wire; the Twenty-eighth constructed lines opposite them in the open swamps. These they built during the night, after first making roads for the horses out of logs placed on the boggy ground. Papa wrote: "When all was done, we lived in comfortable bunkers behind solid earth walls, about 500 meters away from the Russians in the woods."

For both sides, information was at a premium. The Russians had retreated. But now the Germans were bogged down, literally, in the swamps. Terrorists (otherwise known as partisans) were striking here and there behind the lines. Who were they? What would they do next, and where? The partisans were invisible, living in underground forest bunkers or disguised among the civilians in villages. Often, they were the local people themselves, who had at first been friendly to the Germans but were now growing increasingly intolerant. As could be expected, the inevitable recriminations and attacks from those not wearing a uniform, to mask their identity, made everyone an enemy and escalated to produce suspicion and hatred. A vicious circle between Germans and Russians had begun.

One day the command brought their unit two "spies"; orders were to shoot them. Lots were drawn, and Papa, then age nineteen, pulled the short stick, the unlucky draw. In his book of 1937, *Von den Fronten des Krieges und der Wissenschaft (From the Fronts of War and Science)*, he wrote:

My conscience was not bothered when in the fight at the front I had to kill—provided there was the same chance for friend and foe. On the other hand, to shoot defenseless people—even if they deserved to die—filled me with loathing and disgust. Everything in me rebelled . . . but the military discipline is cruel. I knew that I had to obey and that it was my duty to do as I was ordered. The only thing I could do was to make it quick and final.

In this same account Papa described the two prisoners in detail—one was a dark-haired broad-shouldered man; the other a youth wearing a light-colored cap. After their verdict was read, the two hugged and gave each other the Russian brotherhood farewell kiss, and when the sixteen shots of the firing squad Papa had organized rang out as one, the young man's cap flew up into the air. "This demonstration of their extraordinary strength and will, that tamed the natural fear of death and allowed them to look fate directly in the eye without blinking, was the only light on this dark day," Papa wrote.

I do not know what I would have done if as a teenager I had been ordered to murder two men point-blank. But Papa was a product of a historical context in which obeying authority was the highest virtue. As Bertrand Russell wrote, "Germans before Hitler had an exceptional sense of public duty and service to the state seemed the obviously right moral ideal." But Papa's description of this execution horrified me. He seemed to feel that he was showing me how he had triumphed over his soft inner core to become a better man. Why else did he make it public—unless he thought it gave him bragging rights? It was only long after his death that I began to think that maybe his real goal had been to convey the true brutality and horror of war. Recent technological advances allow us to kill large numbers of people from a distance without confronting our acts. I wonder whether any of us would use these industrial killing tools if all of our victims were lined up in front of us and we were forced to look them in the eyes? It would be possible only through the mechanism of hate—

that, and our huge capacity for denial and self-delusion. It's all part of our innate human biological heritage, which helps us to justify what we do, and still maintain our sanity, and sometimes to even think we deserve medals for it.

Shortly after the execution of the two Russians, Papa asked for and received a month of furlough. On his way to Borowke through Berlin, he found "Mek," his old classmate from the gymnasium, who was recuperating at home with an arm shattered by an artillery shell. A party was arranged immediately, and among the guests Papa found Elly Below, a classmate on whom he had had a crush. He wrote:

> The spark of love ignited almost immediately when we saw each other. I danced with her all night and invited her to come to Borowke to spend the time of my furlough with me. She accepted and we left Berlin together the next morning. My mother received Elly with great friendliness. I showed her my paradise—the forests, the fields, the meadows. Sometimes we walked around. It was the month of May, and the beauty of Borowke was breathtaking. The fruit trees were in full bloom, the extensive grain fields were a bright green color, and above were the larks, rising up into the sky with their beautiful, uninterrupted song of rejoicing.

At the end of the month Elly accompanied Papa to Berlin, where he boarded the military train for the Russian front that left each day from the Zoologischer Garten station at midnight. He climbed in and opened the window. Elly walked along the train as it started to pull out of the station. Soon she was running to keep up, then she jumped up onto the footboard, and he opened the door and pulled her inside. The couple got off in Königsberg (now Kaliningrad, Russia); hired a horse-drawn taxi; and drove to the Baltic Sea, where they rented a cottage close to the shore. "It was here that I experienced for the first time in my young life the sweetness and depth of two lovers' final surrender." Papa wrote, adding: "I was born in the nineteenth century, and brought up to obey the strict moral rules of that time. It was clear to me that . . . I had to marry Elly."

After his return to the trenches in the swamps of the Tirol, he wrote to her and proposed marriage. The answer that he got was no. She explained that she had a friend who was stationed in the trenches of Verdun in France and that if she wrote that she had betrayed him, he might sacrifice his life. Papa became quite distraught. Perhaps he had some premonition of the slaughter to come, in which half a million soldiers on each side would die there in the trenches. Perhaps he had his own death wish. In any case, he applied for a transfer to pilot in the newly established Luftwaffe (air force). Airplanes, which up until recently had been used to observe troop placements and movements, were just coming into use as combat weapons. They were equipped with guns but no parachutes.

Papa was assigned to the pilot school based at the Junker plane factory in Koeslin (now Kossalini, Poland), where thirty others were also in training. His teacher, named Stockhausen, used an instruction plane with two steering wheels. One wheel was for the actual steering, and the student held the other to get a feel for the process until he could eventually do it alone. The Junkers at that time were unstable flying contraptions derisively called *Kisten* (boxes). They were "head-heavy"—that is, they tended to fall to the ground headfirst—and so they were also called "flying coffins."

After passing his field test—during which the pilot had to remain airborne for at least half an hour; reach an altitude of 2,000 meters (about 1.2 miles), and make a smooth landing at the end—Papa was sent to support the army fighting the Italians in the battle of the Isonzo River. There he reported at the Flug Park, where "flying marriages" were made. Each plane had an "Emil" (the pilot) and a "Franz" (the observer). The Emil could shoot straight ahead "through" the propeller (the machine-gun firing mechanism was synchronized so that the bullets would miss the blades, of course). The Franz sat directly behind his Emil and served as the plane's "eyes," ever alert for the enemy's single-seater "hunter" planes, and he directed his machine gun to the rear or the sides. It was a life-or-death proposition to have the right flying mate.

Papa had already examined the crop of potential partners at a party where pilots and observers mingled to prospect for "mates," and he had excluded "all those who were busy passing the time drinking alcoholic beverages." That left only one possible candidate. In order to consummate the flying marriages, pilots demonstrated their skills in the air while the Franzes watched. After Papa's flight, the nondrinker and prospective Franz, Walter Grotrian, approached him and asked, "Would you like to fly for me?"

"*Abgemacht!*"—done!—said Papa. "I myself had the intention of making that suggestion to you."

This marriage was to last much longer than Papa's first real one, which occurred around this same time. While attending the theater during his pilot training, he saw a young woman, and when their eyes met, he "was instantly in love." Again. So much for Elly! (whom he never forgot, incidentally). However, from all accounts and from reading his memoir, I gather that Papa soon decided that his bride was "spoiled" because she had independent views. They quickly divorced. He wrote: "My character, as it developed in the war and following, was imbued with an unstoppable thirst for freedom that made me absolutely useless for the 'yoke of marriage.' "

Soon after Papa's much more successful "marriage" to Walter Grotrian, the Heinrich-Grotrian team received orders to fly over the lowlands of the Po River and to land on an airstrip at Istrago. "During this flight, along the front lines, we saw the first air battles in our life. We saw two big Italian aircraft shot down in flames by a German hunter plane. It gave us an indication of what to expect."

At the Istrago landing strip, a small, lively-looking officer who was decorated with an Iron Cross stepped up briskly and introduced himself as "Oberleutnant Hirsch, adjutant to the leader of the Field-Flier Unit Seventeen," to which they were now assigned. Then they met the commander of the unit, Captain Grauert.

The fliers of Unit Seventeen at the time of Papa's arrival had two pilots, Sergeants Noeding and Ricker; and their observers, Lieutenants Briese, Moesnew, and Hemprich. All were later shot down and killed in France. Brüggendiek, who joined later, lost an arm in a crash, and Walter Grotrian took a bullet in his lungs, but survived to fly with Papa again. Papa was the only one, though shot down twice, who was never wounded.

Once, when he was testing the motor before the start of a mission, his observer (whose name he had forgotten by the time he wrote of the incident) said:

"Gerd, we must and shall win the war!"

"What makes you so sure?" Papa wanted to know.

"Because we are fighting for a better cause. We fight for right and justice, not for gain and money as do most of our enemies."

Papa answered with a non sequitur: "There is only one right and one 'justice' in our world, 'survival of the fittest.' "

The first time he told me this story, these words sounded profound to me. I did not understand them. But when he repeated them to me half a century later, after I had become a professional biologist, they sounded circular and empty. They were also frightening, as they fit perfectly with the social Darwinism that came right out of the tool kit of Nazi propaganda.

Gradually the German front lines advanced over the alpine mountains, into the lowlands of northern Italy to the fortified Mount Grappa, which was held by Italian infantry. The German general staff ordered an attack on the fortress, and Unit Seventeen was called to support the offensive. Papa wrote that as they approached the slopes of Grappa,

> I was stunned by their steepness. . . . Here and there I saw troops of German infantry trying to climb up these murderous walls, under fire of Italian cannons and machine guns shooting down on them from the summit. I felt a burning pity for the boys caught in such a hopeless situation. . . . I flew up to the summit and attacked with our machine guns—but one of theirs got me into its sight and blew a burst of bullets into the middle of our airplane. Several bullets pierced the gasoline tank against which my back was leaning. Within seconds my uniform was drenched in gasoline. Within several more seconds we might explode, and we'd go down in flames.

This story was one that Papa told me and Marianne at bedtime, and now, eighty-five years after it occurred and fifty years since he began telling me bedtime stories, I still recall how he would interrupt the flow of the story at the point where he was doused with gasoline and ask us the question, "What did I do next?"

"I don't know, what did you do?" we would answer, and he would tell us that to prevent the plane from catching fire and exploding, "I switched off the motor."

Now I am amazed that he managed to survive. I marvel that he, a beet and potato farmer, ended up finding a solid rank among scientists and kept himself alive through two murderous wars. I used to attribute it to a large measure of good luck. But after reviewing this story, I can now see that there was also a large measure of good judgment. Here was a mere child—younger than my oldest son, whose days are filled with classes in English composition—piloting a plane over mountainous terrain, looking down into the enemy's guns. He had no parachute. Turning off the engine meant for certain that they would crash. Burn or crash, crash or burn.

That could easily have been the end of the Heinrich-Grotrian team. But because the motor wasn't running, the gasoline did not ignite, and they managed to land safely in a streambed. They wandered for two days, during which time they were declared missing in action, before they managed to return to duty with their unit.

Papa and his observer had worked out a communication system. One pound of a fist on Papa's back meant that Grotrian had sighted one enemy fighter plane approaching; two pounds of the fist meant two. Most of the fighter planes were agile one-seat "hunters" whose daily missions were to go out looking for prey to shoot down. One time Papa felt a fist pound, then two, and then three. Looking back, he saw Grotrian swinging back and forth wildly with his machine gun, trying to keep all three pursuers at bay. Then there was a frantic pounding on his back—the gun had jammed. They were surely doomed. Yet Papa clung to a wild hope, trying to execute impossibly steep dives with simultaneous sharp curves in all kinds of unpredictable directions, to give the machine guns trained on them a more difficult target. Their evasive maneuvers meant, though, that they had to quickly sacrifice altitude.

Again they pulled through. They were not hit, and they had maintained just enough altitude to elude their pursuers. The only injury was a sprained ankle that Grotrian incurred from the violence of being thrown around the plane. Apparently, he was not wearing a seat belt.

In the spring of 1918 Papa and Grotrian were transferred to the front in France, while Captain Grauert was promoted and transferred to the general staff. Papa was in charge of the squadron and soon received a call from Grauert telling him that the German infantry was setting up a fortified position. French reinforcements were being rushed to the site by railroad. The Luftwaffe had received orders to interrupt these trains. The problem was that there was a continuous cover of rain clouds just fifty meters (about 150 feet) above the ground. An approach by plane would be suicidal (radar for blind flying had not yet been invented). A call for volunteers had gone unanswered. Grauert declared that this lack of enthusiasm violated the honor of the German officer, and he added that although he now belonged to the general staff, he would fly himself.

"Would you like to be my pilot?" he asked Papa. If there had ever been any question about Papa's ability as a pilot, this surely stilled it. His old boss would have chosen only the best. For a couple of seconds Papa was silent; then he answered, "Yes, sir." Walter Grotrian had to cede his Emil to the captain, and given the mission, there was little chance of getting him back. Unknown to Papa at the time, his response would seal Grauert's respect, which would in an unanticipated way save Papa's life in the next war. He wrote in his memoir—at age eighty-seven, the year before his death, "Of all the incredible, great favors fate has granted me during all my long life, the acquaintance with Captain Grauert has been one of the greatest. He was a man of unshakable honesty and unlimited courage, and of never-failing good judgment in his decisions."

In sizing up the mission, Papa decided that flying for 100 kilometers over enemy territory at fifty meters above the ground meant certain death. Flying blind in the clouds therefore held the better chance for survival. It would be Grauert's task to maintain the height by altimeter and, by accurately measuring their speed and time traveled, to calculate the precise moment when they should strike down through the clouds to hit the railroad.

The two men performed their task with perfection. They burst through the clouds exactly on target, dropped their bombs, and then received the anticipated hail of bullets. On their return they almost crashed. Their plane had received forty bullet holes.

A few days later Papa received a call from Baron von Richthofen, the legendary Red Baron, asking if he'd join the baron's personal unit. Papa wrote in his memoir, "That was a very much honored call and my decision was not easy. However, I was so much enmeshed into my own unit that I could not just leave, and I remained loyal to it until the end of the war." "Besides," he once confessed to me, "I wanted to live a while longer."

By the summer of 1918 Russia had been defeated and disabled by civil strife. The war had shifted to the front in France, where the now gigantic struggle had escalated into barbarous and horrific chaos. It was described later by Erich Maria Remarque in his book *All Quiet on the West-*

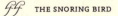

ern Front, about his experiences with a group of bewildered young German soldiers during the last desperate days of World War I. Papa called it a *Materialschlacht*, a battle of materials, with continual innovations in armor, armaments, and firepower: "There were six planes of theirs for every one of ours, and where the numbers lacked, we had to compensate with performance. We flew three times per day against the enemy, and at night when our nerves started to calm, the British would bomb us."

One of the many inventions used in this war was a new Junker's plane called the *Infantrie Flugmachine*, or the *Blech Esel* ("tin donkey"). It was so heavily armored with metal that it was thought to be impermeable to bullets. The downside was that it was so heavily armored that it could fly only about 100 meters above the ground. Papa and his new observer, Brüggendiek (Grotrian had been wounded), flew one of these "tin donkeys," crisscrossing above the trenches to map enemy machine-gun positions.

According to Papa's description:

As far as the eye could see, the surface of the earth seemed to be visited by a terrible catastrophe. In places the ground was covered with smoke and steam. New and mighty steam balls puffed up toward the heavens as all around the fountains of smoke and steam and masses of soil of the grenade [artillery shells] explosions spurted upward. Into this hell, we had to descend. The pressure waves of the countless grenades that passed over, under, and all around us threw the airplane back and forth as if smashed with giant fists. Simultaneously the machine-gun bullets rattled like hail against the armored machine, as though one were throwing peas nonstop against a windowpane. We raced close over the ground—and when we saw French steel helmets at the bottom of an earth funnel we rattled our machine guns at them, and if we recognized German soldiers, then Brüggendiek drew their positions into his map.

On one of their missions over this inferno, Papa felt his observer pounding his back. He looked in the mirror and saw Brüggendiek bent over with his hands up over his head, hot water pouring down on him. A bullet had smashed the generator, and the motor would soon stop. Papa made a quick turn, hoping to get back to their lines. He aimed the aircraft diagonally to a trench so that the landing gear would be ripped off and they could slide to a stop on their belly rather than flipping over. (Otherwise, with landing gear on, the plane would have catapulted and smashed to pieces on the rough ground.) It was only a short way down, and they bellied out right over the trench barely behind their own lines. As planned, the landing gear ripped off and they slid harmlessly beyond it to then jump out of the plane. Within seconds the enemy field artillery had already zeroed in and shells were coming in, demolishing their *Blech-Esel*. With this near-death experience behind him, Papa asked for, and received, a two-week leave to visit his mother in Borowke. While there he became ill with diptheria and was bedridden as Germany fought increasingly desperate battles against the Allies. Then, on November 11, 1918, Germany signed an armistice with the Allies and the war ended.

PAPA LIVED WITH HIS WAR EXPERIENCES FOR THE rest of his life. He remained friends with many of his fellow soldiers, such as Walter Grotrian. The soldiers on the other side received the same respect for having survived their common ordeal. I think he felt a strong kinship and empathy with all combatants, bordering on affection. He had much admired the Russian Cossacks whom he had fought as an uhlan. He learned their dances and wore their costumes when he performed. He repeatedly told me of the valor and respect the British had for the Red Baron, as demonstrated by the lavish funeral they gave him. Much later, in the 1970s, when Papa worked on ichneumon collections at the Smithsonian Institution in Washington, D.C., he found companionship with an old Frenchman who owned a motel in the city. This man let Papa stay at his motel for a dollar a day after the two discovered that they had both been pilots, though on opposite sides, at the same sector of the front. I found among my things a letter that Papa wrote to me after watching the movie *The Blue Max* on television:

> It was an English production about the German airforce in World War I—two evenings, each time two hours. There were a few blemishes of minor importance in it, but, as a whole, the British have done us justice. The pictures of the battles and mainly of the dogfights were very realistic. This movie was a great experience for me.

It has stirred my emotions deeper than anything else during the last long years. I was young once more, in the midst of it all, full of love for my country, ready to sacrifice my blood and life at any hour. And with this I gained the greatest freedom a human being can achieve. For the sophisticated, realistic young man today, heroism is just stupidity. In a way this may be true for every kind of idealism. But then—I would prefer a world of stupidity. It is more rewarding. I myself would certainly be much poorer without those years of knighthood in the sky, which have formed me.

In November 1918, while Papa was still bedridden, one of Borowke's tenant farmers paid him a visit. "He was a young man in uniform, and on his chest was fastened a strange, red band. He told me that a revolution had broken out."

The war was over, but the problems that caused and fed it remained, and many new ones had been spawned in its wake. The Treaty of Versailles imposed severe punishments against the losing Austro-Hungarian side. There was also hyperinflation. Many people found that their life savings were not worth even the price of a postage stamp. People took home their daily pay in a wheelbarrow. Papa went to Berlin to visit his father and saw rioting in the streets. The mob scenes disturbed him. He must have wondered what he had fought for, especially since the war was lost. But he could not imagine that "the war to end all wars" was only a prelude to the worst war that has ever been.

From the eighteenth century until the end of World War I, parts of what is now Poland had been occupied and divided and divided again by the Austrian, German, and Russian empires. Borowke, situated in the northwestern part of Poland, near Danzig (now Gdansk), was in the German section. But the Treaty of Versailles, which marked the end of the war, essentially created Poland. In the previous 150 years, Poles had been second-class citizens under various occupations. Thus, when West Prussia (where Borowke is located) became Poland, the 1.5 million Germans there suddenly felt *themselves* oppressed. Papa, whose loyalties and responsibilities were strongest to the unchanging landscape of his home, Borowke, did not care to immigrate to Germany. Unfortunately, he had been born in Berlin; this meant that he was German and as such might not be welcome on the piece of earth he felt bound to as his home. Fear-ing that he might be expelled from Borowke, he traveled to Toruń to plead his case for Polish citizenship before "an elderly, friendly gentleman with a full whitish beard and mild-looking eyes" named Brejski who was the governor or *wojewoda* of the local government district, called a *pomorze*. Pleading for leniency, Papa, who spoke Polish, explained that his ancestors of both lines had lived in this area for centuries. The *wojewoda* shook his hand and sent him on his way saying, "Do not worry. You will receive citizenship in our country."

Papa and Oma ran Borowke together and had all that they needed in terms of food and materials from the land, but they also needed some cash to pay the hired hands who lived on the estate. Papa had some money in a bank in Berlin and decided to retrieve it. But there were tensions due to the border changes, and the roads into Germany were tightly controlled to prevent smuggling and were guarded by militiamen. Papa managed to board a train for Berlin, where he withdrew all his money from the bank, fastened the bills to his ankles, and pulled his socks up and his pants down over them. He then took the train as far as the new Polish border. From there, he hiked thirty kilometers through the night back to Borowke. Now he would be able to pay his help.

In the chaos after the war, Poland and Germany were hit with a crime wave that regressed into anarchy. Theft and robbery were common. Papa noted that onions were in short supply, and not understanding why nobody planted them, he planted fifty acres. "This turned out to be a mistake" because "as soon as the onions reached usable size, their existence became known all over the country. Every night dozens of people arrived on foot and in carriages, stealing the onions in large quantities." They even cut down the estate's trees for firewood and cookstove wood. He fought intruders nightly, firing his shotgun at looters to keep them away. Papa had been pleased to discover the nest of a crow high in the crown of an old pine tree. But when he came later to check on the babies he could not find the tree. Instead, he noticed a patch of green moss on the ground that seemed out of place. Lifting it, he found the stump of the missing pine tree underneath. Later, while following a dirt road, he realized he had not seen a beautiful old oak tree that he used to behold with pleasure whenever he passed it. Again he discovered its stump cut off at ground level and covered with moss.

Slowly the wave of anarchy subsided, and Papa could once again think about his future. He had once considered becoming a physician, like his father, but in his mid-twenties it already seemed to him "late in life" to begin that pursuit. He also considered furthering his interest in zoology with formal studies, "to follow what I had started since I came out of the crib." But he claimed that his command of the Polish language was then too poor to study there, and Borowke was now within the Polish borders, so he could not cross into Germany to study. He also felt obligated by his responsibilities to his ancestral family home. His wartime aversion to the "yoke of marriage" notwithstanding, he married anyway, but only with the up-front stipulation that love would be "free" and not exclusive. This marriage would turn out to be a lifelong relationship, though not a monogamous one and not one that would be without pain and anguish to others.

His new wife, Anneliese, was the daughter of a lumber merchant from nearby Bromberg (now Bydgoszcz in Poland). He met the family while purchasing lumber to build Margarethe a cottage. Franz Machatchek, the lumber merchant in Bromberg, lived about fifty kilometers away from Borowke, and to arrange the purchase Papa arrived at their home at noon, as the family were sitting down to their meal. He was invited to join them, and he later wrote, "It was the gayest company I had ever met." Franz and his wife had four daughters and two sons. Apparently this time Papa did not fall in love "instantly"; he visited the family several more times before he realized he had fallen for Anneliese. His intended already had a boyfriend, however, a Polish lieutenant who visited her daily. One day Papa arrived to find her not present in the living room, and asking for her was informed that she was upstairs with her beau. Papa later wrote, "Jealousy stabbed my heart like a red-hot knife."

He was determined to wait for his rival to depart. It was a long wait. Suddenly, in a fit, Papa grabbed a dull knife from the table nearby and sliced it across his hand, leaving a deep gash that bled profusely, then he left the house clutching a handkerchief to the wound. This antic was perhaps more dramatic than dangerous, but apparently it did the trick. I assume it made Anneliese think that he was madly in love with her and would endure sacrifices and pain for her. In November 1924, Anneliese Machatchek became his wife; and on Christmas day the next year, 1925, she became the mother of my half sister, Ursula.

Anneliese apparently did not consider herself attractive—she once told me that she was the "ugly duckling" of her family. I do not know if her own assessment was correct, and perhaps it was influenced by the fact that soon after they were married, Papa started an affair with her sixteen-year-old sister, Liselotte. But Anneliese had many charms. Papa leaned on her as he would on a strong mother, as someone who competently ran the household and entertained guests, and as an intrepid and capable coworker on his expeditions into the jungles. For as long as he lived, she would be his confidant and companion. And when his physical passion for her waned, she encouraged him to have other lovers, presumably because she did not want to lose him entirely. Papa felt he should not be constrained by conventions, but I don't think he thought much about the consequences, either. He lived in a world where children could be (and were) easily relegated to nannies. Indeed, later Anneliese periodically took care of me as if she were my own mother.

Papa would eventually treat Anneliese as though she were *his* mother, and I believe that in many ways she was indeed a lot like Oma. They were close, and I think Anneliese modeled herself after Margarethe. Both were extremely tolerant of Papa and both sank their roots much too deeply into Borowke. The apparent resemblance later became even more close, to my mind, when long after Margarethe had died, Anneliese came to be called "Oma." I remember Anneliese as kind and generous and the only person who always signed her letters to me with, "I love you." I remember her as thin and frail, with a high prominent forehead and wispy gray hair. She had a mole above her mouth (as I do), and it seemed to me she often had a lit cigarette at hand.

Long after Papa had been forced to divorce her and marry my mother in order to take me, his only son, with him when he immigrated to America, Papa confided to his mentor, Stresemann: "She [Anneliese] had with much female intuition recognized the strengths and weaknesses of my character. In selfless love she fulfilled my every wish and freedom. So I thank her for a long row of the happiest years of my life, and more than that: a togetherness with her, which, like the phoenix bird of the Andes, reawakened my old zoological passion to live it again."

Stresemann, in his letters to Papa, often referred to Anneliese as *die schöne Helene* ("the beautiful Helen"). In my mind, however, I associate her with the "expeditions"—Papa's wonder trips to far-off lands. He talked to me about them when I was a kid, and they were to me what television shows are to many kids now. And now I take you on Papa's and Anneliese's maiden voyage, in which she awoke his zoological interests after they had necessarily lain long buried during the war, to remerge like "the phoenix bird of the Andes."

four

Eggs, Birds, and a Panther

Only the curious have, if they live, a tale worth telling at all.

ALASTAIR REID

FOR SOMEONE WHO IS MOVED BY THE BEAUTY OF nature's creations, bird's eggs have much to offer. The colors of their shells range from chalk-white to sky blue and dark ocean blue, to luminescent green, and to shades of yellows, pinks, and browns. While some are immaculate in the purity of their colors, those of other species are decorated with odd assortments of bold blotches, squiggles, spots, and lines. One might think their shocking visual beauty to be the product of a creator who used them as canvas on which to unleash his or her creative genius. However, each color pattern has an evolutionary explanation.

After the war, Papa and Anneliese took excursions into the country around Borowke during breaks in the agricultural work. They explored the fields and swamps and woods and took up the common Victorian hobby of searching for birds' nests and collecting eggs. Usually the entire clutch of eggs was taken from or with a nest, so as to induce the bird to lay a replacement clutch, usually within a week—a principle we exploit in chickens to keep them laying an egg a day, year-round. Collectors would note the location of the nest, the number of eggs, and their condition on data slips that were included with each "set."

The insides of the eggs were blown out through a small hole in the shell, and thereafter the shell, with careful handling, would last indefinitely. Even the eggs of the now extinct elephant bird from Madagascar, when found, look as fresh as the day they were laid about 1,400 years ago.

Many prominent ornithologists from both Europe and North America began their careers by collecting eggs, gaining an education in the ways of birds through an activity that involved patience, perseverance, and scientific reasoning. The draw of oology, as this egg collecting was called, was not only the beauty of the eggs but also the challenge of finding the nests, making it an adventurous hobby that took you outdoors into the fields and forests and that gave bird watching a concrete purpose. It was a purpose packed with challenge. Through aeons of evolution, birds have evolved many tricks to keep their nests safe; for most species, "safe" means hidden. (Very few are really safe, though, because predators evolve counterstrategies to find them.) The nests are often camouflaged in such diabolically clever and species-specific ways that to find them one must follow the birds themselves through swamps and woods. Many species do their best to mislead any followers. For Papa, this was good sport. The nests of the large birds, such as hawks and eagles, were less hidden than inaccessible, and climbing to reach them was to Papa yet another physically demanding sport.

Egg collecting has long been abandoned as a nature pursuit, and most of the private egg collections have ended up in public natural history museums where they and their associated data continue to be valuable in a variety of studies, especially those involving totally unanticipated uses in environmental monitoring. They provide baselines for analysis of the distribution and abundance of toxins. Most famously, they were used for studies of eggshell thinning (which was traced to DDE, a metabolite of DDT) that caused the extirpation of peregrines, eagles, and pelicans. This induced Rachel Carson to write her influential book *Silent Spring* in 1962, and this book in turn initiated the era of environmental awareness (and a prohibition against egg collecting). But right after World War I, the future problems and the need for data to solve them were still undreamed of.

Most early natural history investigations followed the Victorian model of people like Henry Walter Bates, Alfred Russel Wallace, and

Richard Spruce—all unsalaried entrepreneurs who collected for museums and did it for the sheer joy of discovering and documenting nature's beauty and diversity. I believe Papa modeled himself after these people. Excellent experimental biology was in the offing, done by academics at universities, but this was not Papa's model. At the time, the naturalists' reputation shone brightly, and Papa by inclination and necessity gravitated toward it. He wanted above all else to go out into the wild. As the American naturalist John Burroughs said, "Take the first step in ornithology, procure one new specimen, and you are ticketed for the whole voyage."

Papa and Anneliese soon found that there were simply not many species of birds' nests to be discovered locally. During the winter they studied their bird book, *Die Naturgeschichte der Deutschen Vögel (The Natural History of German Birds)* by E. G. Friedrich and learned that many interesting species lived in the wild natural areas at the mouth of the Danube River. This area was known as Dobruja. Apparently, no area of the European continent was so generously populated by dream birds: eagles, falcons, vultures, pelicans, ibises, spoonbills, rails, ducks, and herons. Papa wanted badly to explore there, and so, on April 25, 1925, he and Anneliese set off carrying tree-climbing irons and ropes. Papa, who had still not lost his childhood love for collecting ichneumon wasps, also brought along two insect nets.

The Dobruja's Danube delta is known for its remote swamps and extensive reed jungles. The river itself is bordered by a strip of elevated land where members of a religious sect (the Lipowans), refugees from czarist Russia, built their huts and lived by fishing. They were known for their trustworthiness, their intolerance of tobacco, and their intimate knowledge of the forests and swamps. Papa hired one, named Wassily, to serve as their guide. Entering the swamps of the Dobruja, he and Anneliese found a surprise: this forest of ancient oak trees and beeches opened up into meadows covered with flowers that attracted many insects. There they found and collected many rare ichneumons that he had not seen before, to the accompaniment of birdsong.

Papa had never before encountered the bird of legend, the raven. But in this forest he heard one raven's deep sonorous croaking. He spotted the nest and could hear the rasping calls of young ravens begging for food. When he climbed up the beech tree to admire the half-grown babies, to

his joy Papa found two eggs in the nest. Shaking them, he heard the sloshing sound of their liquefied contents; they were rotten. Despite their pu-
trid smell, he pierced a hole into each end of the two eggs, applied his
mouth to one end, and blew the liquid contents out the other. The pair
would become, he claimed, the "pièce de résistance" of his collection.

At the time, however, an eagle nest up in a tall oak produced the biggest surprise. After an arduous climb, Papa managed to get up to it, but
then he faced the real challenge. The nest was very wide, and although he
was under it, he could not look or reach into it. He managed to tie himself
into the tree with a rope and, using a stick, rolled a precious egg to the
edge of the nest. As he reached over the nest edge and felt the egg with his
hand, it seemed to be strangely oblong, rather than "perfectly" round, as
most raptors' eggs are. He was even more startled when he finally got a
look at the egg, because it was not decorated with the characteristic dark
chocolaty markings. Instead this egg was—Papa blinked disbelievingly—
a bright pink, like raspberry punch! Turning the egg in his fingers, he
then found the explanation: in clearly legible letters he read two words
printed on it in English: "Too late!" He took the painted hen's egg anyway, intending to blow it out as a memento and a story for his collection,
but again he was foiled: the egg was hard-boiled and he had to settle for
eating it. A British collector rumored to be also working in the area had
beaten him to this particular trophy.

After Papa and Anneliese left the forests, most of their time was
spent in the guide Wassily's small flat-bottomed boat. Day and night for
over a month, this craft became their home. Wassily propelled it leisurely
through the swamps with a long pole, and they lived on the fish he
grabbed out of the water with his bare hands. Every time he heaved a fish
into the boat, Wassily said in Romanian, "Acuma esti oprita." Papa thought
he was reciting the name of the captured fish, and noted it in his diary.
Later he learned that the words meant, "It is forbidden." It was spawning
time, when fishing was not allowed.

Wassily took them to a colony of white herons, egrets, spoonbills,
night herons, ibises, and cormorants all nesting close together. Of the
many eggs, Papa considered the deep blue-green eggs of the ibis to be
the prettiest. But for the sake of thoroughness, he was also interested in
obtaining a uniformly white pelican egg, and so he was led to a colony of
ten to twenty nests in reeds along the borders of open water covered with

the blossoms of yellow and white water lilies. Papa took a pelican egg, being careful not to disturb the nest further (from the big birds he took only one egg, rather than the whole set). Noticing that Papa wanted no more eggs, Wassily raised his pole and destroyed the nest and then maneuvered the boat from one nest to the next systematically smashing the eggs of the whole colony. Papa, furious, tried to stop him, but Wassily was adamant—he had been instructed to demolish any eggs of pelicans he encountered, because pelicans eat fish.

Papa never mentioned how many kinds of eggs they found and brought back, but he did talk of the many ichneumons they collected from the flowers of the umbelliferous plants growing in the forests surrounding the swamps. Maybe the eggs as such were not the main prize. Papa said, "Of the many research expeditions I would make to distant lands in later years, none gave as much satisfaction and as many happy experiences as this one. My highest expectations were not only met; they were exceeded. They created in me a longing for remote, distant places." Hearing his stories, I felt the romance of poling through the reed jungle of the Dobruja. These images, which are still in my mind half a century later, created my own longing for swamps and exploration in the wild places of nature.

After coming back home to Borowke, Papa wrote up a comprehensive account of his ichneumon collections, which resulted in his first scientific publication on the subfamily Ichneumoninae. Additionally, he wrote up his bird observations, and with the latter "in his pocket," he went to see Dr. Lutz Heck, director of the Berlin Zoo, and confided his passionate desire to take part in an expedition to explore and collect specimens for the zoo. The director replied, "So, like you have many others sat before me. It's not so simple! And besides, your work is ornithological."

Papa's hopes were dimmed. Nevertheless, as he left the office, he handed Dr. Heck a copy of his account of bird observations in the Dobruja.

Months went by. One day Papa received a phone call from Erwin Stresemann, who was then the curator of birds at the Museum für Naturkunde in Berlin. Stresemann, who was also president of the German Ornithological Union (and who would later be described as the most influential ornithologist of the twentieth century), said that he had received Papa's manuscript from Dr. Heck, and he judged it worthy of publication in the *Journal für Ornithologie!*

On February 16, 1926, Papa wrote to Stresemann. He thanked him for accepting his accounts of the Dobruja birds for publication, then continued:

> Soon it will be a year since my trip to the Dobruja and with every day my desire to travel grows. I love nature above all else, especially the extensive swamps and forests that have remained untouched. You should expect to find less in me of a scientifically exact theoretician than one in whom the eagerness to explore is closely allied with the love of the practiced hunter and sportsman capable of surmounting difficulties. My position with regard to science here in Poland, as elsewhere, has become difficult, but I hope nevertheless to set out again in the next years. I would be happy to receive suggestions from you of an area where exploration would be of interest, and to receive specific scientific goals.

Stresemann answered swiftly, and suggested that Papa travel to collect birds in Armenia. Papa replied enthusiastically, "Armenia also promises, aside from the ornithological catch, something in which I have a great biological interest: the ichneumon wasps that I'm presently working on."

He continued:

> If this expedition should come to fruition, it would have to be thoroughly prepared. . . . I'd need about a month after arriving in the country to get oriented, and I'd stay until the season of the ichneumons is concluded. Once the starting date is determined, I would immediately learn the language, study the land, its people, and the birdlife. Last, but not least, would be the need to resolve the pecuniary matters. It is essential that I take a preparator [taxidermist] with me, because [one] cannot have the time and peace to do the observations and the collections while also preparing the skins.

Almost a month went by, and there was no answer from Stresemann. Finally on March 16, 1926, Papa wrote to him again. Apparently Papa had taken Stresemann's suggestion about Armenia as marching orders and was surprised not to receive specific instructions immediately.

Already in the 1920s, puzzles of animal speciation were being worked out in birds. Unlike ichneumonids, birds are convenient "model" organisms for all sorts of theoretical problems, because enough is known about their distributions and regional variation to suggest where to look for answers. Stresemann told Papa that the Great Spotted woodpecker, *Dendrocopus major*, found all over Europe, extends its range even to Persia. However, in Persia there are two similar forms. Were they two subspecies (*D. major poelzani*, with a dark brown ventral side; and *D. major syriacus*, with a white ventral side)? Stresemann thought it possible that the two forms were separate species, but if so, they should live in different habitats, since each species is adapted to occupy a specific ecological niche.

Papa was eager to try to settle the matter of the woodpecker, and so he wrote Stresemann to make plans for an expedition to Persia, rather than Armenia. "Here it is not possible to do much because of the work on the farm. . . . I did take the opportunity to practice making a museum skin, a woodpecker, since it is this group that our project is especially concerned with. I will send you the result of my efforts for criticism. I am not pleased with it myself, but I don't know where the errors are." (Actually it was Anneliese who learned to prepare specimens. She trained herself to become an accomplished taxidermist and therefore was an invaluable contributor to later expeditions.) After some delays Stresemann answered and finally made the Persian venture possible by promising to buy skins for the Berlin museum.

Mek, Papa's boyhood friend, was now a medical doctor. He had survived (though wounded) four years of war in the field artillery, and had not forgotten Papa's schoolboy integrity. He agreed to accompany Papa and Anneliese, and the three left for Persia in January 1927. With their combined strength they decided that they could "get the devil himself out of hell." By that time, Ulla (conceived in the Dobruja) was two years old, and she was left under the care of her grandmother, Oma, the ever-steady presence of Borowke.

The fauna of the Persian ichneumon was almost totally unknown. That provided an added bonus for the expedition, if not the primary one for Papa. There were no research grants to be had, and Papa had no professional position, so he had to raise capital to finance the expedition.

Together he and Mek pooled 8,000 reichsmarks, the new currency in Germany (about two months' wages for a factory worker), but they were still 4,000 short. All their expenses would ultimately have to be covered by the sale of what they brought back. They would have to meet their own costs and take their own risks.

Before leaving, Papa went to thank Heck for having introduced him, through his Dobruja manuscript, to Stresemann. Heck mentioned that there lived in Persia a rare, light-colored subspecies of the panther, *Felis pardus tulliamus*, similar to the snow leopard of the Himalayas. He, as the zoo director, wanted one. "There is not one of this cat in all the zoos of Europe," and to help rectify the situation Heck provided them with the additional financial support required to get started, along with the admonition that the best thanks they could offer him would be to bring him back a live panther—and if not that, then at the very least a picture of one. Heck lent Papa an expensive camera worth over 1,000 marks, to take on the expedition. Papa confided, "I have always tried very hard, when there were hopes put on me, to not disappoint." I think that was true, whenever something scientific was at stake.

Papa, Anneliese, and Mek departed by rail from Berlin via Moscow to Baku on the Caspian Sea, and then continued from Baku by boat to the Persian port of Enzeli, which had just been renamed Bandar-e Pahlavī, after the newly chosen shah of Persia, Mohammed Reza Pahlavi. This was a small town with one small hotel, the Grand. Papa described it as a hellhole that was up to par with a European hotel "only in price." Here they stayed a couple of weeks to get organized. Mek, according to Papa, "always an ardent student of the opposite sex of any nationality or social status," soon found a woman who helped him pass the time. Papa also got lucky. He made the acquaintance of a young, well-to-do Persian named Ali Dadashoff, who helped make their arrangements. Dadashoff hired six horses to carry their supplies and also a *sharvand-dar*, or groom. His name was Mohamed Aga, and he became "Mandaga" for short. Papa said he was "trustworthy, indefatigueable, friendly, and always jolly." Papa had learned to speak Farsi and also tried to learn the customs of the land. He met Mohammed Uli, who informed him that on Ramadan, the Muslim month of fasting, nobody is allowed to eat or drink between sunrise and sunset. But he offered a remedy: the German expedition could make "the night into day"—work at night and sleep during the day.

60 THE SNORING BIRD

After the baggage was split among the six horses, Papa, Mek, Anneliese, and Mandaga began their long march through the forests to the summit of Pish-Kuch. They encountered many denuded hillsides, where charcoal had been harvested. The results were, Papa said, "Like a caterpillar plague—humankind has chewed on the happy green forest to leave a total baldness and an ugly desert."

As they approached higher elevations, they encountered dense shrubbery, and then blackberry thickets. Papa told a native they met along the way that he was searching for panthers, and asked if there were any in these woods: "Yes, many," was the eager reply. Papa had not yet learned one other Iranian custom (also an American custom, it seems to me): that it is polite to answer so as to bring happiness to the listener, "without regard to the truth."

On March 12, 1927, Papa wrote to Stresemann: "As you see, we are still in the same 'nest.' Allah willed it. [Mek] Dammholz and I are both down with fever. Luckily my wife is able to care for us. . . . Tigers and leopards are no longer found here, but there are supposed to be some still near Astarabad." Six weeks later the men had recovered from their fevers, and Papa reported from the Elburz Mountains:

For four weeks we sit, literally, in the middle of wooded mountains. Our tent is in deciduous woods at a spring, about 800 meters [2,400 feet] in elevation. There is no human habitation for miles—only forests, forests without end. In theory it is a little spooky, but in actuality, one feels safe as at home. The climate is crazy. Two-three days the sun burns, then one sits six days in clouds, and masses of water stream down and bring with them an unpleasant wet coldness. The consequences: first I came home one evening after the hunt with a lung infection, which (again!) made me incapable of fighting for fourteen days. And now, for the past five days, my wife lies sick in bed. We don't know what's wrong. As soon as she is well we will go—either up to the snow, or farther to the east. Despite all difficulties we—the healthy—have this month produced 200 mostly well-made skins. I myself have spent sunup till sunset on my feet, hunting like a wild man up and down the steep hills . . . while Dammholz makes the skins also from sunup till sunset. I am astonished that till now we have not encountered a single bird which is not also found at home. One has the impression that you go to Persia to meet all your relatives?! Maybe you could pass this letter on to my mother. I'm too exhausted to write her separately.

A week later, on May 5, 1927, he wrote that they had just received the first mail—one letter from Stresemann and one from Margarethe. "We are now all healthy, and hopefully acclimated to this difficult weather. I have very thoroughly explored the area where we have been staying. There should not be any bird species that has escaped me." The team packed and moved their camp a couple of days later.

Two weeks later, Papa again filed a report to Stresemann, a long and detailed letter on bird life. They had just made their new camp in Pish-Kuch, still in the Elburz. Papa always had time to write when they first arrived at a new camp, because it was a time of easy hunting. "Yesterday," he wrote, "I reached the last, from here, reachable mountain section. Up there I finally got the field lark and also the redpoll—a very cute little bird, but quite shy, and to get two specimens cost hours. I think you will hardly be able to imagine what extreme efforts the securing of this little mountain bird is connected with! After we break camp next we will go slowly again to the coast, and then farther eastward again into the mountains. . . . Still no sign of the panther."

A village chief and a mullah came to visit the camp. Papa shook the hand of the first, and then that of the second also—"the last with a malicious pleasure because I knew that he would consider bodily contact with a Christian as a befouling. But the mullah came prepared for such a possibility: a boy with a water jug stepped forward and poured the water, before my puzzled eyes, over the defamed hand that I had held."

Papa described another surprising Muslim practice he encountered in Persia:

During the holidays of Shah Hussein-Wah Hussain the people engage in a fanaticism of imagination that borders on insanity. Huge crowds of men clad in white shirts down to their knees, with knives and long curved daggers, were hopping wildly and using their weapons on their own heads so as to bleed profusely all over themselves. The crowd, and the frenzy, grew throughout the day, as those who fell to their knees were dragged off. Others bloodied their naked backs with whips. The mullahs wept.

If one had behaved similarly on a street in Berlin or New York, this person would be deemed stark raving mad. But here thousands were carrying on as though it were normal. Papa, who had seen his share of frightening spectacles since becoming a soldier at the age of seventeen, wrote that this was "the most frightening scene that I had experienced in Persia." Was he contemplating human irrationality and anticipating what could happen anywhere, maybe even at home?

But he also realized how bizarre, and ridiculous, he surely seemed in the Persians' eyes: "It must have been a strange sight, as we three—and especially me, with my wavy, black, full beard and short pants—armed with butterfly nets, were springing in great leaps hither and yon over the fields chasing insects." Once, as he and Anneliese were catching insects in a field, two men stood nearby and watched them in wonder. Eventually one approached and asked, "Why are you catching bugs?" Papa had an answer. "I'm a doctor and I catch them to make medicines from them." They seemed satisfied with this, but then asked: "Do you know how to distinguish the useful ones from the others?" "Of course," he replied, so they watched closely for a while to see which bugs he and Anneliese chased and which they ignored. The two men withdrew, talked, and came back to show Papa a tiny insect. "Do you know this one?"

"Yes—easy." Papa didn't know the Persian word for "louse," so he scratched himself. The two men grinned.

As the team climbed higher and higher into the mountains, Papa wrote:

Hour by hour the forest trees become taller and more ancient. Mighty beeches, maples, elms, and alders are in the majority. Once in a while there is also a tree species that reminds us that we are not in a woods at home. . . . The leaves are not yet unfolding and in the gray fog on the rainy days they all look extraordinarily knobby with bent or twisted boles and hanging branches all covered with moss and lichens so that they look truly spooky.

Luckily there are sunny days once in a while, and precious minutes are those when such a day is announced at daybreak by a flaming red glow. At those times the whole dawning forest is filled with a huge birdsong symphony, as occurs in the northern but not the tropical forests. From all sides—from the trees as well as the brush and the ground—sounds the high rapid cackling of the wren, then the exulting beats of the finch join in, the uninterrupted delicate and soft chatter of the redbreast, the loudly fluting blackbird, and the affable sound of the various titmice.

High up on the mountain, the next camp was established in a stand of hawthorn bushes alongside a brook that trickled down from the snowfields above. Papa, Anneliese, and Mek saw bumblebees flying from flower to flower on the vegetation mats just beneath the snow. The bees' furry pelts were white, like snow, decorated with black bands and touched with red or yellow.

Pure water was scarce here. Papa took a drink from one mountain stream only to discover to his shock that someone upstream had recently defecated. He wrote:

> One of the Mohammedan commandments is to wash after relieving oneself. To do this the religious go to water . . . so all Persian water is polluted. Mohammed's law was meant to serve cleanliness and hygiene but its consequences now are contrary to the postulates. It's the same as a poisoning and in many cases the cause of terrible epidemics, especially cholera and typhus. This is another example of the tragic contradiction between the purpose and the letter of the law, between living thoughts and dead dogma.

Whether or not the water was potable, the mountains were a cool pleasure compared with what came next, as the team hiked with their horses 200 kilometers down to and then along Persia's Caspian Sea coast. "The heat in summer was indescribable, and our march on totally shadeless paths was hellish," Papa wrote. "Here it was so hot that even grasshoppers could be close to dying of thirst." When they finally made it to a village, Papa drank some tea in the local teahouse and found it "a pleasure of such intensity that I will remember it perhaps a lifetime."

Hordes of mosquitoes tormented the team at the next campsite. Papa reported that after a sleepless night, he spied hundreds of them perched on the inner walls of the tent, and he noticed that unlike the culex, the mosquitoes he was familiar with in Poland, these insects had the elevated abdomens and hind legs peculiar to the malaria-transmitting Anopheles. Soon, all three Germans were infected, even though they had been dutifully swallowing their quinine tablets, the only known antidote at the time. Nonetheless, they continued their march. There was no other option. It was sparsely populated country lacking motorized transportation. At one point Anneliese sank down behind a large rock and asked Papa to let her die there, but he and Mek got her up and they struggled on, finally reaching the city of Astarabad. Here their loyal Mandaga found them shelter and a *hakim* (doctor). Papa described Mandaga as "the best helper I have ever had on my numerous expeditions." The *hakim*, "a man of unusual kindness of heart," took them into his house and did everything he could. He collected herbal medicines and brewed and applied potions. Anneliese and Mek soon recovered. However, Papa was infected with a particularly virulent kind of malaria and he continued to decline, despite taking huge doses of herbals and quinine.

One day, while in a half stupor, Papa heard Mandaga talking about a *palang*. The word caught his attention. He had used it many times during the past months in talking with herdsmen and farmers. It meant "leopard."

Leopard! He was awake now.

"Mandaga, what is this about *palang*?"

"Sir, in this town there lives a military officer who owns a living panther."

"Go to him right away and ask if I may buy it!"

The next day while Papa was without fever (malarial fever is cyclical), Mandaga came to him, his face beaming in happiness, and he set a half-grown leopard on the bed. It was a gorgeous animal, with blue-gray eyes and the typical leopard spotting on a pale yellow background: the long-sought panther. But disappointment followed as quickly as hope had arrived: Mandaga admitted that the officer (a major in the Persian army) who owned this leopard had declared that it was not for sale. He had sent it only to be admired.

A day or so later the major himself paid a social call. He casually asked about the prices of all of Papa's possessions—guns, camera, tent. . . . Then he left.

Days went by, and then the major gave Mandaga an amazing bit of news: he had decided to give Papa the panther! Papa, however, was by this time somewhat familiar with Persian customs. While they were up in the Elburz, a man had appeared at their tent offering a few eggs and wrinkled apricots. No one was interested in purchasing the wares, but when Papa came back from hunting, the man was still there. Papa turned away, but Mandaga said, "The old man wants to make you a gift!" Papa immediately felt shamed. He took the eggs and the apricots and thanked the old man profusely in all the Persian that he could muster. Still the man didn't leave. Finally Mandaga explained to Papa that a big person is expected to give to one more humble, and the "bigger" one is, the greater should be the value of the "counter-present" that one gives.

Papa knew that the major expected a counter-present for the "gift" of the panther. But he couldn't give away their guns; under Persian law, they had to be accounted for on leaving the country. Heck's camera was far too valuable; nor was it his to give away. So, Papa instructed Mandaga to thank the major profusely, and to tell him that he could not offer him a suitable counter-present.

A few days later the major returned. This time Papa's high fever was back, but he sensed that the meeting would be a crucial one, so he marshaled his strength. He had a table set up with a samovar to serve his guest the customary tea. Then the customary small talk ensued. Given his fever, the difficulty of conversing in a combination of Russian and Persian, and the long, seemingly pointless yet important discussions of the weather, winds, and so forth must have been unbearable. Finally the major got around to mentioning, as an aside, that the tent, the guns, the binoculars, and the camera would be a suitable counter-present for the panther. Papa patiently reiterated the prices of all these things, and the major politely disputed the correctness of the prices, while Anneliese was declaring in the background, *"Unverschämtdich, schmeiss den Kerl raus!"* ("Outrageous! Throw the bum out!") Papa's mind raced feverishly as he tried to figure out how to rescue what now seemed like a hopeless situation.

He summoned his strength and said: "Major, if you give me the panther as a present without demanding a counter-present, then I will see to it that there will be a big sign with your name on it by his cage in the Berlin Zoo. Millions will read this sign, and you'll be famous."

The answer to that proposal was instant: "Sir, if you give me your camera and tent, I will also put a sign with your name in front of them as the friendly giver of these presents."

At that point Papa informed the major that he was too sick to negotiate any further. The tent he could offer, but more than that he could not, and would not.

The major blinked, the deal was sealed, and that evening they exchanged their presents. From then on the *palang* became "Peter." He moved in and settled at the foot of Papa's sickbed.

The malaria attacks became worse. Papa realized that he would surely die. He arranged with Anneliese to be buried in Astarabad, saying that he could "become one with the earth" just as well in Persia as at home. Even if help could be summoned, there was no medicine other than quinine, of which they had plenty, but which was ineffective against this strain of malaria. Finally, in desperation, Anneliese found a small telegram station near Astarabad and sent a message to the German ambassador in Tehran. It seemed like a pointless thing to do, because even if the ambassador sent a physician, it would take him at least two weeks to cross the Elburz Mountains. And even if one did come, what could he do?

On the second day after the cable had been dispatched, Papa said he heard an improbable sound—of a plane. A Junker? He wondered if it was a dream brought on by fever. He thought of the air battles he'd fought over the Somme. But the steel "song" of the plane only got louder and louder, and then it stopped. Moments later, Dr. Höring, the physician of the German embassy in Tehran, was standing before the bed. The doctor said that they had come to take him to the hospital in Tehran. Then he added: "We have a new remedy for malaria. It is called plasmoquin, and this new drug kills the blood parasites of the tropical malaria variety where quinine fails." This new medicine had just been invented by the Russians, who had gotten it into Tehran.

Just before Anneliese's telegram arrived in Tehran, the pilot had landed in that city with a load of anti-cholera medication for fighting one of the periodic epidemics that had broken out in the country. This pilot, Herr Mossbacher, was just preparing to return to Germany when the ambassador, Count Frederick Werner von Schulenburg, asked if he would risk making a landing in southern Persia to save an explorer, Herr Heinrich, who was deathly sick. The pilot replied that he would, but only if the embassy would guarantee covering the costs that could be incurred in an accident during the risky landing in the hinterland. As it turned out, the ambassador could make that guarantee. A rich businessman named von Cramon was visiting from Germany, and had been present at the conversation. Herr von Cramon had spoken up, "That can't be anybody else but the son of my old doctor, Hermann Heinrich." He volunteered to shoulder all responsibility for the possible loss of the airplane.

Papa later visited the ambassador after his recovery, to express his deep gratitude. "No fairy tale of a thousand and one nights can be so fantastic, so colorful, as at times life itself," he wrote.

Seventeen years later, in 1944, Count von Schulenburg and about 500 others were hanged on butcher hooks, tortured, and killed by the Nazi SS. Von Schulenberg had, along with General Erwin Rommel and other Wehrmacht (army) officers, been active in the Resistance, which was exposed after the attempted assassination of Adolf Hitler on July 20, 1944.

FROM PAPA'S BED IN A TEHRAN HOSPITAL, HE could look out onto a sunny garden. One day, he saw a woodpecker on an old apple tree. It looked to him like a bird he had collected in an almost treeless valley of the Elburz. Papa, who had his trusty gun with him, got up out of bed and shot the bird.

It was indeed the same kind of woodpecker, and here was proof that this was not a geographic subspecies of *D. major*, the bird of deep forests (which he had also collected on the trip), but a different species altogether. Stresemann had wanted to clarify the relationship between these birds, and now Papa was able to provide the needed evidence. The bird he had just shot in the hospital garden favored a habitat ecologically different from that of the woodland woodpecker. It lived in open woods, like those in parks, rather than deep forests. In current bird books that I have consulted, it is now indeed listed as a separate species, the Syrian woodpecker, *Dendrocopus syriacus*; and Papa wrote that "it was the last bird that I collected on the expedition to Persia."

The Syrian vs. the great spotted woodpecker.

On the return voyage through Bandar-e Pahlavī, Papa again met up with his friend Mandaga, who handed over the much grown but still tame and friendly Peter the panther, whom he had been taking care of while Papa recovered. Papa, Anneliese, Mek, and Peter were to return to Berlin by train via Poland, but immediately ran into trouble because there was some rule about not allowing large predatory animals on a passenger train, even though it was obvious that the gentle Peter posed no threat. Finally, after daylong negotiations and much baksheesh (palm-greasing), one official had a brilliant flash of insight: "We will just not write 'panther' in the freight declaration; we'll just say 'crate' instead!"

And so the crate went clear through Russia, through Baku, over to Moscow, and on to the Polish border. Everyone along the way knew and loved its occupant and spoiled Peter with fresh meat. The train officials, the train drivers, the cook from the dining coach, and especially the Polish customs officials in Stolpce, received him in the most friendly manner, and let Peter as well as the rest of the team's exotic collections pass on through.

At last in Berlin, one day in late September 1927, from the direction of the Kaiser Wilhelm Gedächtnis Church, a convertible came down Budapest Street. The panther perched in the front seat must have peered curiously at the gaping city throngs as Anneliese, Mek, and Papa escorted their spotted guest to the Berlin Zoo, where he was received "with open arms" in the office of Dr. Heck. Peter was a showpiece there for many years until he and the zoo and most of its inhabitants were obliterated by bombs in the war that was soon to come.

five   Ichneumon Hunts

I found it and named it
Being versed in taxonomic Latin;
Thus became godfather to an insect and its first
Describer—and I want no other fame.

—VLADIMIR NABOKOV

A FEMALE ICHNEUMON WASP, PROTICHNEUMON *pisorius*, is on the most important mission of her short life: finding a caterpillar. Most probably she is hunting for one that blends in with the foliage or bark where it hides and is almost invisible to human eyes and to the birds that have hunted these caterpillars relentlessly over millions of years. Birds hunt by sight; the ichneumons hunt mainly, but not exclusively, by scent. Not any caterpillar (or its pupa) will do. Like the over 6,000 other species of the Ichneumoninae—all of them parasitize Lepidoptera (moths and butterflies)—this particular *Protichneumon* will search for something very, very specific. For her, it is the caterpillar of a green sphinx moth that is artfully camouflaged like a leaf. This ichneumon has a bright reddish-brown abdomen. Her thorax, legs, and antennae are black, but they are adorned with splashes and rings of lemon yellow. She is beautiful and she is large—at over 2 centimeters long, she is one of the largest ichneumons occurring in Europe.

There are a huge number of species in the Ichneumoninae, because

Female ichneumon wasp.

almost every moth and butterfly in the world has at least one of these wasp species as its parasite. Like birds' eggs, most ichneumons are strikingly colored in combinations of reds, yellows, blues, and browns, and black and white. Their color variety is endless, and so are their shapes and their body sculpturing, which Papa used to identify and describe them. As some moths and butterflies are rare, the ichneumons—their parasites—are necessarily much rarer still. Beyond the subfamily Ichneumoninae lies the rest of the family Ichneumonidae, which includes probably well over a million species in numerous subfamilies. These other members of the family parasitize a wide variety of insects and their larvae, including Lepidoptera, among others. Although most species are rare, collectively they play a vital ecological role in preventing or reducing caterpillar plagues that can, and occasionally do, defoliate forests.

Wasps are relatively ancient insects. They first appeared about 150 million years ago in the Jurassic period as contemporaries of the emerg-

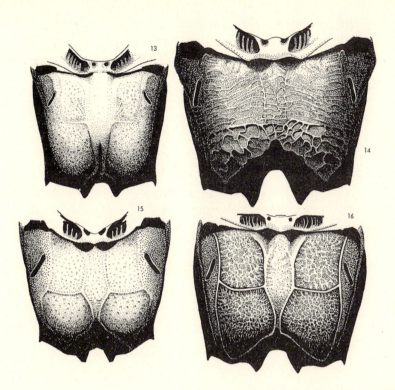

The sculpturing of the thoraxes of four different species of ichneumons. (Figures 13–16 in volume 1 of Papa's *Ichneumoninae Stenopneusticae of Africa*. These drawings are by Erich Diller.)

ing dinosaurs. Bees, which derived from wasps, did not appear until the Tertiary period, after the demise of dinosaurs and the rise of flowering plants.

Each ichneumon wasp has a narrow window to the world. The males have their antennal scent receptors tuned to the females of their species. The females in turn are tuned to hosts in which to lay their eggs. The wasps do not react to variables that do not affect them. A female flies resolutely in search of her host with her front legs tucked snugly against her body, like an aircraft's retracted landing gear, and her two sets of hind legs trailing behind. Her main caterpillar-detector is a pair of thin, black-, white-, or yellow-ringed antennae that she stretches out ahead. They are studded with tens to hundreds of thousands of olfactory receptors narrowly attuned to the small number of chemicals that identify her specific host or the kinds of leaves it feeds on, or both. Capable of detecting even just a few molecules of the scent, she flies upwind, zigzagging if necessary to relocate caterpillar scent.

When she finds her caterpillar, she may land on or near it and vibrate her antennae even more rapidly, all the faster to encounter and sample more scent molecules. She touches her intended victim with the tips of those antennae, and then drums them along its back. The caterpillar has no idea what is happening and may hardly notice as the wasp straddles it with her six spindly legs, bends her abdomen at her narrow waist, and thrusts her sharp syringe-like stinger—her ovipositors—through the caterpillar's thin cuticle, or skin. She will next inject a single small white egg into the caterpillar's body by propelling it through the hollow canal of the inserted stinger. (Polyembryonic parasitoids take the process to

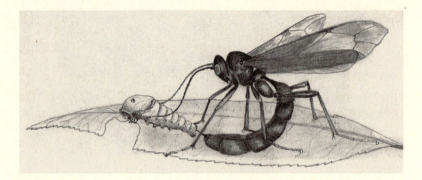

Ichneumon female in act of laying her egg in a swallowtail butterfly caterpillar.

exponential extremes; in a trick of physiology rivaling that of the movie *Alien*, a single ichneumon egg can yield 700 or more individual tiny wasps.) Then she flies off, having completed her mission, while the caterpillar feeds on, oblivious that its fate has now been sealed; it will now "hatch out" a wasp (or many wasps) rather than a moth or a butterfly. In most cases the egg within the caterpillar soon hatches into a small white grub. First the grub feeds on the caterpillar's body fat. Then, it feasts on the vital organs, until little is left of the caterpillar but the skin. And just so, little things have large consequences. This is why, technically, ichneumons are called parasitoids rather than parasites. They are essentially predators that eat their prey from the inside out, unlike, say, a flea or a louse.

The discovery of parasitoids (and parasites that spread disease) was perhaps a shock to naturalists of the Victorian era, who were motivated in their studies by the belief that they were unveiling the grandeur of God's creation. In the year after he published *On the Origin of Species*, Charles Darwin wrote to his friend Asa Gray, who was a leading American biologist, "I cannot persuade myself that a beneficent and omnipotent God would designedly have created the Ichneumonidae with the express intention of their feeding on the living bodies of caterpillars or that a cat should play with mice."

We now know vastly more than Darwin did 150 years ago, and there is no more need to posit an intelligent and caring designer of the universe than a sadistic one. We know now, for the first time in human history, a breathtaking creation story that ties us all to the same earth by the same roots, regardless of who and where we are. It takes us to the beginning of the universe, 14 billion years ago, to the beginning of the atom. Only the story of evolution—a coherent narrative that unites many dozens of empirically proved theories—provides a scenario that has the power to explain, without exception, every one of the millions of organisms that exist on earth. This is not to say that science is superior to religion. Religion is superior to science in that it can be a balm to master pain, evaporate fears, empower conscience, and rationalize away doubt.

The specifics of how the many different kinds of wasps other than ichneumons have evolved from solitary to highly social animals (such as the familiar yellow jackets and hornets) have interested many scientists since Darwin. Most wasps (thirty-seven of the fifty-two families) make their living as solitary parasitoids, and it is therefore likely that the life cycle of ichneumons is the least derived—i.e., the ancestral or "primitive"—type. From examining the variety of forms and habits of all the fifty-two families of wasps (of which the ichneumons are only one), we now can write a tentative scenario of the wasps' evolution from parasitoids to the social, i.e., the most derived.

The trajectory leading from grub-eat-host parasitoid to social being probably began when some wasps provisioned their young in a hidden place, with several prey items to feed on. These prey needed to be immobilized, yet still kept alive to retain freshness. That became possible when the wasps injected their prey not only with eggs, but also with a chemical acting like the anesthesia we now inject into patients to put them under during an operation. Once the prey were chemically incapacitated, hiding them in specially prepared nests became ever more necessary so that birds would not eat them. The provisioned nests gave an additional advantage—they permitted the laying of numerous eggs in one place and allowed longer time periods over which food could be brought in. This in turn led to overlapping of generations all in one place, and that set the stage for social life, where a dominant female lords it over several thousand of her offspring. The ovipositor now was usurped for an additional function. Social wasps, having heaps of riches that would attract predators, needed to be able to mount a defense against more than birds, and they evolved poisons from their prey-tranquilizing chemicals. These (along with appropriate atavistic behavior necessary for delivery of such poisons—as anyone who has ever bumped into a hornet nest can verify) were specialized to produce pain as a deterrent to large nest-robbers. Immobilization of prey now became unnecessary, because there were enough nest members to help provision the young with fresh food continuously, and the larvae then were fed chewed-up prey in direct mouth-to-mouth delivery.

LET US RETURN TO THE *PISORIUS*, WHO ON THIS sunny day flies through a forest of slender birches, over ground covered with moss and sprinkled with wildflowers. Like a beautiful little ghost, she would normally pass completely undetected, leaving no trace. But this is an unlucky wasp. Although her general name, ichneumon, comes from the Greek word for "tracker" or "hunter," this time it is she who is being tracked and hunted by one who has long been specialized for this task. Papa, seeing her land on a leaf and seeing the alertness in her palpating yellow-ringed antennae, moves closer, and carefully and slowly raises a net in his right hand. He holds it steady for a second as the wasp pauses, then swoops the net through the leafed branch and over her. In a seemingly absentminded gesture that I would make use of much later, Papa swings the net once more around his head to make sure the wasp is swept to its bottom. In a continuation of this graceful gesture, he then turns his wrist and flips the net over its metal bow to collapse the netting and close off the opening. Now he lifts the net to the sun to examine his captive. The wasp, buzzing against the white gauze near the bottom of the net, starts to crawl upward. Holding the net with his left hand, Papa grasps her in a fold of net and reaches in with his right hand to retrieve her. She tries to sting him, but her ovipositor barely scratches the skin of his finger and her caterpillar-tranquilizing venom produces only a mild irritation. Papa puts her unceremoniously into a small medicine bottle filled with small paper strips, cotton, and a couple of drops of ether. Within a few seconds, inspiring the ether through the tiny spiracles on each side of her body, the wasp stops moving.

Papa undertook ichneumon hunts around Borowke whenever possible, often in the company of various female companions. They would pace about twenty feet apart along the edge of the woods bordering a field of rye dotted with blue cornflowers and red wild poppies. Reaching their carriage at a secluded spot in a clearing, they would picnic on homemade bread with butter and sausage from the farm. Afterward they probably reclined somewhere on soft moss and admired their catch, as Papa did with me decades later. Then they mounted the carriage, slapped the reins on the horses' backs, and retraced their way back to the Borowke homestead. As they drove, they would have chased up basking peacock butterflies and tiger beetles, the horses' footsteps clip-clopping and the metal-rimmed wheels swishing along the sandy path. Approaching home, they passed under the arching branches of rows of horse chestnut trees, which in spring were resplendent with small white, yellow, and pink blossoms that soon rained down like snow. Margarethe often sat here painting as her pet deer rested nearby.

Papa then usually retreated into his study with the day's catch. The walls were lined with cases of his precious zoological books, full of hand-painted plates of moths, butterflies, beetles, and birds. Along the upper edge of the walls, extending all the way around the room, was a fresco, painted, of course, by Margarethe, and depicting the local flora and fauna. Settling at his desk, Papa took up his forceps and, opening the first medicine bottle, carefully removed the strips of paper, the cotton, and finally the *pisorius*. When all were laid out in front of him, he could begin sorting.

In addition to the wasps, he also collected various other insects for his own interest and for friends and colleagues with whom he had ongoing exchanges. Arraying them, each on a thin sheet of cotton, in cigar boxes, he included a label with each; a tiny white card-stock tag on which he wrote the date and locality of the capture in neat, almost microscopic handwriting, using a tiny quill pen dipped into a well of black India ink. India ink is indelible, and only ink that would last through the ages was acceptable. The catch was then covered with a sheet of paper, and then another layer of cotton. Thus packed, the insects required no other immediate attention. They quickly dried and were preserved. Later, he could retrieve the specimens, lay them on moist sand, and cover them with a jar to hold the moisture. After twenty-four hours he would retrieve the softened insects to be mounted, pinned, and labeled for the collection in glass-covered cases.

Papa's prize catches were given special attention immediately. The wasp would be secured to a sheet of pressed peat moss using a pin passed through the center of the thorax. The pins were made specifically for this purpose and could be purchased in different sizes, depending on the size of the insect. The appendages of a freshly caught wasp could be moved, as they were still supple and loose. After the wasp dried, it would be as brittle as a flake of baklava and could not withstand handling.

Papa would spread out the wings and legs by tugging and pulling them very gently into place with a pair of small thin stainless steel forceps. Then he built a scaffolding of about ten to fifteen additional insect pins, by which he fixed the positions of the wings, legs, and antennae. The antennae and front legs were set to reach ahead; the second two pairs of legs pointed to the rear; and the wings, two on each side, were joined together and spread to the sides like a butterfly's. (Later, to save space, he developed a new style of having the wings pointing up and back.) Only about five or six specimens of each sex of any one species made it into Papa's collection. Once in, a specimen stayed. And is there still.

A day or two later, after the specimen had dried, Papa pulled out the scaffolding of pins; the wasp was removed from the prep blocks; and a data label was created and attached to the bottom of the pin holding the insect through the thorax. Thus the precious wasp was transferred to its final resting place, a spot in a row with other members of the species, within a nearly airtight glass case.

As I was growing up, I found it hard to understand why Papa would lavish so much attention on the pinning and spreading of his wasps and their preservation. Later it became clear to me that for him, each specimen was not just a piece of data. Robert Carlson at the Smithsonian Entomology Division, where Papa once worked, told me that Papa "was so emotional about his ichneumons. He really loved his animals." Indeed, my most common nickname at Borowke was "Pise," presumably coming from the aforementioned *Protichneumon pisorius*.

Late in his life, on the farm in Maine, Papa was still hunting ichneumon wasps and spending hours examining them in excruciating detail, matching the specifics of their exoskeletons, color patterns, and dimensions with those of specimens in his own collection and with those in museum collections all over the world, as well as with descriptions in the international literature. It appeared to me that he spent an inordinate amount of time meticulously spreading the wasps' wings, legs, and antennae into precise positions and writing a tiny tag for each and every wasp—all for no obvious purpose that I could discern or appreciate then.

But there was a purpose, an ethic, even a morality to what naturalists of his time practiced. After they cut their teeth collecting eggs, beetles, or butterflies, they went on to become ornithologists or entomologists. Some collected insects of the very "difficult" groups such as flies—or ichneumons. Mastering these groups required extraordinary dedication and patience. One man who had perfected this craft and may have been a *Vorbild* for Papa (a role model who would inspire him and keep him on his path) was Baron C. R. Osten-Sacken. The passion of this Russian-born nobleman was crane flies, which are often mistakenly called "giant mosquitoes" and engender horror stories, though they don't bite. There are about 11,000 species of crane flies (a couple of thousand more than the total number of bird species). They are one of the smaller groups of the order Diptera (the flies), just as Papa's Ichneumoninae are a subfamily of the family Ichneumonidae (the ending "-nae" indicates a subfamily; "-dae" connotes a family), which is only a tiny section of the order Hymenoptera (wasps, bees, and ants).

Despite his narrow specialty—just one small group of the flies— Osten-Sacken became known as the "father of American dipterology," and during his stay in the United States in Washington as the Russian consul general (1859–1877) he was the only dipterist in America. He was succeeded by an American, Charles P. Alexander, who collected all varieties of flies and described 10,800 new species of them by the time he was ninety years old (in 1979). Both men, like Papa, had picked the right group—they never ran out of new species to discover and describe, even in their old age.

They took their work seriously. Osten-Sacken devoted over 100 pages of his memoir, *Record of My Life Work in Entomology* (1904), to his disagreements with colleagues about what he called his "most important discovery": enunciating the value of bristle arrangement in fly taxonomy, for which he invented the term "chaetotaxy." (It means "arrangement of bristles on an insect" and is still an important taxonomic tool.) One of his detractors, a Professor Brauer, pointed out that the doctrine of distribution of bristles had been put into practice in 1873 by Professor Joseph Mik and Dr. Hermann Loew, long before Osten-Sacken had "invented the learned term chaetotaxy."

This fracas, which seems humorous to us now, illustrates the systematic attention to and importance of minute detail, which extended even to the calligraphy used on labels. The dipterist Hermann Loew could crowd fifteen lines into an inch with his exacting pen, in longhand script; and Osten-Sacken, despite his quarrel with Loew, considered him to be "not only among the heroes, but also the martyrs of science."

Method in science, then as now, was often enshrined as though it were an end in itself rather than a means to discoveries. One senses the stern seriousness with which the practitioners regarded their science. Terms like "duty" and "strength of will" crop up repeatedly in their work—and these were words I had heard Papa use many times. For example, Osten-Sacken accuses Loew of not "doing his duty" when he mistakenly named as a new species a fly that he should have recognized as an existing one. Papa seems to have followed Osten-Sacken's example of exactitude, and he tried to instill it in me.

He taught me the proper protocol for collecting beetles when I was about six years old. I could not, for example, use just any pin to secure my specimen. It had to be one of the long narrow insect pins of just the right gauge fitted to the size of my beetle. Specimens could not be pinned haphazardly; they were to be secured only through the right wing cover, about a third of the way down its length. The legs had to be teased to the sides and held there with scaffolding in some semblance of order until they dried. Most important, each specimen had to have a data label attached to the bottom of the pin below the specimen, and that label needed to have the date and location of capture written in India ink. Finally, the specimens had to be arranged in orderly rows and columns in the collection box, with name-tag labels pinned at the front of each column of a species. At the time I gave these procedures no thought. I knew that this was the way it was supposed to be, and so I did it. When I was a teenager and had already spent quite a lot of time hunting ichneumons, I think Papa assumed that I would continue his tradition. I remember once sitting across from him in his study at our farm in Maine as he prepared the day's catch. He told me that in addition to my lifelong hobby, I needed to choose my *Vorbild*. I think he was intimating that it should be someone very like himself. I did in fact train to become an entomologist, but I would have been incapable of concentrating again and again on such minute details of body sculpturing, color, and dimensions. For many years, Papa's obsession with what were (by then) academically very obscure details, the same details he had attended to for decades, seemed bizarre to me. Yet decades later, I began to appreciate that the habits instilled in me a sense of "doing it right." Excellence for its own sake.

The researchers of Papa's generation felt blessed to be at the forefront of science, an endeavor noble enough to call forth their uncompromising zeal. Papa also wanted to make discoveries of new species. Heeding the advice he'd received at age sixteen, he knew that only the ichneumons, the unknown insects, promised science, and so he pursued them all his life. In *On the Fronts of War and Science*, he wrote:

> When I close my eyes I see the dream—small pictures of my imagination—the demonic small shy creatures who in secret are busily flitting in the foliage thickets of distant forests—I see pairs of white-ringed antennae vibrating in nervous haste, I see red- and black-banded animals with a white spot at the tip of their abdomen, blue-black bodies with yellow spots or those with black- and white-banded bodies, appearing and disappearing. My imagination re-

produces the thousands of pictures of ichneumon wasps that I have hunted with indefatigable passion and it conjures visions of such that I have never seen but hope to catch, or discover, someday. Visions of ichneumons lure me into the hazy future, and I follow them—who knows where to, still.

six

The Snoring and Drumming Rails

I did not tether myself. I let go entirely and went,
  I went into the luminous night—
And I drank of potent wines, as only the
  Valiant of voluptuousness drink.

—FROM "I WENT," CONSTANTINE P. CAVAFY

WHEN PAPA RETURNED FROM HIS EXPEDITION
to Persia in 1927, his marriage to Anneliese had cooled and he had started
an affair with her sixteen-year-old sister, Liselotte Machatchek, whom he
hired as his secretary. Ulla, handed over to Oma during the Persia expedition, had become "Oma's little girl" and was just two years old. Papa had
met his benefactors' expectations by collecting many rare and interesting
birds for Stresemann, and he had brought back the rare panther for Dr.
Heck at the Berlin Zoo. Of course for him the lasting treasure of that expedition was his collection of ichneumon wasps, which clearly inspired
him, because from 1927 to 1930 he kept his secretary busy dictating
twenty-one papers based on his collections from Dobruja, Poland, Germany, and Yugoslavia.

Persia awoke in Papa a realization of the extraordinary variety of species, something that can really be appreciated only by someone who is
collecting in many areas and immersed in the details of discrimination.
He and Anneliese—with whom he apparently kept his promise of friend-

ship, as she kept hers to him—and now Liselotte as well, were restless and eager for new opportunities and adventures.

Thinking that formal studies might enhance his career, Papa at the age of thirty-two spent a year (1928) in Berlin studying zoology and taking a course in botany, leaving the management of Borowke to an administrator. But academics did not suit him, even though he had been a star student at the gymnasium almost two decades earlier. I suspect he was restless and not used to taking instructions from anyone, given the intervening life of adventure. He turned his attention instead to acquainting himself in person with European ornithologists and entomologists. He approached them, and through common interests, doors were opened to him. He visited and spent days with the ornithologist and pastor Otto Kleinschmidt at Wittenberg; he was the originator of the concept of the circle of races, which I learned about in graduate school. (*Rassen-Kreis*, such as that of herring gulls around the North Pole, where different contiguous races are capable of interbreeding, except at one end of a circular distribution where they treat one another as different species and do not interbreed.) Papa visited the British Museum of Natural History in London to see his friends and fellow ichneumon specialists Geoffrey Kerrich and J. F. Perkins. While in England he also visited Lord Rothschild at his castle, where a green carpet of lawn sloped down to the Rothschilds' famous Tring Museum. There, Papa conferred with the museum's two curators. One, Dr. Ernst Hartert, was a renowned ornithologist; and the other, Karl Jordan, was the world's authority on fleas and tenebrionid beetles. Lord Rothschild personally guided Papa through his collection of some of the greatest zoological rarities in the world, including the only existing egg of the extinct giant auk, and a specimen of the rare Wallace's rail from Indonesia—a species that Papa would soon go through much travail to obtain. Stresemann and Sanford were both particularly interested in the relatively unexplored Malayan bird fauna, in the area where the "Wallace line" runs between Celebes and Borneo. This area is the meeting point of Indo-European and Australasian animal faunas, and was the place where, arguably, the theory of evolution was first articulated.

The British naturalist Alfred Russel Wallace had collected in the Malay Islands for eight years in the mid-1800s and had brought back 125,660 animal specimens, including 212 new species of birds and 900 new species of beetles. Wallace noted a curious continuum among bird species; and one night in the field when he was lying in bed sick with fever, he experienced a eureka moment. He wrote down his sudden exciting insights on the origin of new species, based on his lifetime of observation and collecting.

Wallace's concise twenty-page manuscript elaborated what we now call the theory of evolution. In 1858 he sent his manuscript to his idol, a more senior naturalist named Charles Darwin, for evaluation. Darwin had already written thousands of pages that skirted the same topic of how species came to be, but he had no theory formulated for the mechanism of speciation. Wallace provided the key. Darwin, who had pondered the question for decades, now saw it clearly and succinctly articulated, as though it were his own thought. Within a year he wrote *On the Origin of Species*. His influential friends helped see to it that their colleague at Cambridge got the credit for discovering the concept of evolution, while Wallace, the amateur collector, was given a consolation prize: credit for the discovery of the "Wallace line."

Papa would have the good fortune to roam in Wallace's Indonesian paradise, and the path was paved by this meeting with Sanford at Stresemann's house. Papa wrote (in *Der Vogel Schnarch*):

It is a warm summer evening in the suburbs of the city. The evening meal has just ended in the villa of the Berlin zoologist Professor Stresemann, and he has just settled himself with his guests in the study. One of these guests, Dr. L. C. Sanford, the well-known American ornithologist and passionate promoter of the collections of the American Museum of Natural History in New York, is leafing through a scientific work of more than biblical dimensions—"Birds of Celebes by [A. B.] Meyer and [L. W.] Wigglesworth" can be read on the title. Suddenly the American scholar turns to me, and while his right forefinger thrusts several times in succession to a colorful illustration, he says slowly and with emphasis: "This bird you must get. . . ." In this moment there shoots through me an intense pleasure, because I immediately understand the meaning of those laconic words: Dr. Sanford has decided to accept the recommendation of my honored maestro, Professor Stresemann, and will entrust me with the execution of a zoological expedition into the wilderness of Celebes!

The illustration Sanford referred to in Meyer and Wigglesworth's book was of a rail named *Aramidopsis plateni*. It is a species of flightless rail that had been caught for the last time thirty years earlier in the virgin forests of north Celebes. Sanford was interested in the bird because he had a fondness for rails, and because this particular rail had been described on the basis of only one specimen. *Aramidopsis* was thought to be extinct, or at best very, very rare. Finding it would be Papa's mission. Although Sanford wanted *Aramidopsis* above all else, he also hoped Papa would come back with Wallace's rail, *Habroptila wallacei*, known from only two specimens collected by Wallace on the island of Halmahera in the Moluccas. (Halmahera is only 100 miles from Celebes—or Sulawesi—and adjacent to Ternate, where Wallace had his great insight.)

The contract for the expedition, signed by Papa as well as Stresemann and Sanford, stated that Papa should stay at least eight months and bring back a minimum of 3,000 bird skins. This time he would receive an advance. He would receive $5,000 to cover expenses for food and travel for himself and two others. Even in 1930, $5,000 to finance the whole expedition was not much. Papa wanted to employ two bird skinners, one cook, and porters for carrying equipment and supplies.

On March 16, 1930, accompanied by his tried and trusted crew of Anneliese and Liselotte (young Ulla, now just four years old, had again been left home in Borowke, in the care of Margarethe), Papa landed in the harbor of Makasar on the southern peninsula of Celebes. Papa wrote to Stresemann immediately: "I obediently report a happy landing on the 'Eilund' of Celebes. I used the journey over to think." The result was a four-page letter in which Papa wrote of his plan to go first for about five months to the Latimodjong Mountains, with their several peaks over 10,000 feet, where he hoped *Aramidopsis* might be found amid the dense forests.

The trio, filled with enthusiasm and eager to begin their mission, immediately visited the Dutch governor of the island, and looked up Paulus Moninka, the native hunter who had captured the only *Aramidopsis plateni* three decades earlier. The old man thought that he remembered the bird as being from "around the Minahasa." Papa thought that surely Moninka was senile and forgetful; professional ornithologists had long assumed *Aramidopsis* to be a forest rail, most likely found in the vast virgin jungles in the high mountains, not in the coastal swamps.

After hiring two helpers—Tiga as cook and Salong for general work—the party, with nineteen horses carrying the equipment and food, started its march from the village of Kalosi. In the evening, after traveling through hot, dry grasslands, they stopped at the village of Pasui to visit the residence of the *aru*, or prince, to ask for his help in procuring porters. Fifty were standing at the ready the next morning. They took over the loads from the horses and began their ascent into the Latimodjong Mountains.

"We entered an untouched, inaccessible, virgin forest of a kind I had never seen before; not a tree had ever been cut by human hands," Papa wrote in *Der Vogel Schnarch.* "One man had to work at the head of the column to slash the lianas obstructing the way with a sharp machete."

At an elevation of 6,000 feet the long spiny lianas disappeared. The trees became shorter and grew in fantastic shapes as a consequence of the enormous cushions of moss growing on them. The party had entered a zone of constant rain that was enveloped in thick clouds. The ground was soggy and covered with thick carpets of moss, alternating with patches of dense low plants. Papa considered this a good possibility for a place where *Aramidopsis* could have remained hidden and decided to camp there. The porters were paid and asked to return in a month, when the march to the higher elevations would continue. The hunt began.

Ten years later Papa wrote about this hunt in an article for the *Journal für Ornithologie:*

When hunting birds in the moist and cool forests of the high mountains, one very seldom hears a single bird sound during many hours of wandering. One gains the impression that the woods are dead and devoid of birdlife. Suddenly one hears faint peeps and chirps coming from somewhere in the tree crowns, and if one is at home here and has learned the voices of the birds, when one comes closer one recognizes the calls of a half dozen or more different species. One then sees the lively small birds flutter from twig to twig and wander from tree to tree. The previously barren forest seems suddenly filled with colorful and cheerful birdlife. For a few more minutes it swarms, chirps, and peeps all around and then the forest slowly sinks back to its previous silence: the wandering band of small birds that we had met has passed over us. On countless occa-

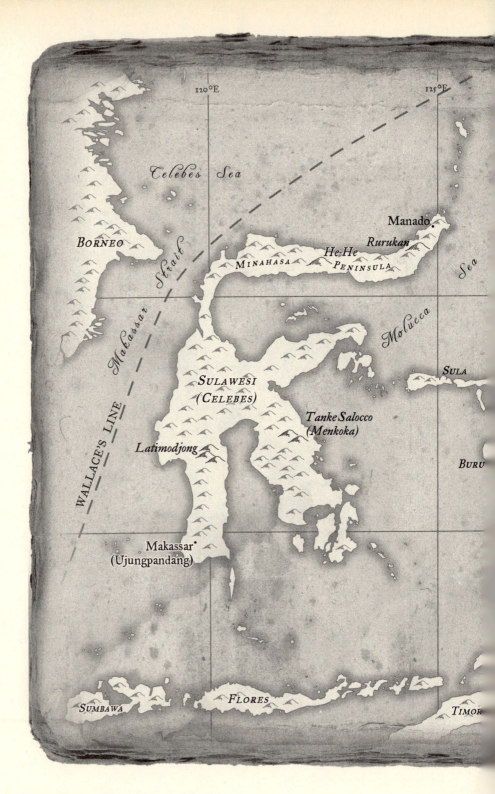

Showing the area of the rail hunt in Indonesia.

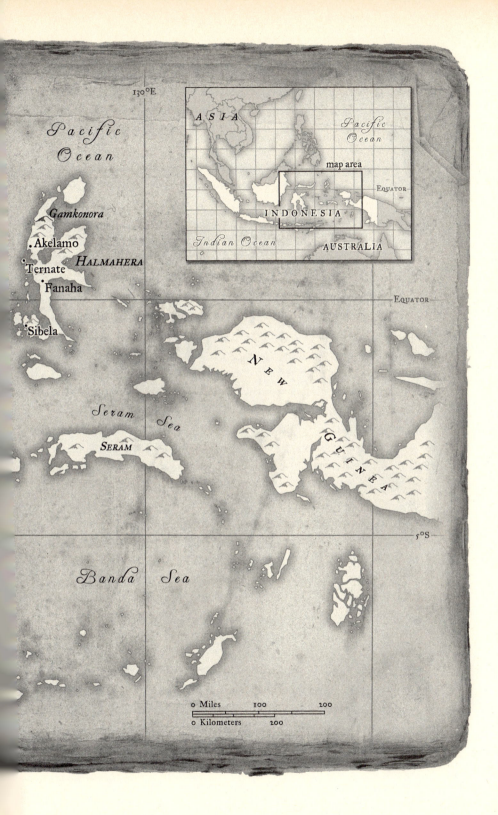

130°E.

*Pacific
Ocean*

ASIA

*Pacific
Ocean*

map area

EQUATOR

INDONESIA

Indian Ocean

AUSTRALIA

Gamkonora

Akelamo

Halmahera

Ternate

Fanaha

EQUATOR

Sibela

Seram Sea

N E W

SERAM

G U I N E A

5°S

Banda Sea

0 Miles   100   200

0 Kilometers   200

sions I have had the opportunity to observe such bird wanderings in the Latimodjong mountains, as well as in the Matinan and Mengkoka mountains. They form a thoroughly typical feature of the high elevation forest.

Papa went on to explain that these bird swarms were probably not found at lower elevations, because the small species of Passeriformes that live in tree canopies are absent. He also wrote that the flocks reminded him very much of the winter flocks of mixed species, including various species of titmice (chickadees), tree creepers, and nuthatches found at home in Europe. He had seen these flocks in Celebes in June and July, but also in November through January, and he therefore speculates that in Celebes they are not restricted to certain times of year. He also discussed at length how some bird species have distinctly different voices or dialects in different parts of their geographical range.

Papa wrote Stresemann several long letters filled almost entirely with discussions of birdlife. The camp in the cloudy moss forest had proved to be spectacularly productive, with Papa collecting *Mizza sarrasinorum*, a honeysucker the size of a starling; and also a flightless snipe, *Scolopax celeberisis*, known from only one specimen. Both were on Sanford's list of most desired species. But when Stresemann wrote back, he reported that Sanford had visited him again and still wanted Papa to make *Aramidopsis* a priority:

> As soon as you get one *Aramidopsis* telegraph us, since Sanford's good mood depends on it. Sanford is pessimistic, since it has dawned on him that there are two women on the Heinrichi expedition. He seems to have had negative experiences with the female sex, and it is up to you to thoroughly cure his misogyny. Best, by reporting to him that your wife caught the *Aramidopsis* in her bare hands.

Almost three months later, on October 11, and still in the Latimodjong, Papa wrote, "Rain, rain, rain every day. But surely no *Aramidopsis*!" Stresemann answered: "Dr. Sanford recently wrote to me that, as before, he wants *Aramidopsis* above all, and that you should not leave Celebes, before his dream, *Aramidopsis* for the New York museum, is fulfilled."

calligyna ♀

simplex ♀

picta ♀

The females of the three subspecies of
Heinrichia calligyna, as described by
Stresemann.

The expedition's results continued to be quite substantial, despite the lack of *Aramidopsis*. The collection that Papa sent to Stresemann in Berlin contained five new species of birds, two of which Stresemann named in honor of Papa: *Geomalia heinrichi* and "the one with the beautiful female," *Heinrichia calligyna*. (The males of this species are plain dark blue, but the females have a pleasing pattern of blue, russet, and dark brown with a white eyespot.) The catch also contained a new genus of rodent, a small ground squirrel with a long snout and a short bushy tail (later named *Hyosciurus heinrichi*).

But no *Aramidopsis*. So higher they went.

The porters appeared punctually on the appointed day, and in only twenty-four hours the expedition reached the summit of the Latimodjong. There, they entered another world altogether. The porters, who had never been this high on the mountain, shivered in the cold, and as evening approached Papa's concern was that they should have a comfortable shelter for sleeping. He had them dig a ditch. Dry grass was spread out on the bottom, and a canvas was laid over it for a roof. They were on a plateau covered with scattered oak trees, and the only flowers were blue forget-me-nots, like those in Europe. They found few birds other than a thrush (*Turdus celebensis*), similar to the American robin.

The many weeks of hunting for the rail appeared to be futile. Most of the previous expeditions to Celebes had explored the lowlands, but the American mammalogist H. C. Raven had explored and found a wealth of new forms at higher elevations, and that is why they had concentrated their search there. Was *Aramidopsis* perhaps a local species on another, more isolated, mountain?

The team decided to leave the Latimodjong and climb the He-He. In the misty, mossy regions of the He-He, they found the same bird species as on the Latimodjong, but also a new species of flycatcher, which Stresemann named *Cyornis sanfordi* in an attempt to appease and please the expedition's anxious benefactor. On December 9, as they were leaving the He-He, Papa wrote:

> We have lived for the last two weeks only for *Aramidopsis*. I have showed dozens of people pictures—none knows the curious beast. Set nets. Set more traps. Three hunters have spent all day crossing all the thickets. We went out at night with lights, and we took only a

few birds in order to not scare off the lovely rail by the noise of shooting. In short, we tried everything, yet still without effect.

Meanwhile, Papa had received a letter from Stresemann ten days earlier advising that the group take a rest (Papa had written long letters describing frustrations, sickness, and mental stress) and also reporting that Sanford had sent $2,000, even though he again expressed concern about the women's being a possible liability. Stresemann commented: "If the expedition should end prematurely owing to the bodily failing of the two women, then that would be a sort of triumph for Dr. Sanford and a blemish on me, since I had set great confidence in them. May you spare me this worry!"

Stresemann need not have worried. Both women wrote to him personally expressing their enthusiasm for the expedition. Contrary to Stresemann's fears, I believe they were the pillars of strength and determination. Having myself served as an expedition taxidermist, I can vouch that they did the hard part—the daily grind of sitting at a table, often tormented by insects and the weather, while painstakingly skinning and stuffing. Papa could do none of this. He was totally dependent on them, and would never have ventured out if it were not for them. Furthermore, with women he was the boss—and probably acted that way.

On Christmas day Papa received a batch of six letters from Stresemann, and in answer reassured him that *Aramidopsis* would be the primary goal. The next attempt to find the mysterious rail would be in the region called the Minahasa. But, "the civilization of the Minahasa, from all information that I have, is frighteningly advanced and with it, deforestation, so there seems little hope of success. Still I will try it. If I'm unsuccessful, then I'll carry the costs myself. *Aramidopsis* has become a compulsion with me, exactly as it appears to be that of the honored Sanford!" He detailed travel plans to the Rurukan Mountains in the Minahasa and mentioned having exceeded by 100 the 3,000 skins specified by the contract. He had used up 5,230 rounds of ammunition—he was apparently a very good shot. He asked for more ammunition, and the bottom line was still his all-or-none concern with the rail; he ended with the words, "To meeting again in one to one and a half months with heaven-high jubilation, or depressed to death."

On reaching the Rurukan a week later, Papa was tending toward the latter: "The civilization of the Minahasa exceeded all fears. Asphalt roads crisscrossed this 'jungle province of Holland.' " And, two weeks later, with rainy season at its peak, "I get at best one to two birds per day, and then usually not the ones wished for, so one grows despondent after awhile."

Papa decided to break camp only when he had thoroughly explored all the forests of the Minahasa:

> Only then can one with relative confidence say that *Aramidopsis* no longer exists. When six good hunters, fired up with enthusiasm by high prices for rewards, have for months hunted in all forests without sighting the animal—only then is it probably not there. You cannot know how much worry all this gives me and how nervous this hopeless thing makes me. Please, for my nerves, leave me time and patience to look thoroughly and be prepared that the wished-for result may not come true. The "by-product" of this *Aramidopsis* hunt promises not to be too valuable. Expect to receive only all the remaining rails, owls, pitas, and thrushes. . . . C'est tout—
>
> For the *Aramidopsis* program there is at least the rest of February, maybe even half of March. Nothing helps. Halmahera will in any case be carried through. To leave here before all has been tried would for me be unbearable.

In mid-February 1931 Papa wrote to Stresemann, speculating about all the places where the bird might be. "I have to admit I've become very skeptical about our further efforts." At the end of the month he wrote, "*Aramidopsis* truly does not exist here in the Rurukan." And then he speculated: "Could *Aramidopsis* perhaps actually be a swamp rail, not a woodland rail, as Sanford had presumed?"

AFTER AN ENTIRE YEAR OF FUTILE SEARCHING, papa finally decided to search instead for Wallace's rail, *Habroptila wallacei*, which Sanford also wanted. They began their search on Halmahera. Halmahera is situated between New Guinea and Celebes and is in form a miniature Celebes, with four long arms reaching into the sea. It also has a volcanic mountain, Gamkonora, which Papa expected to be covered with jungles all the way to the top. Instead, they found dry sparse woods, surely not a rail habitat.

Might *Habroptila* be in the well-populated lowlands? It seemed unlikely that this rare rail would be near civilization, but if so, then perhaps some local hunters might have seen it. Papa made a sketch of the black bird, with its bright red bill and red legs, from the skin he had examined in the Tring Museum. He showed this sketch to native hunters, asking if they knew it and had a name for it. If so, then its habitat or "address" could be found.

Wallace's rail, and the various native names that Papa thought might refer to it, eventually crossed off one by one.

Papa's picture acquired dirty fingerprints, and a gradually increasing succession of names—iiki, *waringiti*, *gabura*, *goortongo*, and *soissa*—were scribbled on the back of it. Each name led to a wild-goose chase (literally, as most informants confused the bird in the picture with ducks, common rails, or pigeons), and one by one the names were crossed off.

On May 10, 1931, Papa wrote to Stresemann from Halmahera:

> We sit in our swamp camp, half eaten by mosquitoes and gnats. I was led here by the reports of the natives, who after studying the beautiful Habroptila picture agreed that it was "iiki," who is here, but rare. After a few days I was on the track of iiki and knew his strange call, which one hears at the most seven times per week. Shirt, boots, socks, and hat I leave hanging on the nail so that when I hunt I myself appear to be half swamp rail. Finally got "iiki." Exactly as in Celebes. Nobody knows *Habroptila*. It is miserable! I am again in despair. Where shall I go when every track, every hold, fails?

Of all the names that people had given Papa, only one had not yet been eliminated. This was *soissa*. One day a native hunter who loitered at camp told Papa, "*Tingal dimana banjak duri*" ("He is at home where there are many thorns"). He claimed that *soissa* lived near Akelamo in a narrow sandy coastal strip of fully flooded sago swamps where it would be difficult to find an island big enough even to put up their tent. It might be another false lead, but it was their last hope.

It was pointless to wear boots in the swamp water, because every step would have been audible for a long way. It was also senseless to wear a shirt, because the long thorns from the sago palms would soon tear it to shreds. To hunt this presumably shy bird, it would be necessary to remain stoically quiet, while being pierced by thorns and bitten by swarms of mosquitoes. Last but not least there were land-living leeches. If one paused even for a moment, the leeches came from all directions, seeking a blood meal.

For a long time when I was younger and heard Papa tell of his adventures, I thought that this was an exaggeration meant to entertain. Now I'm not so sure. Shirts were not dispensable, especially in the wilderness and on a very tight budget. Furthermore, as I studied the photograph of the thickly bearded "wild man" with the bush knife in his right hand and the shotgun in his left, I saw plenty of formidable thorns. In the picture they looked, if anything, even more threatening than he had described them.

Days of hunting in this "green hell" went on and on. Neither Papa nor any of his hired hunters saw or heard anything that could have been *Habroptila*. It continued to rain almost constantly, and the team was nearly flooded off its little island. Had *soissa* itself been driven away by the high water of the rainy season?

One day a small old man came from the village of Fanaha, about a day's walk to the south, because he had heard that an *orang blanda* (a white man) was crazy with the wish to hunt the *soissa*. This man claimed to have seen its tracks in the mud, set snares, and caught and eaten the bird. "Are there more *soissa* there?" Papa asked, thinking the man worse than a cannibal. The old man allowed that possibly there might be more, deep in the swamp. Papa offered a reward for catching one or providing the critical clues that would lead to its capture.

A few days later, the man returned, very excited, in a pouring rain. He had heard the *soissa* the evening before, he said, and could lead Papa to the spot. As if possessed, Papa immediately got his hunters together and asked the man to lead them there.

The thickets at Fanaha were even denser than those at their camp at Akelamo. Sharp finger-long thorns lay embedded in the mud. But the hunters entered the thickets at once. Papa hunted among palm bushes where the water was two to three feet deep in places, and he saw a large dark body submerge—a crocodile? Perhaps. But one thing was sure: at the end of the day, no one had seen or heard the *soissa*.

The next morning Papa instructed all of his six men employed for the search to split up and hunt for the *soissa*. All day he saw and heard nothing, but as he returned in the evening, Papa thought he heard a drumming sound at some distance out in the swamp. What could drumming possibly mean in this wilderness, where there were no people for miles?

Slowly the other hunters straggled back to camp, empty-handed. The last to return that evening was Salang, whom Papa considered his best and most trusted employee. With shining eyes Salang exclaimed, "*Ampar dapat!*" ("Almost got it!") And then he told of how he had suddenly seen what could only be *Habroptila wallacei*, the black bird with red bill and red legs. It had stood next to a sago palm, picking at something on the ground. Salang had lifted his gun to shoot, but with lightning speed the bird had disappeared. "Only his voice remained—it sounded exactly like a drumroll. . . ."

"We were drunk with excitement," Papa said. "Until late in the night, all the hunters chattered among themselves, and as I fell asleep, I [and presumably Lotte and Anneliese as well] could still hear Salang mimic again and again the drumming call of the rail." They were so close now. Whoever got the bird first was to yell.

The next day the men gathered where the bird had been heard, and they hacked lanes through the thickets where the ground-hugging bird could be seen by a vigilant hunter. They also built low walls of twigs and palm fronds with holes at ground level spaced several yards apart, where a bird might slip through, and at these spots they set snares. At dawn the hunters dispersed to hunt in the swamps nearby.

The next morning as the hunters were checking the snares, Papa heard a jubilant yell. They all knew what it meant, and they crashed through the thickets toward it, to find a smiling hunter lifting a black rail with a bright red bill and red legs up into the air: *Habroptila wallecei*. There were congratulations all around. They then caught its mate, and Papa immediately instructed the others: "Order the carriers and a boat, we leave these terrible swamps tomorrow!"

When the hunters stepped into the canoe to leave, they again saw the same little old man who had led them and who had remembered the reward owed him. Papa handed him the promised coins and said: "A funny call your bird has—he sounds almost like a drum." "So it is," the old man answered, "and from that the bird has its name, since *soissa* means in the language of my tribe nothing other than drum. . . ."

Papa was exultant, and quickly wrote to Stresseman from Akelamo, District Oba, Halmahera:

Dear Herr Professor,

The two last days were extraordinarily eventful. The hunt for *Habroptila* was an immensely hard piece of work; one could call it martyrdom. But all the details later. Today only the event: a pair of *Habroptila* are well prepared and drying. I would have liked to get a third for the Berlin Museum. . . . *Habroptila* lives, naturally, not in the forest! I am solidly confident that before me no European has ever seen this rail alive, for that requires such a degree of toughening and such demands on oneself as I cannot so easily attribute to others. [Presumably Wallace got it from one of his hunters.] *Hab-*

roptila is shielded by the awful thorns of the sago swamps, and on top of that is exquisitely rare. In this thorn wilderness, I walked barefoot and half-naked for weeks.

It was with immense satisfaction that Papa, Anneliese, and Liselotte then left the sago swamps to embark on the voyage back to Makasar, and, they hoped, home. Anneliese had come down with a heavy fever on the day the *Habroptila* were brought back to camp, and despite prophylaxis and visits by a doctor who showed up in a motorboat, she remained sick with "double tertiana malaria." Papa had contracted a nasty foot infection in the *Habroptila* swamps and told his friend Stresemann, "You are now reading the last report of the 'expedition Heinrichi,' which in four days completes its work. Tomorrow we leave Sibela." He reported that his toe had turned violet and he had opened it with a razor blade, finding a thorn inside—"a memento of the *Habroptila* hunt."

In Makasar Papa found a euphoric letter from Stresemann awaiting them. Their collection, he said, contained "fantastic discoveries." And new financial support was promised if they only would continue. Richard Archbold, an avid American mammologist who was interested in obtaining mammal skins for his collection, chipped in another $10,000 to keep the expedition going after the first funds ran out.

In the letter Stresemann also elaborated about a caprimulgid (a member of the "goatsucker" family, along with nightjars and whip-poor-wills) that Papa had thought was the well-known *Caprimulgus celebensis*. In fact it turned out to be "a wonderful new species of nightjar—fabulous luck!" *Eurystopodium diabolicus* became its species and genus name; satanic nightjar and devilish nightjar became two of its common names. But in the most recent books, it came to be called Heinrich's nightjar.

Today you can find it, specimen number 3972 of the expedition, in the American Museum of Natural History in New York.

A few years ago, while visiting my alma mater, UCLA, I stepped into the office of a friend and colleague, Jared Diamond. The topic of the Indonesian birds came up, as Jared, along with Ernst Mayr, was then working on a book about the biogeography of the birds of northern Melanesia. When I mentioned my father, Jared got up, opened a drawer, and to my great surprise pulled out a reprint of a paper by himself and D. K. Bishop, "Rediscovery of Heinrich's Nightjar" (1997). Bishop and Diamond wrote

Heinrich's nightjar.

that the forested region where Papa had found the bird (on March 5, 1931, at the base of Kelabat volcano in the Minahasa), had since been cleared and replaced with coconut plantations. The bird was sighted and identified (presumably because of the unique small white spot on its wings that can be seen in flight) sixty-six years after its first and last recorded sighting, in a small piece of forest that remained undisturbed and protected as the Lore Lindu National Park. It is still positively known only from Papa's one (female) specimen. It is now illustrated in two recent books on the nightjars, where we learn that it was collected by G. Heinrich.

PAPA WAS SO BUOYED UP BY THE NEWS OF THE DIS-covery of an entirely new species of nightjar that he, Anneliese, and Liselotte felt renewed enthusiasm for continuing the expedition for a little while longer. He made immediate plans and began writing a long letter to Stresemann. As I read this letter, I had the impression that it had been suddenly interrupted, as with a jolt, because there is a long line scrawled clear across the page and underneath it is written:

Now comes a surprise—new report. On Halmahera I developed the conviction that *Aramidopsis* also is not a forest rail, and therefore I presumably looked for it in the wrong places. Now on my return I had a three-day stay in Manado and visited the Minahasa. There one of my native helpers delivered the news that *Aramidopsis* has been seen. . . . I think I would get her, if I could devote six to eight more months to the task. I am ready to again start on this demanding hunt, but only after we return to Europe. I only tell you this so that you can talk of plans with Sanford.

Papa's letter crossed in the mail with a letter from Stresemann that said:

Maybe news has reached you of the catastrophic economic conditions in Germany. Even in your paradisiacal seclusion this might force you to abolish your wish for a speedy return. Your opportunities for lectures have worsened extraordinarily, since most businesses and societies have no more money and will not be able to honor their contracts. But otherwise welcome back home!

Papa didn't have much choice. They needed to return home to recoup their strength and acquire new funds—but he still had several more days to hunt while waiting for the boat.

Their exhausting search had lasted nearly two years. Though *Habroptila wallacei* had been found, *Aramidopsis*, the "holy rail," appeared unattainable. To pass the time Papa was now hunting bats at night—one of his favorite activities, as he continued to have a small private collection of bats—and while using a flashlight to search for a bat that had fallen to the ground, his hunting partner called, "*Tuan, ada ularitan besar*" ("Sir, here is a big black snake"). Papa saw an extraordinarily long snake that, in the beam of the light, proceeded to raise itself slowly up in front of him. In the light the snake looked shiny, like polished metal. Papa's mind quickly calculated the possibilities: The snake doesn't look like a poisonous viper; it looks more like an adder; not poisonous—except for *Naja*, the cobras—but those aren't in Celebes—even Governor Caron told me so personally in Makasar at the beginning of the expedition. So there is no reason not to grab this snake and catch it to add to the collection of live snakes and birds (most of the birds died in passage from an outbreak of avian flu, as reported on May 8, 1932 in *Hamburger Nachrichten*) for the Berlin zoo.

Papa told his assistant to keep the snake in the light as he put his gun in his left hand to keep his right free to grab the snake behind the head. He stepped forward—the snake raised itself still higher in obvious alarm. "*Ati, ati—ular mara*" ("Careful—the snake is angry"), the assistant warned, just as the snake lunged forward. In the same second Papa raised his gun and fired without taking the time to aim, but the snake collapsed into a writhing mass. It had been hit by only a single pellet, but that pellet had hit its only vital spot, the head. Angry at himself for having been provoked and even panicked enough to shoot and kill this beautiful animal needlessly, spoiling the opportunity to bring it back to the zoo, Papa grabbed the still wriggling snake and shoved it into his knapsack to take to camp and surprise Anneliese and Liselotte.

A few minutes later he stepped into the tent. The two women had just finished their work. One was reclining on her camp cot writing in her journal; the other was cleaning her fingernails. "What do you have?" They wanted to know, expecting some kind of bat. "Today something special!" Papa said, as he reached into the rucksack and threw the snake onto the bed. They screamed. "No fear," he said, "it's only a harmless adder."

"I don't believe it," Anneliese said. "It has a definite mean and creepy look."

"OK," Papa replied. "We have a little alcohol left. Let's pickle it and get it identified."

And so it came to the zoological museum in Berlin. To everyone's great surprise it was identified as a new species of cobra! It would be named *Naja celebensis*. Cobras are among the deadliest snakes in the world. They are aggressive, and one bite is usually fatal. Papa wrote, "Only then did I learn that my life had hung on the proverbial 'silk thread.' " And it was certainly not the first time, much less the last.

As they were making the final preparations for their departure home, and not having anything else to do, Papa went once more on a hunt. He wrote near the end of *Der Vogel Schnarch*:

With quiet, strong cuts, my bush knife slices here and there through the tendrils and vines that block the way. Carefully, each footstep gropes for a place where it can descend soundlessly. Carefully, inch by inch the body pushes, sometimes erect, sometimes crawling, through the thicket. I have to fight for soundlessness in every step by force of will. This is the stalk in the jungle thicket! There—what is that—a sound in front of me somewhere in the endless chaos of plants, a rumbling that resembles the sound of the black-warted pig—no, more like a deep snoring! In this second, I freeze into total immobility, and as the sound is repeated once more, I sink down inch by inch, submerging into the plant thicket. It hums again, first farther, then closer, and closer. Minutes pass—or hours?!—I do not know any more. I consist only of two alert ears and two eyes that in terrible tension try to penetrate the thicket. There—there in the distance—something moves—a couple of blades of vegetation move, a leaf vibrates. . . . My hands clamp the shaft of my shotgun . . . again it moves . . . a reddish bill? . . . The shot rips through the silence . . . Have I hit and—what is it?! I leap in bounds through the thicket aimed for the spot. There lies a bird, a bird with black and white stripes under the wings and with a strong, partially red bill—*Aramidopsis plateni.*

He had called Wallace's rail the drumming rail because of its voice; and he now named this one by its voice as well: the "snoring bird." And he continued:

After years spent poking all alone through the solitude of virgin tropical forests, I have acquired the habit of occasionally talking aloud about whatever occupies my thoughts. While I was wandering back to the tent with, in my rucksack, the most priceless catch that I have ever hunted or will hunt, I recalled the memory of that early summer day in Stresemann's house, the moment of the contract of this hot and long battle for this zoological rarity. I murmured to myself again, again, and again: "Doctor Sanford, I've got your bird!"

Papa finally got the real prize: *Aramidopsis plateni,* the rail that Sanford wanted "über alles in der Welt," and on almost the very last day of the expedition. However I could find no letters referring to it. I presume he chose to deliver the bird personally, as a surprise.

Papa must have recounted the moments of getting the rail a million times over, in his head, for the remaining years of his life. Certainly he told us the story many times, bringing the hunt to life as he relived the excitement of finally attaining the quarry he had struggled so hard for. This was, for him, the crowning moment of his career as a collector. But from my perspective it seems odd that *Aramidopsis*, and to a lesser extent *Habroptila*, assumed such mythical importance. Papa and his party discovered a whopping six new species of birds, of three new genera that had never been known to exist until then. Yet these new species generated little excitement, relative to the two already known rails.

But rails are special. They are an embodiment of the wild and inaccessible nature hidden from human senses. They are otherworldly birds, because they live where few people venture, and where those who do go seldom get to see or hear them. They very rarely take flight (both *Habroptila* and *Aramidopsis* are ground dwellers), and they spend their time stalking on mud amid dense vegetation. If lucky, one may know of them from their voice, which is usually heard only at night. There is always more than is seen, and they come to dwell in the imagination to join the bestiaries that may grow, thrive, or die there.

Throughout all his meanderings in search of the rails, Papa had persistently collected ichneumon wasps. The collection would grow steadily, with many completely new and fascinating species of wasps. Two years after his return, a 263-page manuscript, "Die Ichneumoninae von Celebes," emerged from its host expedition. However, Stresemann would henceforth refer to Papa as *die Ralle* ("the Rail").

Responsibility for publishing the results obtained from the birds brought back from Celebes fell on Stresemann. Between 1931 and 1938 he would write a series of nine papers titled "Vorläufiges über die ornithologische Ergebnisse der Expedition Heinrich 1930–1931" ("Preliminaries Regarding the Ornithological Findings of the Heinrich Expedition") in *Ornitholische Monatsberichten* (*Ornithological Monthly*). Finally, in 1939, he allotted a total of 461 pages in the last consecutively printed installment in the leading ornithological journal of the time, *Journal für Ornithologie*. In 1941, this manuscript fell into very good company, alongside the works of the Nobel laureates Niko Tinbergen and Konrad Lorenz. That company also included other leading lights in the revolution of biology that marks this century and that will never be repeated. These were

Label on the snoring rail specimen.

Otto Heinroth, Ernst Mayr, and Gustav Kramer. In the correspondence between Stresemann and Papa I found discussions about color plates of the new birds. Unfortunately, the war curtailed finishing this work and these plates never materialized.

It is interesting that although Papa wrote all the biological observation in this large work, he is not listed as a coauthor, as would be de rigeur these days (when it is not uncommon to see dozens of names as authors of a single publication). As already indicated, Papa did later write a popular travelogue of the expedition: *Der Vogel Schnarch*, from which I have quoted. A premier American ornithologist, Margaret M. Nice, reviewed it in *The Condor*, calling it a "masterpiece of bird study." She remarked, "The undaunted courage of all three participants [in the Celebes expedition] and their dogged determination to achieve the allotted task gives an example of devotion in the service of science which may well serve as inspiration to the rest of us." It did to me.

Three years after Papa's death I traveled to New York City with a friend and with my older daughter, Erica, to the American Museum of Natural History to view the *Aramidopsis* specimen. My contact there was Dr. Mary LeCroi. When I called her, she said, "I know your father's birds well." Now she met us at the elevator and invited us up to the fifth floor, to the collections where the public is never allowed. We walked through a room filled with tall sealed cabinets that only research scientists would ever have access to, past original bird paintings by Louis Agassiz Fuertez and John James Audubon. When we reached the "rare bird" cases, Dr. LeCroi scanned the labels, unlocked one case, and pulled out one of dozens of drawers containing rows of birds, most of which hadn't been exposed to light for perhaps decades. Then she pulled out the rail, and handed it to me. It looked fresh—Anneliese or Liselotte could have prepared it the day before. I noted the label attached to its feet, as well as its rich but muted browns, russet, cream, and black, that I would render in watercolor. One side of the label indicated: "Heinrich-Expedition 1931. Skin number: 6999. Locality: Foot of Mengkoka Mountains in Southeast Celebes. Date: 20 January 1932. Gender: Female (♀). Iris: light brown. Feet: gray. Bill: red, tip and back black." The opposite side of the label listed stomach contents: crabs. And there was the name, *Aramidopsis plateni.*

My mind reeled at the thought that this bird, which to the eye did not look very different from our common Virginia rail, could have had such a profound influence on the imaginations of Sanford, Stresemann, and Papa, and ultimately on me. In my mind's eye I pictured Papa, bare-chested in the thorn-swamp thicket, bush knife and gun in hand, and triumph written all over his face. It was a momentous occasion, maybe even more so than he realized at the time, because with this bird, Papa had earned his stripes as a collector.

ON SEPTEMBER 15, 1932, NOW BACK IN BOROWKE, Papa wrote Stresemann, addressing him as *Teuerster Impressoria* ("Dear pillar, or masthead"):

It is raining and the sky is dark and not one sun-loving ichneumon will be out, and I therefore write. For the first time since [May] 1930 when we started the Celebes expedition I now have ten days vacation to hunt ichneumons and it fills my time. Hardly a half year has passed since the dark, wet, and cold primeval forests, and already my consciousness pushes all the pain away, and brings the beautiful blossoms into double grandeur. Already I see myself in the future and in wilderness and I believe I know one thing: I will not tolerate staying in Europe for long. I will find ways and means to get rid of my responsibility to the estate, and will look for opportunities for new deeds.

seven   A Switch from Light to Dark

PAPA'S FIRST CAR WAS A 1919 MERCEDES WITH A gold-plated gearshift mounted on the outside, so that you had to roll down the window to change gears. It had a long running board on each side. This vehicle came to him while he was destitute—the Great Depression was under way in Germany—and Borowke was falling into neglect. But as fortune would have it, while passing through Berlin, Papa happened to meet a man named Reimer, the owner of a glove factory who also collected zoological rarities. Herr Reimer gave the car to Papa with one string attached—that he use it on his next expedition.

Papa bought a trailer to tow along behind the Mercedes and after a few adjustments drove from Berlin to Borowke, a straight-line distance of 200 miles. The journey, mostly on sandy roads, took two days. With such a fine vehicle, most of Europe was now within reach of his insect collecting net, and on April 15, 1935, with Anneliese and Liselotte again as his faithful companions, and Ulla again at Borowke with his mother, Papa left for the Rhodope Mountains of Bulgaria. The director of the zoological museum in Warsaw had indicated that he would purchase bird and mammal skins, and Papa thought that the Bulgarian ruler, King Boris III, would be accommodating as well, as he was a member of the German Ornithological Society. Boris wasn't much help, but the trip was ultimately financed by contracting to sell specimens they would gather not only for the natural history museum in Warsaw, but also for those in Berlin, New York, and London.

Leonard Sanford's misgivings notwithstanding, Papa had figured out that having another European male on his expeditions was not desirable—no man, he said, would be as compliant with his leadership and decisions as Anneliese and her sister were—but for this trip he made an exception. Dieter Radke, a nineteen-year-old from Schneidemühl (on the Germany side of the German-Polish border) was a tall, gangly, outgoing youth who had impressed Papa during a social hunt at an estate near Borowke. Friends from all over Europe had been invited, and during that hunt Dieter had shot two foxes while nobody else got any. After that, Dieter was thought to bring good luck, and on a hunch Papa asked him to go to Bulgaria. Dieter's luck would follow him wherever he went, and decades later at the Archbold Biological Station at Lake Placid, Florida, he pulled a shiny blue ichneumon wasp out of a spiderweb—a new species, to be named *Protichneumon radtkenorum*.

Like other German young men of his time, Dieter had been unemployed, available, and eager for adventure. Besides, he had few other options. His father had been relieved of his job, so there was no money to continue the son's education. Political violence was also brewing. At the beginning of 1932, Dieter stood on the balcony of his home and saw protesters with red banners decorated with hammers and sickles, and others with flags emblazoned with the swastika, battling each other in the streets. He heard the communists' "Internationale" and the "Horst Wessel Lied," the Nazi anthem. Only in Borowke and the surrounding country was there little change, it seemed.

Much later, Dieter wrote in his diary of the trip to Bulgaria:

> For an adventurous young man like me, the next year was the fulfillment of many boyhood dreams: hunting in the mysterious woods and mountains of the wild Balkans, meeting interesting people and living in castles and tents. We crossed mountain chains, pulled through swamps and deserts. We always had trouble starting the

Mercedes on cold mornings, so we had a supply of charcoal with us. It was my job as the young man of the expedition to get up two hours earlier than anybody else and warm up the motor. It was the most free and adventurous and happy year I ever had.

The passage above was written in retrospect, decades after he and his family had settled in Florida, and after he had served as driver and companion to Papa on trips to collect ichneumons in Newfoundland and other destinations in eastern North America. I expect some of the happiness that Dieter remembered and that he, Papa, Anneliese, and Liselotte experienced on this expedition resulted from what they were escaping and perhaps from premonitions of what was to come. These were Hitler times. But on the day of their departure, spring had begun, the birds were singing, and the leaves on the trees were fresh and new.

East across Poland and south through Czechoslovakia and Yugoslavia they drove in the Mercedes, until they reached the Romanian border, where customs officials refused to let them pass. The officials were suspicious of this quartet of Germans with guns and boxes of ammunition in their trailer. After a day of hassles Papa was on the verge of exploding at the Romanian bureaucracy, but the rest of the crew wisely restrained him while the necessary permits (which he did not have) were issued from Poland.

After crossing the Danube at Giugio, they were finally in Bulgaria. This country, together with Austria and Germany, had been on the losing side in World War I, and Bulgaria's vast Dobruja and Danube delta had as a consequence of the war been awarded to Romania. Dieter remembers the people of Bulgaria as "subdued and gloomy," and sometimes "openly hostile" to their neighbors, Romania to the north and Greece to the south. The travelers noted that Turkish Muslims in Bulgaria did not mix with the Christian Orthodox majority. That is, the people were alert to differences, and when this strange group of Germans pitched their first camp near a village close to a river, curious onlookers came immediately. Two women and two men from some foreign country living in tents and cooking simple meals over open fires were an attraction indeed.

Aside from his other chores, Dieter was to hunt birds, and in the Rhodope Mountains he saw strange ones he had never seen before: rose starlings, green bee-eaters, yellow-black hoopoes, and blue-black crows. Papa again saw the species of Syrian woodpecker that he had shot from his hospital bed on his earlier expedition to Persia. According to Dieter, Papa "followed the woodpecker immediately, through groves, fields, and finally to a cemetery, where the woodpecker was picking at a tree—he shot the bird right in the middle of the cemetery, and was promptly arrested by the village cop for disturbing the peace of the deceased."

Papa found many new species of ichneumons on mountain summits, where they were hidden under moss and apparently in hibernation. However, in a letter to Erwin Stresemann, who was visiting the United States at the time, he said he considered this trip "training" for a true discovery expedition, and asked, "Please do not forget my wishes! I am hoping that in the course of the summer in America you will get a contract for me!" Anneliese added to the letter, "It is beautiful here, but not very interesting. Have the Americans still not the money for New Guinea?" And Liselotte added: "I want to go to New Guinea, too!" Perhaps they were hoping to collect birds of paradise. Instead they got two live eagle-owls.

One day a herdsman approached Papa with two young eagle owls (similar in size to the American great horned owl), and Papa bought them. The two young owls thrived on a diet of the animals that Papa caught (as usual, he trapped small mammals as well as hunting birds and ichneumons) and they kept him busily trapping mice—they swallowed squirrels whole. The owls, perched on two poles in front of the tent or riding in the trailer behind the Mercedes, became camp fixtures.

A year later, they headed home. Progress was slow. They had one flat tire after another: by Dieter's count, "almost 200" in total. On reaching the Polish-Romanian border, they were asked by a customs officer, "What do you have in the trailer?"

"Just some collections for the museum in Warsaw," Papa said

The official seemed satisfied, but just at that moment, out from under the canvas cover of the trailer came a deep, loud, penetrating sound that might have given him pause.

"*Ooh—hoo-hoo*," one of the owls, or *Uhus*, as this kind is called in German, had hooted. Papa reacted quickly by stepping hard on the accelerator. The group proceeded through Poland and back home.

They, and the owls, made it safely to Borowke, where Margarethe had in the meantime managed the farm and taken care of Ulla, now eleven years old. Ulla would soon assume the chore of feeding the owls their daily supper of mice, and she later wrote to me:

> Ruth Michelly and I were working and helping with the harvest that summer and we and a few other youngsters soon made a game out of "who can catch the most mice for the *Uhus*." We eagerly tried to outdo one another in catching the mice found under straw ricks in the grain field. My problem, being dressed in suntop and "short" shorts only—how to get the dead mice home. I found a piece of string and started stringing the mice (dead of course) as I was catching them, onto a necklace. I hung the thing around my neck. So far so good and not really anything funny. Well, one day we—with horse or horses—came galloping into the farmyard, I in front with long hair blowing in the wind and mouse necklace swaying all around me—I probably had some thirty mice on my string at the time. I noticed a group of visitors—Polish military—standing almost center in our yard. Needless to say I was going to show off my horsemanship. I came to an abrupt and dusty stop right in front of them and dismounted with bravado to greet them. Well, they looked at me as if I was not something of this earth, and for a moment they seems to have lost their speech. Finally one of them pointed at my necklace. "W-wha-what is that?" he asked. "Oh that—for supper," and I walked away. I can still see their faces, and I smile.

I suspect Ulla may have first told me the story when I was small, because I once used the same "costume" for trick-or-treating, on our first Halloween in Maine, and I can also remember the shocked expressions when people opened their doors to offer me some candy.

The *Uhus* were eventually donated to the zoo in Poznan, Poland. Papa, Anneliese, Ulla, and Liselotte continued their idyllic life at Borowke where they were largely insulated from the events in Germany, which I suspect they cared little about or thought peripheral to themselves. They made short ichneumon-collecting trips to southern Poland, Transylvania, and the Carpathian Mountains, and Papa petitioned Stresemann to be sent farther. Only poor Dieter had to face reality; he got drafted into the German army.

There may have been more than scientific reasons why Papa wanted to collect ichneumons and go on expeditions to the tropics. The National Socialists had come to power in Germany by solemnly promising to bring work and food to everybody. Their slogan was "Honor, bread, and justice," and their agenda was war. On February 27, 1933, Germany's most beloved government building, the Reichstag, was set ablaze. Only one arsonist, a young former communist named Martinus van der Luppe, was apprehended, but the Nazis blamed the fire on the communists collectively. A newspaper in Berlin declared, "This act of incendiarism is the most monstrous act of terrorism so far carried out." Hitler, then the leader of the National Socialist (Nazi) Party, called a news conference and declared, "This fire is the beginning of a great epoch in history . . . a sign from God" to declare war on the terrorists and their ideological sponsors.

Though some believe that the Nazis themselves set the fire, Hitler instantly exploited the event. The very next day the Nazis used the terrorist act as an excuse to invoke Article Forty-eight, a presidential decree to suspend civil liberties (freedom of speech, inviolability of property, freedom of association) in an emergency. Temporarily, they claimed. Of course the article was never revoked. The number of "enemies of the state" simply kept increasing as more people were hunted down, and the rights for which the liberal movement had fought during the previous century were demolished. The Weimar Republic was destroyed, and the full dictatorship of Hitler became established. He immediately started a giant propaganda campaign, and sought and got support from industrialists to prepare for war; and by the end of the year, about fifty concentration camps had been constructed and opened to receive all those ever more numerous "enemies of the state."

I DO NOT KNOW WHERE PAPA STOOD ON THE issues at that time, during the rise of the Nazis. I can only guess in retrospect. He always tried to impress on me to stay clear of the crowd, and any crowd whatsoever was suspect. He was not a born democrat. He did not believe "the people" could govern themselves by reason, and I think as proof he would point to emotional mass demonstrations. I think he was uninterested in anything as transient as the changing human tides of politics. He had his ichneumons. He had his enduring Borowke, and at Borowke at that time everyone lived in a seemingly impenetrable bubble.

In the 1930s in Germany and Poland there were other such pockets of oblivion, willful or otherwise. In January 1936, Erwin Stresemann (who like Papa was focused on birds and probably also looked without interest, if not with actual disdain, on the political scene), was concentrating on trying to get Papa to New Guinea. He asked the American millionaire Richard Archbold, who was interested in tropical mammals, to use his private plane and fly Papa into New Guinea. Stresemann even went to New York and met with Archbold, but the plan fell through. Papa then told Stresemann: "I'm ready to go anywhere."

It seems that Stresemann did pull together funding for an expedition to the Philippines for Papa, because two weeks later Papa wrote to thank him. "I have enthusiastic response from my coworkers in Borowke," he said; "they are ready to resume the battle of the rain forest." However, another letter from Papa to Stresemann, dated May 3, 1936, indicates that this expedition also fell through:

I am sorry that nothing came of the Philippines project. . . . I feel damned to have to be devoted to growing potatoes. My only elixir for living is the strong will to start new field biology in the coming year, no matter where. After one reaches forty years of age one sees over the top of the hill and finds that one has no more time to waste.

The letter is also signed by Anneliese, and by Liselotte, who says, "I am totally unemployed, and . . . holding still in place without doing anything does not become me."

Papa knew that Stresemann and his backers would base their funding for any expedition on the expectation of bird and mammal skins, but for him, as always, the main attraction was ichneumons. A report he received from a Swedish entomologist, René Malaise, soon whetted his enthusiasm for an expedition to Burma. Malaise had invented a new type of insect trap. The trap could be set up in the forest and was capable of collecting vastly more insects than a person with a net could. Malaise had collected in northern Burma and sent Papa all the ichneumons he captured. Papa was highly impressed by their variety and at once realized that only a small part of the Oriental fauna was known. He set his mind on collecting more from that region and made a trip to Berlin to study the Burmese birds at the museum and discuss their appeals to collections with Stresemann. No easy feat, Stresemann had written to Papa: "The Americans give no money and we at the Berlin Museum are not in position to purchase Burmese skins, even though we need them."

Papa responded that he could at best scrounge up 1,500 marks, and he asked Stresemann to see if Sanford at the Museum of Natural History in New York "might give $5,000, or guarantee purchase of at least 3,000 birds." He continued: "I do not want to stay here much longer. I need to be able to forget the sadness of this world here"—presumably he meant the deteriorating political situation (which remained unmentioned in the correspondence), or his own inability to escape potato farming. He went on, "This year has passed as useless despite all efforts. There lay in my cabinet the correspondence for New Guinea, the best maps of that land and the book of its birds, as well as Haschisuka's book of Philippine birds, the special maps of the Arakan district, Colonel Ruder's *Burmese Self-Taught*, and the list of Burma's birds. I have barely the strength to continue the studies. Please give me your opinion as to the chances of this whole thing."

Stresemann responded in a letter addressed "*Teurer, den Yankee's zu teurer Forscher*" ("Dear valued and too-expensive-for-the-Yankees researcher"), "My last suggestion for you: write to Ernst Mayr." (Mayr was Stresemann's star student, who had received a doctorate at age twenty-one and was then sent by Stresemann on an expedition to New Guinea, and who had gone on to the bird division at the American Museum of Natural History in New York. There, Mayr practically pounced on a huge bird collection from Walter Rothschild's Tring Museum that Rothschild had to sell to raise money to pay off a mistress who was blackmailing him. Mayr made his reputation as one of the world's leading biologists using data derived from this collection.) Papa did write to Mayr, whose reply from New York on September 8, 1936, was curt: "I hope you will be able to carry out your plans without our help."

Then came a small ray of hope. On April 17, 1937, Papa learned that Sanford would probably offer $1,000 for birds from a Burmese expedition. That modest sum was enough to tip the scale. It was "official" support, which would help with obtaining permits from the British administration of Burma. Financially it was extremely meager fare, and success would depend on obtaining skins for later sale to other outlets.

To get the expedition rolling, Papa paid a visit to his friend and cousin Otto Heinrich, who was a banker and lawyer in Berlin. Otto's resourceful teenage daughter Marlis helped Papa find a suitably large and sturdy automobile to take to Burma. They decided on a Stutz limousine, which was at the time one of the most luxurious cars on the market, "comparable to a Cadillac." Papa claimed that this vehicle would impress dignitaries in Burma and thus open doors for the expedition. He and Marlis drove all over Berlin gathering equipment for the expedition, which they stored in Otto's basement. Marlis's efficiency and helpfulness in procuring passports, permits, visas, and recommendations impressed Papa, and he invited her to join the expedition to Burma, giving her a job helping with preparing skins. Thus in the spring of 1937 Papa set off with Anneliese, Liselotte, and Marlis, along with the Stutz, aboard a ship headed for Rangoon, Burma (which since 1989 has been known as Yangon, Myanmar).

It might seem unusual for one man to head into the wilderness with three women, and it probably was. One was officially his wife, although their relationship was platonic at this point. She knew of Papa's ongoing relationship with Liselotte, her younger sister. I suspect Anneliese didn't like it, but she didn't want to lose Papa entirely or give up her home at Borowke. It seems that she did try to deflect Papa's attention away from her sister—attempts in which Ulla, by now twelve years old, may have served as a pawn.

Before leaving for the expedition, Anneliese left Ulla with a newly hired Polish tutor, Hildegarde Buruvna (or Bury). Hilde, as she was called, was the daughter of a German mother and Polish father and fluent in both their languages. Her parentage was then an irrelevant detail, but it would be of major significance later. Hilde had recently received a teaching certificate in Poznán. The training had taken five years and had been "like prison" for her; she experienced it as an "unbearable invasion of personal freedoms." She had just completed her schooling when her biology professor informed her that there was a job opening for her on a beautiful estate, where a tutor was needed for a young girl. Hilde applied, and Anneliese, who came to Poznán to interview her, was pleased. Apparently one of the job requirements was that the tutor be attractive. And according to Hilde, Anneliese told her all about her "charming" husband.

Papa turned out to be even more charming than Hilde had anticipated. Almost as soon as she arrived at Borowke, the sparks began to fly and Papa and Hilde fell in love. Hilde, who is my Mamusha, wrote decades later:

> Gerd had great charm and he used it! Dancing was a wonderful way of being in his arms. In the daytime we undertook little excursions to collect ichneumons and other insects, to look for birds' nests, to sit in the grass and to enjoy each other's company. All of this led to a little bit of petting. I started to get really frightened. I loved this man, but he was married! He told me that he did not live with his wife. She was his wife "only on paper." She was his best friend. I pointed out that, since we were not married, we could not start anything that could go too far. He replied that "we love each other and do not need anybody's piece of paper to seal it." He also maintained that "we do not need to get married as long as there is no child." But he wished so much for a son, and found in me the perfect mother. All of this sounded convincing. I wanted to be convinced. Then Gerd left for Burma with Anneliese and Lotte [Liselotte] for a year. We were sad to say good-bye, and it was understood that I would be waiting for him when he got back. I understood that he too would be faithful, but I was uneasy. When I asked my mother what to do she pointed out "this man is not there alone. He has his wife with him . . . etc. etc." We exchanged passionate love letters, but my conscience bothered me too much and I decided to end the relationship. With a heavy heart, I finally wrote to Gerd that we were through. As an answer, I got an angry letter from Anneliese! "How could you do such a thing! Gerd, upon receiving your letter became very sick, depressed, and unable to work. Don't you know that his work in the jungles of Burma requires all of his energy, that he needs your support . . . ?" She continued, "I also count on you to stay with Ulla and not to leave." She asked me to reconsider my decision to leave her husband.

I have no record of what Anneliese thought of her sister's being supplanted by Mamusha. I suspect, though, that she realized that the new woman could serve as a wedge to prevent her sister from marrying her husband and thus possibly banishing her from Borowke, but only if the new woman could be neutralized later.

Meanwhile, when Papa and his crew entered the mountains where René Malaise had had such stunning success with his insect trap, they followed a brook through a sparse forest and almost immediately saw interesting ichneumons "at every step." The mountain air was cool, and so the chilled wasps were immobile among the leaves on the ground vegetation. Here they set up their camp to start collecting.

It may have been a good area for ichneumons, but it was an unfortunate site for camping. They were in the woods about fifty yards off the road, but the light from their tent lamp was apparently visible from below, on the road that led between the city of Mandalay and the Chinese border. One otherwise peaceful night they heard pistol shots from down near the road. Quickly they doused the lamp. Papa jumped out of bed, grabbing his shotgun, and fired rounds of bird shot into the trees. Then they all lay on the ground, listening. A motor started, and a vehicle moved a short way, then stopped. More shots were fired, and then the vehicle left. Still awake near midnight, they again heard an automobile stop below them. Papa jumped out, barefoot, and immediately started shooting again: "Boom, boom," blasts the shotgun. "Bang" came the reply of a pistol. "Boom." "Bang, bang." While lying on their bellies and listening to wild yelling down on the road, they heard the English words, "I got a shot." English! Papa yelled, in English, "What do you want?" No answer. Then a counterquestion: "Who are you?"

Papa yelled back: "I am the Polish zoological expedition." That ended the conversation.

A second vehicle arrived and directed a searchlight into the woods toward them. At that point, yielding to superior force, everyone grabbed what valuables he or she could and retreated deeper into the woods. From their hiding place they heard voices coming closer; apparently the attackers were approaching the campsite. After two hours there was silence. The next dawn when they returned to the campsite they were surprised to discover that nothing had been disturbed. Later, they learned that the highway was used by opium smugglers, and a special governmental unit patrolled the road to fight drug trafficking. The ichneumon hunters' camp had probably been mistaken for a drug smuggler's hideout.

Most of the other campsites were idyllic retreats far from civilization, but invariably they were transient. As always on an expedition, one day the stillness would be broken when the carriers arrived as appointed to take them farther into some wilderness or higher up into the mountain, and inevitably deeper into debt.

The crew eventually left the foothills of Mount Victoria and with fifty-five carriers and ten packhorses ascended toward the nearly 11,000-foot summit. Already at 1,000 meters they entered a zone of pine forests "where no tree had ever been cut, and probably never will be." They found a recently killed carcass of a wild gaur (a kind of wild buffalo) on which a tiger had fed. They decided to make their camp in that wild area and finish the ascent later.

In a letter to Stresemann written on Mount Victoria on March 24, 1938, at their campsite at near 4,000 feet, Papa mentioned that he had caught two new species of shrews: one without external ears and almost no tail; the other a *crucidura* with a notably long tail and much darker than the species found in the lowlands. He also wrote that he had received an order from the Field Museum of Chicago for $200 worth of rodents and insectivores. "With that I've as good as covered ten percent of my debts."

Up to this point the ichneumon hunting on Mount Victoria had been poor. But one day Papa was very surprised, while stalking through a shady wooded ravine, to see a beautiful wasp, and shortly thereafter, another. That afternoon, he led Anneliese, Liselotte, and Marlis to the spot and they all hunted for ichneumons. The result was so spectacular that for a time they made daily forays to this "ichneumon El Dorado" as he called it. It was the hot, dry season, and the insects from far and wide had gathered here in the cool shady woods.

One evening after Papa returned to the tent, Marlis handed him a cuddly black, hairy animal baby with soft pink humanlike paws—a baby bear! It was so young it could hardly crawl, and "it captured our hearts before we even knew it." Two locals, Chins, had brought it after finding it in a hollow tree where a forest fire had driven the mother away. They wanted ten rupees for it. Papa tried to explain that he could not take care of a bear at any price. The two Chins were angry, having come a long way, and intimated that they would have to kill it. With that Papa changed his mind and kept the little cub.

It was a male. They called him *"Dickie"* ("Chubby"). Marlis became the mama bear, and every night Dickie slept soundly in her sleeping bag. Before daylight everyone in camp would be awakened by loud growls, grunts, and other sounds reminiscent of a cross between a puppy and a hungry pig, coming from Marlis's bed. She would get no rest until she got up and prepared a bottle of diluted condensed milk.

Marlis eventually got tired of sleeping with Dickie as he got older and hungrier; her adoptee was not as hygienic as this city girl from Berlin would have liked. He was now confined to a hollow rhododendron tree at the edge of camp, where he could be shut in for the night. But at intervals he awakened and had loud temper tantrums. Marlis would still have to get up and quiet him, so the others could sleep for another hour or two. Eventually the rapidly growing Dickie grew accustomed to sleeping alone in his tree, but he still came to the tent whenever he felt like it, and he became less discriminating about whom he slept with, not to everyone's displeasure. During one rainstorm he came to Papa's bed and Papa lifted him up and let him crawl in. He lay on his back, kicked with his short legs, and gently took Papa's hand in his mouth, a bear love bite. Then he snuggled up and slept. "In that moment I could not have had a child more dear than Dickie, the little bear," Papa said.

For a time, life in Burma was practically Edenic. The camp on Mount Victoria was set amid large oval-leaved oaks and huge knobby rhododendron trees flowering with red blossoms. Nearby were open meadows over which larks sang in the blue sky "as back home in spring," and the flower carpet over the forest floor glowed with innumerable blue and white anemones. Pale yellow flowers bedecked rose scrub, and in the shady places of the summit forest violet-red primulas and violets bloomed. In his book In Burmas Bergwäldern, he wrote, "What was Latimodjong in comparison with this, God's mountain!"

But alas, paradise was lost—"We soon learned, however, that where there is much light there is also much shade"—mainly because of the biting insects that began to abound in the Burmese summer. "During the daytime the large horseflies attacked the hunters literally by the thousands . . . and sometimes while I was stalking a rare bird and unable to drive them away, they sat upon my face and hands, giving me lessons in stoicism." (This is only a slight exaggeration on Papa's part; in one photograph of just the front of his legs below the knees, I counted 103 very large biting flies perched there.) "In addition," he wrote, "there were always mosquitoes and tiny gnats on my hands and head, be it night or day. Last, but not least, there were ticks in great numbers. The helpers alternately cried, swore, and had temper tantrums. At the end of each day I disappeared to a secluded place near the tent, shed all my clothes and plucked off all the ticks that had attached themselves to my skin."

Then one day the rainy season started, and with the rains came a new pest: land leeches. They approach a person from all sides, dozens at a time, as if drawn in by the magnet that is scent, and attach themselves to the legs. "They sucked blood," Papa wrote, "until they were thick and plump, and then dropped off to the ground." At the end of June, when the crew was still on Victoria after having collected 3,900 skins, Papa's correspondence with Stresemann became somewhat heated. Stresemann had, through his contacts, secured several outlets for the collection, and he wanted the entire collection sent to Berlin to be distributed from there. But Papa insisted on sending it instead to his close friend Piet Hart Nibbrig (general manager of the Bols liqueur company), living in Groningen in Holland, where he wanted the collection to remain sealed and untouched. On his return, he would work over the specimens before distributing them "because in Groningen there is space to sort the collection in peace." He invited Stresemann to join him there. I personally question his stated logic. Surely there was plenty of space at the Museum für Naturkunde in Berlin! My hunch is that he was cautious, if not fearful, of something happening in Germany.

After the crew (and the bear) descended from Victoria and reached the village of Kampetlet, the naturalists received their first mail, and with it bitter news. Otto Heinrich had written to Marlis that she should return at once because there was immediate danger of war in Europe. Papa wrote to Stresemann: "The prophetic words in Marlis's father's letter about the 'dangerous war' represent in my mind the beginning of a drastic change in my world and my life, like a sudden turning of a switch from light to dark."

The crew packed the mammal skins they'd collected and sent them off to the American Museum of Natural History in New York. Writing again from Rangoon on August 7, Papa explained to Stresemann that they were now leaving, and he would personally take the bird collection with him to the Hamburg Freihafen (free port), and from there on to Holland (which is what he did). "Here it is acknowledged," Papa's contract with his friend Nibbrig stipulated, "that the entire bird collection of the Burma expedition is the property of P. H. Nibbrig, of Groningen, Holland." From there the Burmese specimens were eventually distributed to Warsaw, Bonn, Berlin, Chicago, New York, the Phillipines, and the Museum of Comparative Zoology at Harvard University.

Papa left Burma reluctantly. In the last sentences of his account of the expedition he reminisced that they had felt themselves rich and happy in the forests of the mountain peaks, but "we are slaves to a power, which is called civilization, and we must go back, down into the lowlands."

When Papa was back in Berlin, he discovered that Stresemann had arranged for him to give a talk to the German Ornithological Society. But Papa refused, giving some vague reasons having to do with his frayed nerves: "I am presently not able to give a humorous account and anyway I would have nothing to say." In short, he did not feel inspired, even though Stresemann had gone to the trouble to recruit a German businessman named Kurt Rolle to finance making slides for a talk. Furthermore, Papa was remiss to reimburse him. When Rolle wrote to Papa demanding his money, he signed his letter, dated October 3, 1938, "Heil Hitler"—a not-so-veiled threat. It's the first mention I find of the dictator's name in any of Papa's correspondence.

eight   War

PAPA AND HIS CREW RETURNED TO BOROWKE TO
find Hilde, Margarethe, and Ulla still waiting for them. Life at Borowke
had been largely isolated from the festering social malaise that had
spread through much of Europe, yet tensions were mounting. Many ethnic Germans living nearby still had loyalties to their old roots, and some
of them were openly sympathetic with the National Socialist Party. A
group of Germans in the countryside nearby formed a union which called
itself the Young Germans, and which later turned out to be a Nazi avant-
garde. Papa had never been eager to join unions, organizations, or societies of any kind, and he had steered clear of the Young Germans.

In a letter dated June 19, 1939, he complained to Stresemann that he
was not able to get a permit to travel to Berlin. "In Borowke," he added, "I
feel like a peregrine falcon in a cage." A month later he wrote again: "We
who live between the hammer and the anvil see the future with growing
pessimism and are sorry for only one thing; that we are not still on Mount
Victoria."

Papa's increasing pessimism (or was it increasing realism?) that
summer was stoked by a group of young Polish officers who came as
guests to Borowke. They were remembered as "great optimists" who were
fun to have around. However, their joking—about the lower-class rabble-
rouser from Austria who was agitating in Germany to build an army with
the intent of attacking Poland—put Papa on edge. The officers saw this
man as a simpleton who talked and looked like a cartoon character—a
short man with a little tuft of a mustache on his upper lip and a strand of

black hair across his forehead, who spoke of the superiority of tall, blond, blue-eyed people. Ha, ha, ha! Hilarious. They saw no real threat coming from that quarter. But if this blustering idiot ever did have the audacity to attack Poland (presumably to reclaim the part of that country that had belonged to Germany before World War I, which included Borowke), the Poles would be ready for him. The officers evidently did not appreciate the events that had brought Adolf Hitler to power. He had ignited sparks from the coals that had smoldered since the end of World War I, when Germany had been left humiliated and in ruins, with nearly 2 million men dead (about forty times the number of U.S. soldiers killed in Vietnam). Kaiser Wilhelm II had fled, leaving political turmoil, social upheavals, and mass hunger. As a twenty-nine-year-old Austrian corporal of that war, Hitler lay in a hospital near Berlin, blinded by mustard gas. He would later remember and write from prison: "In vain were the sacrifices— hatred grew in me—hatred for those responsible—I decided to go into politics." He said, "The German people must be led" into a mass movement, and he became the first to master "radio talk." Through deceit, slogans, symbols, oratory, and eventually terror, he warred first against intellectuals and liberals, and then against everyone else who might think and therefore oppose him. He spoke of "one people, one nation, one leader" who would stand up to external and internal threats. He saw himself as the messiah and when criticized would say, "Two thousand years ago, a man was similarly denounced . . ."

The liberal Weimar Republic that had replaced the failed Kaiser was impotent against Hitler's words. Its chancellor, Paul von Hindenburg, who had led the victorious battle (but ruinous war) against the Russian forces in which Papa had fought, was now a respected gentleman in his eighties, fond of long naps. Unemployment and inflation were rampant, and most people in Germany had no savings whatsoever left. Hitler preached, "Down with the system." He presented himself as antigovernment, champion of the common man, the "Volk," and yet shrewdly painted himself as an ally of Hindenburg, the hero who was worshipped as a soldier and statesman.

Soon things began to change. Hitler gained immense popularity. Historians have debated for many years how this happened. As a biologist, I view it as a response to particular stimuli. Certain grasshoppers, when subjected to crowding, will change body shape, color, and behavior to become migratory locusts. They become highly "nervous," voracious, and mindless crowd followers. For humans, one of the most powerful transforming stimuli is sensing that we are under attack. We close ranks and identify an "enemy," which requires first of all isolating and identifying "us" as separate from "them." Desperate situations always demand blame, and in combination with chaos, they then provide the opening for an authority figure who promises to restore order. "Liberties" must be curtailed while enemies, real and imagined, are sought. Then the spin begins, and those who follow are rewarded and those who don't are punished in direct proportion to the power the leader has attained.

Joseph Goebbels, Hitler's propaganda minister, exploited a disagreement with the French to withdraw from the League of Nations and go it alone. His "genius" was to link loyalty to Hitler to patriotism. Hitler was the "leader," so according to Goebbels if you were against Hitler, you were against the country, and that was treason. He proclaimed a revival of the Christian faith. Every soldier was to wear a belt buckle that said *Gott mit uns* ("God with us"). Any opposition to the regime was deemed unpatriotic, and dissenting voices in the press and elsewhere were muzzled. Soon other parties were banned because they were deemed anti-German. Still, there was resistance. But when scapegoats were needed, Jews who had positions in public life were thrown out, again because they were "anti-German." Goebbels rallied university students to burn the books of liberals, to silence opposition to Hitler's promised path to greatness.

And at the Nuremberg Trials, at the end of the war, Hermann Göring, Hitler's designated successor, said:

> Why, of course the people don't want war. . . . That is understood. But, after all, it is the leaders of the country who determine the policy and it is always a simple matter to drag the people along, whether it is a democracy, or a fascist dictatorship, or a parliament, or a communist dictatorship. Voice or no voice, the people can always be brought to the bidding of the leaders. That is easy. All you have to do is tell them they are being attacked, and denounce the peacemakers for lack of patriotism and exposing the country to danger. It works the same in any country.

When the German army marched into Austria on March 12, 1938, not a shot was fired. Instead, the German soldiers were showered with flowers. Hitler was emboldened, but he needed conflict to legitimize his power. On September 1, 1939, Papa received an order to immediately bring all the Borowke horses to Sepolno, for the Polish military. Within the hour, Papa sent men and all thirty horses so that the army could have their pick. Horses and riders, now panicked and sweat-lathered, came right back in a full gallop. Low-flying airplanes had swooped in over Sepolno and sprayed bursts of machine-gun fire on anything that moved. They were the Junkers Stuka dive-bombers, equipped with automatic weapons and devices that produced shrieking noises, which added to the terror. Behind them came high-flying heavy bombers; and finally 3,000 tanks rumbled across the landscape with more than 1 million infantrymen on the march.

Papa and Mamusha went down to the main road, to see if they could find out what was going on. There they met a troop of magnificently dressed Polish uhlans, with swords and long lances. These cavalrymen were bravely expectant to meet the German offensive. They were prepared to meet infantry charges, but would instead meet fighter planes and tanks.

The blitzkrieg ("lightning war") had begun. World War II was several hours old. Papa, as an ethnic German, was more afraid of the Poles than the German invaders. He had said that he thought war was "possible" in June, "probable" in July, and "certain" by August. Therefore, a month or so before September 1, he had already built a shelter to hide in and provisioned it with food for several days; he expected that the Poles would kill all the Germans and destroy their property when the German Wehrmacht attacked. He had his blacksmith build tin boxes that could be soldered airtight, and into these he put his most valuable ichneumon specimens, including the type specimens of his magnificent Burmese collection. The main part of the collection was too large to be hidden and had to remain in the house. Later he wrote to Stresemann:

> I could easily imagine what fate would await the German people here. I must tell you how I hid to avoid capture. *Aramidopsis* and the other rails have made me into a swamp bird, and so I escaped to the swamps [those below the Borowke house where the cranes nested]. Amid the willow bushes and sedge grass that covered the moor, I made a hole in the surface blanket of growth and dug out the mud,

pail after pail, from underneath. I carried the mud away, and eventually could dive into the hole. Now picture the hat of sod-carpet that I lay on my head, and you will agree that not even the clever Polish soldier or policeman could possibly find the rail researcher, unless he accidentally stepped on his head. A few hundred meters away in a thicket, I had my expedition field cot, sleeping bag, mosquito net, and rain guard. This was the actual shelter where I could stay until danger approached.

Papa made it sound as though he had copied a rail, and indeed Stresemann was calling him *die Ralle*. However, he had learned from the partisans in Russia about submerging in a swamp—to hide in the day and come out at night. He apparently felt that the rest of the family would not be in danger—he didn't mention them. I think he instinctively knew that whenever political "justice" is administered, it is always those (like him, a highly visible male, educated landowner) with the most potential influence who are targeted. "I could hold this position for a week, but in a few hours the war was over and Poland was in German hands. Those were grand and unforgettable moments." (Mamusha said he was overjoyed because he felt liberated. She, on the other hand, was saddened.)

One of the first German soldiers to step onto the yard was Papa's young friend Dieter Radke, who had participated in his Bulgarian expedition. Radke belonged to Company One, Third Battalion, Ninety-Sixth Infantry Regiment, Thirty-Sixth Division. Radke relates that his regiment was ordered directly to the Polish border near a chain of lakes where the soldiers were happily camped and spending their time catching, frying, and eating fish. But on the evening of August 31 they were all ordered to assemble at a barn and sleep with their boots on and live ammunition around their belts. Shortly after they had fallen asleep, Dieter, who had studied languages on Papa's advice, was called by an orderly to appear before a colonel and listen to Radio Warsaw and translate what he heard. Dieter understood only a little and faked the rest, but was complimented for his expertise and then let go. He may not have understood everything he heard, but he accurately translated the implications of what was happening. That translation was confirmed when, at around two o'clock on the morning of September 1, his unit was ordered to leave the barn and start marching. By three-thirty that morning they were crossing the Polish border; the war was on. They marched through the dawn and morning and by afternoon Dieter felt that the sandy road they were traveling was looking more and more familiar. And then he knew why. It had been pure luck or coincidence that his regiment, out of all those that streamed across the 1,750-mile Polish frontier that day—1.5 million soldiers in all—had crossed into Poland near Borowke. It was, actually, not just a coincidence. It was a miracle.

His head suddenly aflame with memories, Dieter rushed up to the captain, who was on horseback, and explained that he wanted to visit his friends and family. He must have been very convincing, because the captain granted him a one-hour furlough, on the condition that he be back no more than one minute late. Dieter jumped on a bicycle and raced off. When he reached the Borowke farmyard he remembered seeing smoke everywhere. All the farm animals were gone; just a few chickens were scratching the ground. Not one human soul could be seen. He entered the house . . . there was total silence. Dieter remembered that there was a cellar deep under the kitchen to keep things cool, and he opened a trapdoor and hollered down: "Hallo—it's me, Dieter! Dieter Radke!" Margarethe, Anneliese, Liselotte, Hilde, and Ulla joyfully opened the hatch door, and in the few minutes' time that he had left he tried to explain the situation to them—and they told him how afraid they had been to see German warplanes and hear wild shooting in the village. After assuring them that they were safe, Dieter jumped back on his bicycle and caught up with his unit. The captain said, "Radke, if you had been one minute later I would have had you arrested. I'm glad you and your family are OK."

Of course the family was not OK, and the war was not over, as Papa thought. In fact, it hadn't really started yet.

That very first morning the German Luftwaffe's Stukas and bombers wiped out the 500-plane Polish air force while it was still on the ground. Warsaw was soon surrounded, but held out, refusing to surrender until September 27, after 12,000 of its citizens had died. The radio station stopped what had been up to that point a continuous broadcasting of Chopin. A quarter of the city was destroyed, and there was no food left. Hitler was incensed at the Polish resistance and sent *Einsatzgruppen* (special units) into Poland to search town after town of the defeated country and murder local officials, teachers, doctors, aristocrats, Jews, clergymen . . . anyone who he thought might be "terrorists." He had vowed "to kill without pity or mercy all men, women, and children of the Polish race or language," and he would eventually attend to that. In the meantime, there was solemn stillness.

Various detachments of the German army, the Wehrmacht, passed through Borowke after Dieter left, and all behaved honorably. But the army, which to many Germans stood as a bulwark against the Nazis, soon marched away and the Nazi Party, represented by Hitler's personal and political fighting units, the SS (the Waffen SS, or *Schutzstaffel)* and the SA (*Sturm Abteilung,* or storm troopers) took over. The SS commission in Sepolno sifted through lists of names to decide which of the ethnic Germans were desirable and which were dangerous to the "new Germany." Papa was thought to be "Polish-friendly," and he was. His loyalty to the regime was suspect.

One person on the committee sifting out the supporters from the nonsupporters was a man named von Wilkens, the owner of a large estate near Borowke. One day, he whispered in Papa's ear, "I am risking my life—but I must warn you, you are on the list for liquidation. Try to disappear."

Papa immediately took precautions, expecting that a car might drive up at any time, with SS officers who would step out and haul him away to be shot. He still had his hideout in the swamp, where he could submerge whenever suspicious-looking people appeared, but that was no long-term solution. He wrote a letter requesting assistance to Ernst Schäfer, the fellow explorer who had gained fame and publicity as the first European to reach the Dalai Lama in Lhasa. Earlier, Schäfer had been a co-leader of two American expeditions to China and east Tibet to study the giant panda. He was considered a brilliant scholar, but his political record was mixed. Schäfer's expedition had been under Nazi sponsorship; and he had joined the party, apparently in order to receive support for his scientific research.

Nevertheless, he soon also became authoritarian and was considered "arrogant." Willie Pruessing, a family friend, read Schäfer's book *Berge, Budhas, und Bäre (Mountains, Buddhas, and Bears)* and then wrote him to ask to be taken on the next expedition, volunteering as an artist. He was accepted, but first had to join the party, which he did. But he argued politics with his boss, who then arranged through his party colleagues to get Willie sent to what he must have assumed would be the young man's death sentence—the Russian front. Willie survived and would later say, "Only a few people had large enough vision to see to evade, before the door closed." Papa, the pessimist, the realist, was one who had enough vision.

Papa wrote not only to Schäfer, but also, on October 20, 1939, to Stresemann, thanking him for his promise to give help and stating, "I do not doubt that I will need it, and the feeling that I am not totally alone comforts me and means a lot more than you can imagine."

Eight days later, on October 28, 1939, Papa wrote Stresemann a postcard. This is the text of that card:

> *Bawah tulung sama sajah lekas!* As an explorer you understand these words in Kiswahili? Hopefully it will again soon be time for an expedition into the German-African bush. But the "rail king" is presumably the first game that will be hunted. The natives know how to set their snares. Ernst, the *Tibetforscher*, could through his contact with the high chief thereby be of help. Unfortunately, I could not get a pass to speak with him. Ernst, on December 1, will presumably follow the customary. The question of the fate of my Burma catch should be accomplished. Visit Ernst yourself. You have long promised—now is the time!! Or send Ernst Sch. Maybe you can also speak with Pg from Bruthausen. Otherwise I am healthy and alive, and also spiritual! You doubt? Yes, there are still trips to heaven and earth. Tomorrow I return to Borowke, without great pleasure. I wrote some days ago to General Grauert, airman and old boss and observer from the World War. I volunteered for the front. No answer (yet). Do not forget your old, faithful Heinrich.

When I read the postcard, it took my breath away. It was filled with meaning—the postmark, the fact that it was a postcard rather than letter—almost every word is carefully crafted in doublespeak to convey to Stresemann the danger of Papa's situation, and his hopes and plans for getting through it alive, not to mention getting his precious collections out of harm's way. It is a missive of diabolical cleverness spiked with humor, and a glimpse into why Papa, and ultimately all of us, lived through the war.

Fundamentally, the message was that Papa feared for his life. The card was postmarked from Bromberg—this is the only time Papa ever mailed anything to Stresemann from there. Maybe he feared his mail was being read at the little local post office in Sepolno (by then given the German name Zempelburg). Also totally atypically, he wrote a postcard rather than sending a sealed letter. I suspect that this was to deflect attention. Nobody would be so foolish as to send an important message in such an open way. But he had a perfect cover.

The opening in "Kiswahili" is total nonsense.* It sets the stage: "This message is to be translated" from surface meaning to hidden meaning. Papa's reference to the "German-African bush" was an allusion to primitiveness and barbarism. There had been no correspondence between them about any expedition to Africa! "Rail king" is obviously a reference to himself, and in the next sentence about the "natives" knowing how to "set their snares," he is informing Stresemann that he is being hunted. "Ernst" is Ernst Schäfer, with connections to the "high chief" (none other than the boss, the all-powerful and later notorious SS leader Heinrich Himmler). I don't know what he meant by "Ernst, on December 1, will presumably follow the customary." The question of the Burma catch is probably bogus. The Burma catch had been taken care of. But Papa was recently in contact with Stresemann regarding his book on Burma. So, what needed to be accomplished probably referred to a letter from the previous week where Papa had said: "I do not doubt that I will need your help." (He had left no doubt then that the help was not with respect to birds.) In the card he then mentions that he is still alive and healthy . . . and "spiritual"! Here he sends a message through a joke: The "trip" that is here referred to is not a physical trip to Africa or Celebes, but an allegorical illusion to the spirit making its trip from earth to heaven. He then becomes very transparent in saying that he has contacted Grauert to volunteer for the front, which he did. The more deceptive the message, the more it has to be sprinkled with truth.

In his next letter to Stresemann, dated October 29, 1939, Papa wrote:

---

* As this book was going to press I received the surprising news from a book reviewer, John Major, that this sentence is only *seemingly* nonsense. It is not Kiswahili (an African language) but Bahasa Indonesian, instead, and it says, "Send me [*Bawah*] assistance [*tulung*] as before [*sama sajah*] immediately [*lekas*]."

In the meantime news will have reached you from several sources. If I could have come myself it would have been better. The secretary of the Landrat informed me that I can't get a pass, not even if requested by an institute. There is a deep tragedy in these things. I was a second-class citizen for twenty years [under the Polish administration], and am one again under my own people. How seriously I have to take my position, and all the possibilities for the near future, you can judge from the card I sent you from Bromberg. Maybe one is hypernervous—maybe. In any case my and my loved ones' concerns are not groundless.

He concludes, "Say hello to father—who is now eighty—who could for years not get a Polish passport [from Berlin] to visit Borowke."

A week later (November 7, 1939), on his forty-third birthday, Papa wrote a four-page, highly deferential letter to Schäfer, thanking him. Apparently Schäfer had intervened to have Papa's name removed from the blacklist. Papa concludes his letter with, "Please give the Dalai Lama my comradely, thankful handshake." He also said that he would "never forget"—presumably Schäfer's favor. The letter was signed with "Heil Hitler," as was customary in corresponding with public officials (who were by definition assumed to be party members). I suspect he gnashed his teeth while writing it, because he knew Schäfer got government support for expeditions that Papa could only dream of. Indeed, Schäfer was leader of a whole institute near Munich, the Ahnenerbe.

Schäfer's word was undoubtedly important (as would be Papa's word for him as a character witness when he was tried after the war), but I doubt that Papa felt safe, even after the local Nazi boss was asked to take him off the list. People were disappearing, even "good Germans," never to be heard from again. A neighbor near Borowke named Pradzynski and his wife were shot without any trial or explanations. Another neighbor, Komierowski, owner of the Komierowo estate, which had been in his family for centuries, was shot to death while walking through his woods. In this way the occupying Germans "acquired" Polish properties. Any small "wrong move" would be used as an excuse for execution. But the Nazis could presumably not so easily touch anyone wearing the uniform of the German Luftwaffe, so Papa, as he had written in the postcard to Stresemann, did indeed seek out Grauert, his old Luftwaffe boss from

World War I, and through Grauert's intervention, once again joined the war effort. At an age when many men would have been deemed too old to fight, he was forced to go to war to save his life.

I found no dramatic or revealing letters over the next year. One letter was signed "Lieutenant" Heinrich. Most of the letters to Stresemann were "ornithological." Indeed, in one (dated July 8, 1940) Papa wrote to his old friend that he had given a slide show and lecture (about Celebes and Burma) "to all formations of the Luftwaffe in Vorthya—a duty that coincides with my expertise." The letter continues for seven pages, all about birds. He hoped to get home soon on vacation and hunt for ichneumons and birds. For the moment, despite his "international orientation," his "Polish-friendliness," and his "intellectualism," Papa was relatively safe.

Nevertheless, he also took calculated risks. There was a severe shortage of men left for agriculture, and as later correspondence revealed, Papa was sometimes relieved from the "guns" to the "butter" detail of managing the agricultural production at Borowke. One day, when the workers at Borowke assembled to be assigned their work for the afternoon, the blacksmith, Josef Kowalewski, made an unfriendly remark about the new Nazis. The foreman, Bernard Zimnicki, who had a grudge against the blacksmith, reported the remark to the Nazi authorities. Kowalewski was promptly arrested and brought to a detention camp.

Of this incident, Papa wrote in his memoir:

I felt that I had to intervene personally and attempt to free Kowalewski. When I got [to the center], the sight that met me made me shudder. There was a freshly dug ditch filled with corpses, ready to be covered with earth. It was an annihilation camp, this Carlshof. I must confess that I was very scared. I had to force myself to urge my horse to continue further, toward the house of the leader of this monstrous place. I found him drunk. I explained that I had come to get my blacksmith back as he was needed to keep the tools in shape, tools needed for the production of grain and meat to feed the population. He tried to explain, in a slurred speech, that this could not be done. I insisted, and finally told him that I would report him for hampering the production of much-needed food. This helped, and I was able to take Kowalewski with me, saving his life. The leader of the extermination camp later committed suicide.

From what I learned later, I have the impression that these camps were well known, but for self-preservation, they were a taboo topic. The German population, including the army, was deliberately screened from the atrocities that were committed there by the Gestapo and the SS. Members of the special units who ran these camps risked their own and their families' immediate execution if they talked about what was going on, as did anyone who knew and opened his or her mouth.

There were rumors. Occasionally the Nazis released selected concentration camp prisoners under the condition that they give fake accounts of what they experienced. There were heaps of misinformation and propaganda from all sides. Where in this mélange lay the truth? Maybe the BBC in London broadcast some of the truth, because people caught listening to BBC broadcasts landed in concentration camps. Many people were stuck between a rock and a hard place. They might risk their own lives in order to make the truth known and be true to their own beliefs, but such honesty became far more difficult when it meant the certain execution of your wife, your kids, your parents. I was very glad and relieved to learn that Papa never joined the Nazi Party or any of its affiliates. But even if he had, I do not think it would be possible for me to judge him. I cannot know what I would have done under the same circumstances.

By December 22, 1940, Papa had become deeply cynical. He wrote to Stresemann:

> Today we can only speak of dreams—the dreams of many: that we hold on to what we have despite the possibility that all will be lost as the destruction creeps threateningly and reachably closer. Before one noticed, half a human lifetime—for me a difficult but incomparably beautiful one—has passed on which stand as markstones the names *Tetragulla, Geomalia, Heinrichia, Aramidopsis, Habroptila, Tropagon,* and *Lymaticus.* I am unhappy without an open view to the future, but still I am happier than most of my countrymen. I also lay my heartiest wish for a New Year at your feet. May you in your work vigorously stride forward and our two loves, our beautiful collections in which the passionate efforts of the wilderness explorers meet with those of the scholar at home—may they long remain with us.

Papa strongly wished to escape from the insanity he witnessed around him, and so retreated into his own memories of wild natural places. One day while he was in Sepolno to buy provisions he came on a column of haggard men being marched by German soldiers through the streets. One of the prisoners was holding a poster that read: "We are going to Jerusalem." Papa recognized them as the Jewish merchants of the town. On the sidewalk looking on was a German soldier, who turned to Papa and said: "They should be ashamed"—and surely he did not refer to the Jews. Papa said: "I felt ashamed myself to be a German, and not to be able to help."

Papa's relationship with Hilde was now dangerous. She was Polish, and relations between Germans and Poles were *verboten*. In the summer of 1939, Hilde had become pregnant (with me), and she and Papa were unsure what to do. Abortion was illegal and unavailable. Hilde had no family to turn to: several special units had descended on the province of Poznan to "clean" the area of ethnic Poles. Hilde's parents and her sister Elisabeth were told to leave their home. *Raus, raus!* ("Out, out!") They were permitted to take one suitcase. Within half an hour they were packed into a crowded van with forty-five other people, no food, and no toilet facility. After three days' journey, they were dumped in east Poland to fend for themselves. That left her brother Rüdiger, who was in a prison camp; another brother, Jan, who was in the British Royal Air Force; and the youngest brother, George, who worked on a farm. Hilde was not safe at Borowke, and Anneliese and Papa had to find a haven for her. They came up with a bold plan.

Papa had heard about the Nazi SS program of *Lebensborn*. They wanted a "population explosion" of Germans. To achieve this goal, they made it easy for unmarried girls to give birth in agreeable surroundings with nurses and doctors to care for them. One such establishment was in Polczyn (then Bad Polzin), which was far enough away from Borowke that Hilde would not be known. Papa and Mamusha traveled there and on arrival they were carefully inspected for racial purity. Ironically, Hilde (going by the name of Bury rather than the Polish Boruvna) passed immediately, but Papa's skull apparently was misshapen. After some debate and more measuring he passed scrutiny, but just barely.

I was born safely. Mamusha later told me that she tried to "rush the birth," because she did not want me to be born on April 20, Hitler's birthday. I was born on April 19, which, it turns out, is the anniversary of Charles Darwin's death. The name they gave me was Bernd Bury. Shortly after my birth Anneliese came and, as Hilde (whom I will from now on call Mamusha) says with rancor in her voice, took me away to Borowke. Admittedly, possibly my mother had little choice. Although she was officially certified as German by a doctor at the home, locally everyone still knew her as a Pole; and while the neighbors might be tolerant of her, there was a question whether someone might snitch on Papa, putting both him and the child in danger. If she could not come back to Borowke with me and had no place to go, as she claimed, then one might presume that she should have thanked Anneliese for bringing me to a safe place, rather than blaming her for taking me away.

When I arrived in Borowke everyone knew that I was not Anneliese's child, but no one asked questions, and so I became Bernd Heinrich. Hilde was told by Papa, "Of course he is your son. Nobody can change that."

Given the times, it is incredible that Papa, while on military leave, visited Mamusha at her rented apartment and got her pregnant again. This time they could not go to the *Lebensborn*, since, as Mamusha explained, "they had family values and did not approve of one having babies every year." Mamusha attempted to induce a miscarriage by immersion in hot water, risking damage to the unborn if it didn't work. It didn't. She needed someplace to go, and eventually Papa decided that they should turn to the von Sandens, an aristocratic family in Berlin. They were more than just aristocrats—they were also naturalists from Papa's circle of friends. They introduced Papa (and later me) to a very special rare mouse, the tiny *Hazelmaus* (harvest mouse in England). It is a beautiful tawny creature that specializes in eating raspberry seeds and weaves little aboveground nests out of grass, where it hibernates in the winter. The von Sandens kept these mice as pets. Mamusha was taken in by the family and my sister, Marianne, was born in their house on June 3, 1941.

One day a few weeks after Marianne's birth, as Mamusha was changing the baby's diaper, she was handed a letter from Papa. She opened it eagerly. It said: "I have just met a beautiful woman and I'd be a fool to leave her alone." Mamusha was so furious she said she "felt like throwing the baby out of the window." The long-suffering and forbearing Anneliese visited shortly thereafter and told her, "You should be happy for him." Mamusha had trouble mustering that sentiment. Many years later, in Maine, after Papa had told her about yet another new girlfriend, Mamusha said that Anneliese again said, "You should be happy for him." This time, Mamusha told me, "I threw the glass of beer I was drinking right in her face."

As an adult I found it quite difficult to comprehend Papa's self-centered behaviors—behavior that bordered on cruelty. How could Mamusha allow herself to be manipulated so brutally? Why did Anneliese not only put up with him but even work as his accomplice? I corresponded with Ulla about these questions. She wrote:

> He never could feel or see beyond his own nose. And about Mamusha being forced to give up her child to another woman, it was not the only way out. The original goal was a "frame" or cover that would allow her to remain at Borowke. The insults that she put up with in order to hang on to Gerd are unbelievable. It must have demolished her self-esteem completely. She did not love Gerd, she loved all that he represented in the eyes of many. And so did my mother [Anneliese], much to the contrary of what she tried to make herself and everybody else believe. Anybody who stayed close to Gerd for any length of time had to become very manipulative and sort of twisted. That seemed to be a sort of prerequisite of survival with him.

I think Ulla was partly right. Given the times, everyone was looking after his or her own interests, and contemplating how one's actions might or would affect others down the road was a luxury. Papa was probably no more deficient than others in his ability to consider the future. However, for years during World War I he had lived "on the edge," where every day of life was a gift, if not a miracle. Now, he was once again witnessing literally thousands of innocents dying in the cold hell of Russia, where he was stationed. I think such experiences could recalibrate one's perceptions of what is acceptable behavior. Additionally, Papa had on his expeditions been exposed to numerous cultures. He would have asked himself why one way of life should be any better than another. He prided himself on not being bound by prevailing norms. As a result he was free. But his freedom exacted a high price from those who relied on him for their happiness or well-being. A different aspect of Papa's character emerges in his interactions with strangers, especially fellow soldiers.

• • •

IN 1941, WHEN HITLER ORDERED THE WEHR-
macht into Russia, Papa and his unit, the Luftgaukommando II, were moved from Warsaw to Smolensk. The staff took quarters in a large villa about six miles north of that city, and Papa was made head of the *Front-Sammel-Stelle* (front collecting place). It was established in the Hotel Molotov, at the marketplace in Smolensk. Once a week a major of the high command in Berlin would fly to Smolensk and bring a top-secret list of changes in the positions of the air force units in the median sector of the eastern front. It was then Papa's job to inform soldiers arriving in Smolensk from Germany where they could find their units and how they could get there.

As Christmas 1941 approached, Papa received at his field post in Smolensk a number of parcels from home with delicate cookies, cakes, and chocolates. On Christmas morning a young soldier, a boy of eighteen years "at the most," Papa recalled, stepped into his office. He saluted briskly and said that he was on his way to the Moscow front, but he wanted to stop and see his brother-in-law, who was a major serving on the staff of Luftgaukommando II. The boy asked where to find him. Papa told him that the Kommando II occupied a villa near Smolensk. Since there was heavy military traffic on the road, he would easily find a vehicle to pick him up and take him there.

The soldier left, but before dark he was back at Papa's desk. He was depressed and very disappointed. He had found his brother-in-law, who told him that his boots were dirty, that the staff had no place for him to stay overnight, and that he should hurry on to get to his unit. Papa said, "I was furious at the cold, inhuman behavior of his brother-in-law, but did not say anything. Instead I suggested to the boy that he spend Christmas eve with me and proceed toward Moscow the next morning."

Papa lighted a few candles and shared his treats with the boy. They talked of peace and the past, of the future, of memories and of hopes. Early the next morning the boy left for the front. I suspect Papa may have known what awaited him. Not long thereafter he received a letter from the boy's mother. Her son had written about the wonderful Christmas evening he had had with Lieutenant Heinrich, and the mother thanked

Papa for the last happiness he had given her son. The first day after arriving at his unit he had been killed in combat.

Papa sometimes walked to the railroad station that was not far from the Hotel Molotov, to see what was going on and learn about what was going on at this front to which he was sending so many boys. Several times he found cargo trains standing there, which gave off an offensive stench that one could smell even at a distance of thirty yards. It was the stench of decay from the frostbitten feet of hundreds of soldiers lying on straw in the cattle cars. He was horrified and deeply moved. After he saw these trains several times, Papa became convinced that Adolf Hitler's war against Russia was lost. It could be fought and survived not in tight-fitting leather boots but only in felt boots like those the Russian soldiers have always worn. However, such trifles as clothing or equipment could not bother the brain of the "greatest commander-in-chief of all time," as Hitler considered himself. He led by personal inspiration, regardless of empirical facts, regardless of the lessons of history.

As in World War I, the Germans had rushed or been led deep into unfamiliar territory until they were isolated with their supply lines cut. It was an old story—the same situation that Napoleon had faced on his retreat from Moscow. It was also the same situation that the Greek historian Herodotus described in Book IV of his *Persian Wars*, when Darius the Great of Persia in 514 BC met the Scythian warriors on horseback and with ox-drawn wagons, with his reputed army of 700,000 men. Darius sent a messenger and demanded surrender. But the Scythians replied, "Go weep," and they then "retreated" out of reach, to successfully continue their guerrilla war. Similarly, although the Germans advanced swiftly and for the first six months, three years after Hitler had sent the 2 million men into Russia, a quarter of them were lost.

In Smolensk, Papa found himself in a field of rubble that had once been a thriving city. The roof and uppermost stories of a brick building that had stood many stories high had been destroyed, but the lower stories were still standing and served as a prison for thousands of Russian soldiers. Typhoid had broken out and they were dying by the hundreds. After the worst harvests and coldest winters in decades, there was not enough food, and most were starving and freezing. Those who survived sometimes resorted to eating their dead comrades. If the Nazi guards caught them in the act, they were shot dead for being cannibals.

With the overwhelming number of dead prisoners came the problem of how to dispose of the bodies. The ground was frozen deep and hard as rock. Instead of being buried, the corpses were piled up, creating a wall along the border of the prison camp. Papa estimated that the wall was, in midwinter, about 300 feet long and almost a yard high. He wrote that this wall of the dead was "the most terrible and sinister sight that I have ever seen." Could anyone seeing all this, and feeling it, and being alone, brush away a woman who also needs companionship and who provides friendship and warmth?

After New Year 1942, the temperature dropped to minus fifty degrees Fahrenheit with the snow knee-high. The Russian winter offensive was under way and desperate fighting was going on both to the east and to the west of Smolensk. If the Russians conquered the city, Papa felt, "all would be lost" (for, presumably, both the Wehrmacht and him). But good fortune once again miraculously came his way: his unit was transferred from Smolensk back home to Bromberg, Poland, and from there Papa was later released from military duty for the spring and summer in order to oversee the agricultural productivity of Borowke.

Papa's experiences caused him to retreat more and more into his dreams. On March 9, 1942, he wrote to Stresemann:

> *Teurer Meister!*—Those glorious forests in which the proud *Tropagon* and *Syrmatius* make their home, where white orchids bloom on old pines and *Aethopygen* visit the rhododendron flowers, I will never see again. That is for me the most painful thing, since these lands were to me the richest and most beautiful. I need them still to finish my small zoological lifework: "The Opuscula Ichneumondgia Orientalis." As best as I can, I must finish this work in any case—if enough lifetime is given me. In the last month I made my reckoning "with heaven." All was clear: "nothing missed—life's richest gifts in all the wonders of nature and the fairy-tale existence enjoyed"— now only one thing weighs on me: that the many years of hard-won and perhaps only knowledge of the system of Ichneumoninae not be left behind.
>
> One thing is certain: after I leave my official soldierly duties I will devote the rest of my life—if there is one—after the war only and exclusively to the service of zoology. This shall be my goal!

It was a promise he kept. On June 29, 1942, he wrote again:

At the present I lead the most boring, barren, and unimaginable nothing-life that is possible to conceive. After I have finished sitting my ten hours, for leave seekers, discipline cases, etc., I can finally go into my quiet room and steep myself in the details of the body markings of Burmese ichneumons, to discover, dream, and to fix my imagination freely on mountains, to be in secret thickets and imagine the whole beauty of my past life.

He continues, on October 23, 1942: "I spent part of this month at home on the *Kartoffelschlacht* ('potato battle'). This was not so easy, since my battalions consisted of English, Russians, Polish, and Germans. It stretched my linguistic abilities to get everyone fired up to the required work ethic." A few weeks later he wrote, "The potato harvest was nerve-racking. There were few workers and the harvest was very poor (fifty percent of usual). Prepare yourself—a very poor year is on the way!"

A contingent of forty British prisoners of war was assigned to work at Borowke. Papa told me about getting them all together to give them a speech. In it he spelled out their common predicament and how they would all need to work together to live sanely through such insane times. Apparently he lived up to what he preached, because I found a long chatty letter (in Papa's correspondence in the barn in Maine) from one of those British prisoners, George Weatherly, written after the war. It was to Anneliese and was dated May 27, 1946:

I received your letter a few days ago, thank you very much. I am glad to know you are all right. I think all the prisoners from Borowke got to England all right—I have had a letter from little George (you know, the Professor), a few days before I had sent him a letter telling him I had received a letter from Ulla. He said he was very glad that you are all OK. I have had a letter from Lotte [Liselotte]. I still have got the watch she gave me. I also have a letter from Ruth [Michelly, Ulla's friend who drew the map of Borowke]. . . . I have asked at the post office if I could send you cigarettes but it is not allowed. Well, I think that is all for now. Give my best wishes to Erna, Frau Bury [Mamusha], and your husband. Best wishes from Kathleen, and June.

He also says in his letter, "It was the chief's humanistic treatment of the British prisoners of war that made life worth living." George, who had been a bobby, or policeman, in London before the war, and a gardener and cook at Borowke, was one of several prisoners that I vaguely remember. I had spent my first three years at Borowke, raised in the midst of the war. George had shown me hummocks of moss in the woods, where the gnomes of the forest made their houses and got out of the rain by standing under mushrooms. Visions of this magical forest and the tiny gnomes living under the beautiful green moss and stepping under the bright red-capped amanita mushrooms with their white spots stayed with me for years.

Another of the Britons, who worked with the womenfolk in the kitchen, was called Starry. Mamusha remembers him as having lost most of his teeth. Starry would spend hours sitting in the kitchen polishing Papa's guns, so I suspect he had much idle time, and he sought diversions. One of these diversions that the Brits and the women collaborated on was trading cigarettes and coffee for eggs and butter, and making brew by fermenting raisins that they got in their packages from England. Another project was to teach us English, or at least a few select words.

In the kitchen there was constant chatter in Polish, German, and English. One person would translate for another, but words were remembered, as most learning is, by making the context memorable. I remember being chased round and round in the kitchen by one of the Brits—I think Starry.

The key to effective teaching is to be simple and uncomplicated, and Starry was both. Round and round we went, and soon I was picking up the beat, and his words: "Captain Focken Bastard, Captain Focken Bastard." We all knew the first word, which probably referred to Papa—although Mamusha then assumed the other two referred to somebody else. These words, which I picked up effortlessly, were the first of my English vocabulary.

The Brits' legacy lives on at our house in another way, with cookies at Christmas. Every Christmas Mamusha makes a special kind of cookie by rolling a white and a chocolate strip of cookie dough into a log shape, and then slicing off rounds to bake. By chance my wife made some just like that, and I said to her, "Mamusha will like these—they are a tradition in our family." When we offered them to Mamusha she said, "Oh—Caruso cookies!"

What did she mean? "One of the British prisoners used to bake them in Borowke. He sang with a beautiful voice. So we called him Caruso after the great singer. That's why they are Caruso cookies."

After about a year, when Papa's role as potential father was no longer an issue, Mamusha ventured back to Borowke, bringing Marianne with her. She said "things had settled down" and she felt safer. Papa could not so easily be accused of having had relations with a Pole. But if he did still consort with her, he tried to do it elsewhere. On one agricultural leave he left a manager in charge of the estate, and despite the raging air war on the cities, he and Mamusha traveled to collect ichneumons in the country in southern Germany. When they returned, Papa wrote to Stresemann on August 20, 1943: "Borowke is no longer as idyllic as in May." I think he could only have been referring to the change of season. Papa did allow that "one thing remains unchanged: the hunt for bats every evening. That is my greatest enjoyment and my greatest interest." He now had nine species.

Two weeks later, when Papa was recalled to military duty, he left with a premonition that he would never see his beloved home or family again. Before he departed he arranged for his trusted friend, whose life he had saved, the blacksmith Josef Kowalewski, to prepare one of the large harvest wagons in case the family would need to escape westward when and if the Russians approached Borowke.

TEN MONTHS LATER, AFTER NEARLY FIVE YEARS of war, the tide was turning. The Allied assault at Omaha Beach on June 6, 1944, took place from the English Channel; and the American and British offensive from the west soon crossed the Rhine at Remagen. In the south, the French crossed the Vosges River, and with a nearly simultaneous assault on June 22, 1944, Marshall Zhukov came from the east against the German lines in Russia. He had the guerrilla forces come out of hiding from the forests where they had been harassing the German troops by setting booby traps and bombs and cutting off rail lines. The giant trap was sprung as troops of Russians, Siberians, and Cossacks crossed the

Oder River. Zhukov ordered 26,000 heavy guns and screaming Katyusha rockets to be let loose, followed by 4,000 tanks and 1.6 million Soviet soldiers. More than 70 percent of the Wehrmacht would fall on the Russian steppes. In a four-year siege of horrendous carnage, 40 million people died, and an estimated 2 million German women were raped. General Ludwig Beck, chief of staff of the Werhrmacht since 1933, had predicted such a rout. He had reported to Hitler that invading Russia might end even more disastrously for Germany than doing so in World War I did. He predicted a military catastrophe because he knew that Germany did not have the resources to wage a protracted war.

Hitler of course ignored Beck's report. Beck in turn grew disillusioned with the moral decline of the regime and the incompetence of the leaders, and he became a leader of the resistance. For this he was executed in 1944 in Berlin, after the failed plot to assassinate Hitler.

Near the end of the war Papa was appointed commander of a small, little-used air base in Strebelsdorf, near Danzig. Life was still quiet and peaceful there, and Papa followed peacetime customs and paid visits to the country squires in the neighborhood, including the nearby estate Felstow, which belonged to the Kramp family. It was situated off the beaten track in the middle of a forest that reminded Papa of the forests back home. He liked to visit Felstow and the old Kramps, whose son had returned home from military service as an invalid after he lost a leg in combat. In a letter to Stresemann on July 7, 1944, Papa expressed his premonitions about the end of the war: "The events are rushing over us. I was in the middle of the harvest, and got called back to military duty. I tell you this so you will not come, as per [the invitation in] my last letter." On August 4, 1944, he wrote again: "Just now came to us from east Prussia the first wanderer—a mother with three children," presumably fleeing the feared Russians. Papa saw this as a sign of what was to come.

On one occasion around this time, Papa was seated at dinner next to a small, white-haired woman, Else von Bonin, who lived on an estate near Strebelsdorf. She asked for an honest opinion: Will the Russians advance still further into this area? Papa hesitated for a moment, wondering if it was his duty to lie or to give his honest opinion. He decided for the latter and said, "I am afraid that one day the Russians will push back our exhausted and decimated forces and will force us to retreat from here toward the west." When Papa first told me what he had said, I thought: So *what? Wasn't that obvious?* Later I learned a little history and found out that expressing any doubt about the power of the German forces was considered defeatist and labeled *Wehrkaftszersetzung* ("weakening of the defense"). It was a crime for which thousands had been executed. The official word spinning tended to project the opposite of what was actually happening. War was portrayed as glorious triumphs, even during catastrophic setbacks, and the enemy was almost always "cowardly."

Shortly after that conversation Else von Bonin was found dead, from an overdose of sleeping pills. Countless others followed her example.

Papa made plans for the worst-case scenario. If we had to flee Borowke, we were to come to him at the base in Strebelsdorf. But if we lost contact in the expected chaos, we had fixed an alternative meeting point in Berlin. Papa also took precautions to see that his most precious possession, the manuscript about the Oriental ichneumons, would be safe, along with his most valued type specimens, which were already buried. On August 21, 1944, he wrote,

> To all the following mentioned friends and colleagues: My manuscript "Ichneumoninae Orientales" contains the first comprehensive and systematic work of the Ichneumoninae of the Oriental region [primarily of Celebes and Burma]. It is a lifework that has cost the author thirty years of unremitting work and much money for expeditions and trips to study type specimens. The publication of the work was to proceed in 1943 by the publisher Dietrich Reimer in Berlin, who had published my popular books and had personally said he would be prepared to print this purely systematic work. Total war and the prohibition against publishing anything not war-related prevented the execution of this promise.

Papa then said that there were two copies of the manuscript. One, the original, was finished and was welded into a tin box and put in the possession of Professor Max Vollmer in Thüringen; his address was given. The second incomplete manuscript had been sent to Stockholm before the war and given to the ichneumonologist Dr. A. Roman to safeguard. Papa explained that this second manuscript was missing some portions (those he had completed during the war), but if the original was lost, it could serve as suitable backup and still be published.

"I sincerely ask each of my friends and colleagues, whom I let read these lines, in case of my untimely death in the war, to help bring this manuscript to publication." He lists fourteen people, ten of whom are academic doctors, including professors E. Stresemann, Walter Grotrian of the Astrophysics Institute in Potsdam, and W. Ramme of the Berlin Museum. Also listed is his friend "Mek," Dr. M. Dammholz, with an address at a hospital in Tornow bei Buckow.

Then, in a letter to Stresemann dated August 25, 1944, Papa writes four pages of fond remembrances of the old times, and he thanks his *lieber Meister* for everything. Papa makes no mention of the ichneumon manuscript. Is he saying a final good-bye?

For a long time, the possibility that the Russians would invade and we would need to flee from Borowke seemed remote. Papa had almost daily telephone connection with Borowke, during which we always said, "Nothing new."

On January 4, 1945, Papa wrote:

> The months fly by like clouds in the wind, robbing me of the actual content of my life. I work with passion and industry on my airport and the fatherly leading and service to my people, and enjoy life as opportunity offers. Quite amusing, to observe myself. But I cannot delve too deeply and lose myself in the romantic forests of remembrances, or I get caught with an elemental force of homesickness. The horned lark twittered today so happily—as if knowing spring is coming. What a wonderful impression! And if one could relive the swamps and the hunt for *Aramidopsis* then all would again be wonderful.

Then one day Papa could not contact Borowke. That could mean only one thing: the Red Army had arrived. He wrote to Stresemann immediately (January 21, 1945):

> Maybe before too long or very shortly, a wanderer from Borowke with little children will knock on your door. Please be so kind as to help them get shelter and to get them farther—to anywhere, where one may still be allowed to live a little longer. For the young, existence is always worth it. At our age it is only not so when we are

separated from our great chieftain, Nature, or when we have lost our loved ones.

Papa waited and waited. There was no word, no way for him to find out what was happening at Borowke. Two days later, on the night of January 23, 1945, he again wrote to Stresemann:

By candlelight—lack of coal, the electricity is turned off—I want to write to you this night in the hope that my letter will then be taken by any machine and in the further hope that the bird will then bring it to your hands. I have nobody else to whom I can send this.

Two things above all lay weight on my heart:

(1) My manuscript. I just received from Professor Volmer the exact location where it was deposited. Please note it and also give it [the address of the location]—Schmölln/Thuringia, Sösstrl, the residence of docent Dr. Stande, in the cellar—to three others (Hedicke, Ramme, Birkoff).

(2) From home I have received no news since the eighteenth, when the Russian invasion reached [German-occupied Poland]. The realization that my family find themselves under such indescribable conditions with the small kids, and without protection and help on their trek through the ice-cold winter, tortures me immeasurably, especially since I have no idea if they got away in time. I count on being here another week. (Tel. Gr. Boschful 43.) But then I suspect the connection will be severed.

Another two days later Papa wrote again:

No word from my family for eight days—and the Russians are already in Bromberg and near Schneidemühl. I worry myself sick. If you should get any kind of news, please, please try to find a way to transmit it to me. What is one's own fate in this ocean of suffering on whose shores we stand. Had we really at one time been able to study nature and wander through sunny meadows? Those were wonderful dreams of flight, and we awoke from them one day as flight-deprived slaves. With the shield, or on the shield! Ergo: when the swamp warbler sings again we will be either decayed and re-

turned to earth, or saved. And this short-range alternative has much
that is comforting!

Papa continued to wait for us at the base, as a stream of humanity
passed by. Finally, giving up hope, he drove to his superior officer and
asked for and received a few days' furlough to travel to the bombed city of
Berlin to seek news of his family. He wrote to a friend, "I am on a difficult
trip to Berlin to look for my loved ones. There is likely no more doubt that
they have all been left with the Russians in Borowke. Life has no more
meaning for me and if I perhaps live a while longer, then only for the
small hope that I may see them once more."

On the outskirts of Berlin, at the designated contact point, he found
a letter that had been delivered from Anneliese. She had written from
Borowke: "All roads are completely plugged by the military and masses
of civilian refugees." She feared that many hardships and dangers awaited
the fleeing population and had decided not to leave home. However, she
wrote that when the artillery fire was close by, Ulla, now nineteen, and
Mamusha decided to leave and took with them Marianne and me, four
and five years old at the time. She reported that they had taken back roads,
hoping to hide at a remote farm where there was a small cabin belonging
to the family of our cook.

When Papa read Anneliese's letter (which she had written on January 23 but he had not received until February 18), he was horrified to learn
that we had left Borowke only when the Russian army was virtually at our
doorstep. That had been over a month ago, and all this time there had
been no other word from or about us.

nine   The Flight

THE RED ARMY INSTILLED TERROR. THESE WERE people from another world, who reputedly ripped out faucets expecting them to continue to yield water, who washed their potatoes in toilet bowls and defecated in open pianos. Fear of the Russians came from firsthand reports and from propaganda, and also had a historical basis. There had, throughout a history of thousands of years, been two types of people—the settlers and the wanderers. Roving bands of nomads had preyed on the farmers: the Asiatic Cossacks on their horses; the Vikings in their ships; the Mongol horde, who killed, looted, and pillaged without mercy.

"Everybody's constant great fear of being overtaken by the Russians may sound like mass hysteria, may sound even silly and unfounded," says Mamusha, "but—we had known too many facts about happenings to people and places where the communists swept through, and so we did all we could to avoid similar treatment. Women were raped as a matter of every soldier's perfect right; they herded the women into some barn or other convenient shelter, and long lines of soldiers performed; young girls and older women were not exempt. When a man happened to be still on his property he was killed; if his wife happened to be there too, she was raped many times and her husband had to watch the proceedings before they killed him. We did not want to be sent to Siberia; we did not want to be separated; we were told that the children would not be left with us should we be captured."

By that time, Papa's parents were already dead. Opa had been sick and had, against Papa's recommendations, been admitted to a hospital in Bromberg. The patients had been loaded onto a train to try to escape the Russians, and that was the last anyone had heard of them. Opa became one of the millions who simply ceased to exist. Oma had suffered a stroke and was mute and paralyzed. She was not likely to recover and could not be taken along if we tried to escape; nor could she remain behind, to await the Red Army. Under the direction of our doctor and in the presence of the family, she was given a lethal injection and then buried with great pomp and ceremony in the Borowke cemetery, next to her daughter. For the rest of his life, Papa always kept a picture of her near him.

As the Red Army advanced, the German civilian population tried desperately to escape. At Borowke, we were particularly vulnerable targets because we were landowners and "intelligentsia," both hated by the communists. The Bolsheviks were killing anybody—of German, Russian, Polish, or any nationality—who had as much as a high school education. These were the people who were expected to make trouble in their newly conceived proletariat state. The "workers" had less to fear. The resulting exodus of Germans from the path of the Red Army has been called "one of the biggest and bloodiest mass migrations in history." After the initial Russian combat troops, called "front pigs" by both Germans and Russians, had passed, more disciplined soldiers generally replaced them. They could be kind and just and sometimes were even known to execute their own who had taken the opportunity of war to behave like beasts.

However, I'm not sure one could blame the primitive drunken foot soldiers more than the British and American pilots and crews who rained bombs down onto the heads of women and children. It was perhaps justified, but to us it didn't matter. All that mattered was that we get away. We would become part of the exodus, and we would survive by what now seems to me like a fabulous chain of luck. Over a half century later, it still shocks me with joy and makes me aware that life is a gift.

ON THE EVENING BEFORE WE LEFT BOROWKE, Mamusha, Anneliese, and Ulla heard a faint but ominous booming in the distance. I was almost five years old and have no memory of it at all; neither does my sister, Marianne, who was four. Such is the gift of childhood (and the cloak of ignorance): one feels safe in the midst of a maelstrom because the dangers remain hidden. Fleeing seemed like the prudent thing to do for the adults, but the Nazi administration had forbidden it and made any attempts to do so difficult. Money could not be withdrawn from banks, travelers' papers were checked, and roads were supposed to be kept free for the army to move on.

Papa had arranged for Anneliese to depart with everyone in the large wagon driven by our blacksmith, Josef Kowalewski. Our immediate goal was simply "west," but our ultimate destination was the home of a friend of Papa's, some 375 miles away in western Germany near the border with Holland. When it came time to leave, Anneliese could not bring herself to do so. She was, at nearly Papa's age, too old to be willing to gamble on her abilities in the face of anticipated dangers in which strength and endurance would count. She had, in any case, to some extent taken Oma's place and was strongly rooted to Borowke. Liselotte chose to remain with her sister. Both had been my caretakers, as well as Mamusha, and they wanted to keep me there. Naturally, Mamusha wanted to take her children. But it was considered unwise to take both children on this dangerous journey where quick mobility might be essential. "At least save one," Papa had told Mamusha, and "at all cost do not fall into the hands of the Russians." Finally Anneliese relented and let Mamusha go with both kids, provided she take Ulla along. Ulla, who was one month beyond her nineteenth birthday, was old enough to help take care of us. I have few hazy "snapshot" memories of our leave-taking from Borowke: the long strong hug from Anneliese, who had cared for me and loved me as her own child, the snow falling in the darkness of an early evening. As we said our good-byes to Oma's longtime chambermaid Erna, and the others who stayed on, there must have been something in their eyes that caused me to remember the moment when we were loaded onto the horse-drawn sleigh that Mamusha chose as our means of transportation (there was a big snowstorm and she thought the sleigh would get us farther than the wagon). We were bundled up in furs, and then we left, gliding under the great chestnut trees in front of the house for the last time.

I was spared the knowledge that we were abandoning forever the wheat fields where the blue strawflowers and the crimson poppies grew and where the lark hovered high in the air singing its melodies while the reapers working in rows bent into their scythes with sweaty backs. I did not yet know that the dark green potato fields ablaze in white blossoms in June, Margarethe's many beautiful paintings of flowers and butterflies, the barns and the horses and cows, the forests, all were being left behind for good. Papa's collection of ichneumon wasps, which he had begun thirty-three years earlier and which filled half of a big room at Borowke, had to be abandoned. It contained the specimens that he, with the help of Anneliese, Liselotte, Mamusha, and others, had collected in Celebes, Burma, and Iran, and from large areas of Europe.

Papa had fully expected to lose it all, but in a desperate and possibly hopeless gesture, he had picked through his collection to pull out all those with a red label—the type specimens. They were the ones used to make the new species descriptions for publication, primarily of the manuscript on Oriental ichneumons that he had schlepped with him throughout the war, to work on whenever opportunity afforded. There were about 500 in this collection of type specimens; and before he left Borowke for the final time, Papa, with the help of his trusted blacksmith, Kowalewski (who would not divulge their project), had sealed these precious wasps into a double-soldered tin box (one box inside another). He then chose a spruce thicket on sandy soil near the Borowke house, and buried this treasure, carrying the soil away so as not to reveal the site. After the box was lowered into its "grave," as he called it, he covered it with soil, moss, and twigs to resemble the surroundings, much as the *Ammophila* wasp camouflages her nest site. He then made a map of the site that he sent to three of his friends, asking them to take care of the specimens if he died in the war.

My account of the events that followed our departure from Borowke comes from many sources, but mainly from Mamusha and Papa, who would recount the story over and over in the years to come. It is derived from the almost pedantic habit common to both of my parents, and then later to myself, of writing things down. When we left Borowke, at around five o'clock on the afternoon of January 23, 1945, Mamusha took what she could of Borowke with her. She took the family photo album and the Borowke guest book containing pages of notes—memories—that she loved to read through. She continued to make journal entries in this book, and it became a diary that traveled with us to America. In addition, she took Opa's old brass clock. Under their coats, Mamusha and Ulla carried his pistol to ward off German deserters who might try to steal our possessions. And in their pockets they carried cyanide tablets, in case we were overtaken by the Red Army.

Ulla and Mamusha were driving the team of four horses (two were "spares" in case of theft or injury) as we began this fearful journey. "Ulla," Mamusha declared, "our great adventure has begun. We are alone now, alone. We must be brave. Let's sing!" And they did sing, to the pounding of the horses' hooves, and the gliding sound of the sled's runners on snow. Mamusha wrote in the Borowke guest book that served as a diary: "The kids, Pise and Butz [as Marianne was called], are comfortably huddled in the back of the farm sleigh in a large fluffy fur rug. Stefan Pawlowski is riding Schnausi, Ulla's horse, to accompany us a part of the way to reassure Anneliese that we are safe and in good spirits."

Around ten o'clock that night we arrived at the home of the Pawlowski family in a village called Obkaz, forty kilometers to the northwest of Borowke. The Pawlowskis were relatives of our cook, who had suggested that we go there and hide till the Russians moved on and we could continue our journey west, or weather out the snowstorm. After a rather crowded night with eight people in a one-room hut, Mamusha and Ulla went out to the highway to try and find out what was going on.

The scene on the road was utter chaos. German soldiers were fleeing on foot, on small sleds, in carts, and in horse-drawn wagons. All previous rumors about the Russians were confirmed by these refugees. In the panic some soldiers were taking horses from fleeing civilians, leaving them stranded. The SS were stopping people because Hitler had decreed that nobody was to leave. The sound of artillery was growing closer. The roads would rapidly become choked up with traffic and impassable, and Mamusha and Ulla now wondered if we would really be safe in Obkaz. Might it be necessary to go farther, and if so they needed better horses. They could also not count on the snow continuing to allow for a long journey by sled. Mamusha suggested returning to Borowke to exchange the sled for the large farm wagon that we had originally planned on taking, getting new horses and loading up on more food and supplies. The local farmers she talked with said, "Stay put; don't do anything so foolish as go back now!" But the next day, Marianne and I were left at Obkaz while Mamusha and Ulla went home. "It was worth all the trouble," Mamusha wrote, "because there was a love letter awaiting from Gerd."

The mood at Borowke was bleak. The booming of artillery was

closer, and ceaseless. Everyone now knew that the Red Army would sweep in and that soon there would be nothing left. All were busily slaughtering cows, pigs, and lambs; boiling meat; frying lard; making sausage to hide; and eating.

After exchanging the sled for a small wagon and loading it up with food for themselves and oats for horse feed, the two women had trouble getting the best horses since Lukowicz, the farm manager, felt he should try to keep the farm intact for his boss, Gerd (but they did get them: Koza, Rebecca, and Osiolek). As it happened, the Russians killed him the very next day when they arrived, and the property he had tried to protect was taken over by the communists, who still claimed it half a century later.

At this point Anneliese and Liselotte had decided to join Mamusha and Ulla on the exodus. But they wanted "just a little more time" for preparation, and suggested they meet up with us at a specific road crossing. Mamusha adamantly declined: "Absolutely not!" she said. "You are totally oblivious of what is going on. You sit here in the house while all of humanity is marching west and south and overflowing the highways, pushing and pulling and robbing. There is no way that I will be sitting there at the corner of the road . . . waiting helplessly. I want to be left alone with Ulla and the children. You go alone, too, and try to save yourself as best you can." Finally, on January 26, she and Ulla set out to return to Obkaz.

It had grown colder. The road was covered in deep snowdrifts, and the wind was howling. Mamusha and Ulla crept along in their wagon, shaken, frozen, tired, and no longer singing. The sound of artillery thundered from all directions. Finally, after six hours, much relieved, they arrived back at the hut in Obkaz where Marianne and I had remained. Days passed in Obkaz. Apparently the Russian offensive had stalled. Mamusha and Ulla spent their time visiting people in the area and trying to gather information. They took a short trip (12 miles) to the village of Chosnica, where a military post was still operating. Here there were many large trees lying in the streets that people had felled to slow the Russians. Ulla was scouting the village, trying to find novels "with a lot of romance" to read. Marianne and I were kept busy playing with some homemade toys. The farmer, our host, slaughtered a pig and arranged a sumptuous feast. The war seemed to have come to a halt, until one day (February 9), Russian planes flew low over the neighboring village and dropped bombs on the houses. The windows rattled. The ground shook. At night, the bombardment was heard all around, and the next morning we moved out of the hut and into a cellar at a nearby farm. It seemed safer there than in the exposed hut. We waited, and waited, and waited. . . . The sound of artillery continued, but did not get any louder. That evening we moved back into the cottage, but at midnight the noise of the bombardment again became deafening.

"Everything looked better in daylight," said Mamusha, and "we filled our time baking cookies." However, soon the thunder resumed and then all kinds of German military vehicles—tanks, trucks, jeeps—started gathering nearby. A German soldier of Polish nationality tried to convince us that it would be best if we stayed. To take off now in our horse-drawn wagon would be foolhardy, since we'd be in the way of military action and would get pushed off the roads.

Mamusha and Ulla were unsure what to do. But the choice soon became clear. Tanks stopped in front of our window. Mamusha looked out and yelled, "Don't you have a better place to park?" No, they did not. They had chosen our humble hut as their new staff headquarters!

They were a panzer division previously stationed in Kurland (also known as Courland, a region in Latvia), and they were retreating south and temporarily slowing the Russian advance. The staff were a Captain Kelsch; his adjutant, Lieutenant Franz Moertel; another lieutenant; the company chief Hannes Haberkamp; and the company doctor, Martin Alberts.

The officers asked: "What are you doing here? Don't you know what happens to women and kids who come into the clutches of the Red Army?" This was to be a sleepless night for Mamusha (although she claims Ulla slept soundly on a pile of straw under the table despite the constant activity of the officers who kept coming in and out and talking on field telephones). The sounds of artillery never stopped. She went out to observe the fireworks from a hill. When she came back, Captain Kelsch declared, "We have to leave here now, quickly. Come with us!" The panzer soldiers had already explained the danger of staying. Mamusha did not want to leave with them, but Ulla was anxious to go. Finally Mamusha made the decision to come, too. Thinking of Papa's words, "Above all, don't fall into the hands of the Russians," had tipped the balance for her.

"Ulla, wake up!" Mamusha woke the younger woman with a start and then rushed out to grab a package of silver spoons (one of which I still own) that she had brought along to exchange for food later if necessary, and had buried for safe keeping in the manure pile. "Ready? Let's go!"

Marianne and I were yanked out of our sound sleep and packed onto a stretcher in the medical tank with Martin Alberts. Mamusha and Ulla, three medical orderlies, and the doctor settled onto a bench opposite the stretcher in the same tank. We were off—riding in a panzer.

We arrived in Chosnica at six o'clock in the morning and expected it to be the end of our journey. But we'd barely arrived when we again heard: "We must leave immediately. The Russians are attacking." Now escorted by a number of tanks, we rode in a truck, first through woods and then through open heath country, toward Tuchela. On the way we stopped at the village of Reetz, parking for a while under the cover of woods as Russian planes flew low and started strafing with machine-gun fire. The planes circled and returned again and again, all the while being shot at and shooting back. Mamusha wrote in her diary, "We feel confident that our gallant knights will protect us. But the children are afraid." (I have only a very bare recollection of the circling planes now.)

Afterward, we found shelter in an abandoned farmhouse. Marianne and I slept soundly for twelve hours, and then at seven in the morning came the familiar order: "All hands get ready to move." There was time for breakfast, served by the orderlies, who "knew how to find food in this now desolate land," Mamusha said. She and Ulla entertained their new friends, lightening the mood by pretending to be Gypsies looking into the men's futures and prophesying splendid things and love just around the corner. The soldiers laughed, and soon we were on our way, to once again be sprayed by bullets from low-flying Russian planes. The doctor, by now called *Stabi* for "stabs doctor," grabbed me and Marianne and ran for cover, with Ulla and Mamusha following.

One evening we moved into another deserted village. The soldiers captured a pig, slaughtered it, and made merry while cooking it over an open pit. The aroma of roasting pork saturated the air, and everyone was looking forward to the feast. Then we were rudely interrupted by a shrill alarm—the Russians were entering the village at the other end. Everyone dashed for the tanks. We lost the pork, but surely it did not go to waste.

At our next quarters, about five kilometers before Tuchela, we were given a cozy room in an abandoned farmhouse, where we found toys. The doctor showed us how to use them, and his fellow officers joked that if they wanted to find the stabs doctor, they had to ask for the family doctor. Mamusha remembers that they had a big party with much alcohol; that there was lots of singing, dancing, and telling jokes; and that the officers in their black uniforms were "a sight to look at in the frame of flowers [us]." "Everything seemed funny," she declared, and "with much laughter one can tolerate a lot of nasty reality." The reality was that the Kurlanders' tank unit was running out of fuel.

"It was a sad leave-taking," Mamusha wrote in her diary. "We thanked the officers for their kindness. We shook hands, got the children from their beds, and then got into Stabi's car. The driver was instructed to take us to forty miles away, to Schwarzhoefen, a town near Braust, where we arrived at two o'clock in the morning."

From there we eventually managed to get on a truck that took us all the way to the railroad station in Danzig (Gdansk), again in the company of combat troops. It was late February now. We had driven through hilly country showing signs of spring. The highway was crowded with refugees, military vehicles, and columns of British prisoners of war who were, along with their German guards, hurrying west to get away from the Red Army.

Our goal was to try to reach Papa at his air base in Strebelsdorf. We were lucky to get on a train and arrived in Strebelsdorf late at night, anxious to finally see his surprised face. We located his apartment, and with great anticipation knocked on the door. Nobody answered. Mamusha and Ulla were stunned. But as had happened many times before and would happen again later, what at first seemed to be disastrous would turn out to be a blessing.

While we were knocking and calling for Papa in the night, a voice answered nearby, "Captain Heinrich is not here. He is looking for his family." It was Papa's orderly, who let us in and told us that Papa had expected us a long time ago. A week before he had given us up for lost and had then headed to our alternative meeting place in Berlin, to see if there was news or information. There he had received the letter of January 23 from Anneliese telling him that we had left Borowke to stay at Obkaz.

Papa knew that the whole area, including Obkaz, had been overrun

Babelsberg 18.II.45

Mein lieber Thank!

Ich war auf schwerste Wege
für 2 Tage von Danzig in Berlin,
um nach den Meinen fahren zu
können. Es ist wohl kein Zweifel
mehr, daß sie alle im russischen
Bereich in Bromberg zurückgeblieben sind.
Das Leben hat für mich keinen Wert mehr,
und wenn ich mich noch vielleicht
ein Weilchen in ihm halte, so einzig
auf die kleine Hoffnung hin, die
Meinen vielleicht doch noch einmal
wiedersehen zu dürfen.
Um Euch hatte ich große Sorge.
Ich habe mich dankbarlich der Nachricht
von dir in Obien erhalten, daß Ihr
vorläufig in Sicherheit seid. Was bedeutet
der Verlust alles irdischen Hab' und Gut's
gegenüber der Tatsache, daß Ihr mit den
Kleinen noch beisammen lebt.

Ich grüße Euch alle so herzlich!
Euer armer Ged.

Papa's letter of despair.

by the Red Army. After reading that chilling letter from Anneliese, he had
sunk into despair. Feeling certain that his family had suffered horribly, he
wrote a letter to his friend Thankmar Muenchhausen. He declared that
"without a doubt, they have all been left in the Russians' sphere, and life
has for me now no value" (fig. 13). The very next day, however, he received

a telephone message announcing a miracle—we had arrived at his apartment in Strebelsdorf! It seemed impossible to believe that anyone, let alone two women with two toddlers in tow, could have made it through the enemy lines. In three days, Papa was back from Berlin for a tearful reunion, one that choked him up every time he told us about it decades later.

But we were not out of danger. The Russians did not have the sympathy, food, or time to take prisoners, especially German officers. Strebelsdorf would be overcome by the Red Army several days later, and our arrival was now instrumental in getting Papa out just in the nick of time. The high-ranking officer who had brought Papa back to Strebelsdorf from Berlin, and who then joined in the festive mood of our reunion, had been smitten with love for Ulla. He knew the Russians were just around the corner and he had the authority to arrange for Papa to try to escort all of us out of this danger zone. He transferred Papa to Schwerin, far to the west, and "ordered" him to immediately take the family with him on an old wood-burning truck. (Gasoline had long ago become a precious commodity.) This slow, awkward truck was not only on a mercy mission. Its official mission was to transfer meteorological equipment.

Mamusha said she remembered that it took "forever" for the men to assemble the equipment and then start the truck. We were finally on our way—only to be stopped by military police: "The road is now closed," they said. "The Russians have just completed their pincer movement. We are encircled—caught in der Kessel (the kettle)."

We had missed our opportunity to avoid entrapment, perhaps by only an hour or so, because the soldiers on our truck had dawdled. Now the only possibility left seemed to be escape via Gotenhaven, the port on the Baltic Sea. We hurried back to the railway station of Stolp. But, when we arrived there, the trains had just ceased running. Nearby, at Stolp-Reiz, was an air base. Numerous refugees gathered there hoping to fly out, willing to go anywhere.

The next day, March 4, Papa, Ulla, and Mamusha met the pilot of a Junkers transport plane who was to fly a load of women and children to Kolberg, a small seaport. They spent a pleasant evening with the pilot, who agreed to take us all on. He allowed all of us to board the plane the next day, but when we were ready for takeoff, uniformed military police came to check the passengers. They ordered Papa—the only male passenger, and in uniform—to get off the plane. However, the pilot declared, "I am the captain of this ship"—he had the final say as to who could board. Miraculously, the military police left.

We were now set to go. There was only one problem: the plane would not start, we sat inside for a few hours before finally disembarking, when the motor problem was declared irreparable. Finally, on the sixth, a replacement motor had been secured from another airport. But then, the pilot realized that there was not enough fuel to bring us to our intended destination, Tutow and possibly Demmin. The sergeant in charge of the fuel pump at the air base declared that we already had the maximum amount of fuel permitted. Papa, a teetotaler, had enough foresight to have carried along a small bottle of very excellent brandy for just such an occasion. He gave it to the pilot, who negotiated the exchange; we got a little more fuel—enough, we hoped. Now greatly relieved and finally ready to depart, we met yet another setback: another motor was defective.

Apparently what we had boarded was a discarded wreck that had been left on the airfield. We sat and sat and sat in the plane for most of another day; and finally, in utter frustration, the pilot decided that our only option was to risk flying the short distance (about 12 miles) to Vitzka-Strand on the defective motor and attempt to get spare parts there. Despite these mechanical problems, there was a wild struggle to get onto the plane. We made it to Vitzka-Strand flying very low; but while there the weather became too turbulent to fly any farther. We had to spend another day on the ground. We sat around a small barracks with soldiers, and talked and dozed through the night, wakened often by our growling stomachs.

The next morning we got up at five, ready to start. But it stayed at that. We learned that the motor needed further repair. There was little time to fix it, because the German soldiers had been setting the runway with mines and were about to blow it up, to keep the Russians from using it. Somehow, the motor was fixed in time, and finally everyone boarded once again. The plane accelerated down the runway, but then suddenly veered to the side. This time there was something wrong with the rudder on the steering mechanism.

"Orders are orders," said the base commandant: it was now time to blow up the runway. He could no longer wait for this clunker to get off the ground, and the pilot had run out of options. He had to order all of us to disembark: "*Alle raus.*" Women and children, now whimpering in fear, disembarked in the heavy wind of a snowstorm onto the now deserted air base. All stood around in shock. Men swore, women shrieked, and children were crying and clutching their mothers, waiting for the end.

Suddenly, hope arrived in the unexpected form of a truck slated to bring refugees to the nearby Baltic coast. An escape by sea! A steamship had docked to rescue many thousands of the civilians who were caught in the Russian noose.

The panicked horde, of which we were a part, surged forward, fiercely jostling to get onto the truck. There was not enough space for everyone, and this might be a last chance. In the crush of women and children, Mamusha somehow made it on with me. Then, almost inexplicably, Papa yanked us back off, probably because he, Ulla, and Marianne had not managed to get on. Although we were told the truck would return, Papa, the eternal pessimist, did not believe there would be time (he was correct), and was certain we'd be separated. He knew that in this flood of humanity, with no means of communication, there would be little chance to reunite.

We had missed, it seemed, our very last chance by nothing more than a misstep. And so had the other refugees who were left as the truck drove off, leaving a panicked crowd of wailing women and crying children. Then there was stillness. The snow kept falling on the runway under the dark sky.

Decades later I learned that Papa's instinctive act had almost certainly saved our lives. We heard that the ship we had been trying to get on was torpedoed by a Russian submarine and thousands were drowned in the cold Baltic Sea. But at that moment we felt utterly doomed. We sat in fear on the runway. Then we started to trudge, heavyhearted, toward the woods to avoid the mine blasts that were to destroy the runway, and to await our fate with the Russian army. "We were devastated," Mamusha recalled.

We had not gone very far when the pilot of the grounded plane jubilantly yelled, "*Ich hab's gefunden—ich hab's gefunden—schnell einsteigen!*" "I've found it—I've found it—board quickly!" The Junker's crew, making one more desperate attempt to get the plane off the ground, had discovered that the steering problem turned out to be a minor cable dysfunction that could be fixed on the spot. Snatched from the jaws of an ill fate, we were exuberant to be airborne at last. The ground fog now helped greatly, because it shielded us as we flew over land held by the Red Army.

During the flight the fuel was running low and there was a question if we would make it to Tutow. Papa conferred with the pilot, and they decided to gamble on the fuel and try to reach Tutow anyway, rather than land at a nearer base that was commanded by E. A. Behrendt, a friend from World War I who might have provided fuel. It was a fortunate decision, because as we were coming close to that base, we saw black puffs of exploding flack in the air. "Those bastards are shooting at us," cried the pilot. "Don't they know any better?" It turned out instead that the Russians were already occupying the base.

I remember this ride of a little over an hour. It was bumpy, and I became airsick. One of the women had brought along canning jars for preserves, and I barfed into one of them. We joked for years afterward about how funny it was that she would take along empty jars at a time like this. She too might have been amused had she known that Papa had brought along thirty-two rattraps and a copy of a book about a snoring bird!

After landing safely in Tutow we tried to get onto a train in the mobbed station. No luck for three days; the trains were packed. But Mamusha managed to obtain coffee beans from some medics who "were very funny and good company." I suspect they thought likewise of her. After pleading and offering the stationmaster some of the coffee beans, we and the medics finally managed to board a cattle train.

It was time to celebrate. Mamusha put the beans in one of her blouses, and then pounded them with a stone to make grounds. The medics had a pot. Someone got hot water from the train engine, "And then we had a ball! It was the best coffee we ever had in our lives," according to Mamusha.

We reached Demmin (Mecklenburg) in good spirits, and Papa checked in at the garrison there to ask the commander if there was a chance to fly to Schwerin, where he had been ordered to report for duty. Papa said that this commander, who "looked more like an SA than like a Luftwaffe officer," was suspicious. Probably thinking Papa was a deserter, he asked him to wait and stepped out of the room to call the Luftwaffe high command. He received confirmation that Papa was indeed officially transferred to Schwerin. There were no planes going there, so we soon found ourselves once again on a train, settled into smelly hay in a cattle car, traveling farther west.

We arrived without incident in Schwerin, where Papa was put in command of the "Horst" company and he and Mamusha were put up in comfortable living quarters in the military barracks on the base, while Ulla and we kids were ensconced in a haymow in a barn in the village. Papa was put in charge of a group of soldiers defending the airport, and he was also instructed to give the men lessons in National Socialism. Papa replied that he had had no lessons himself, so how could he instruct? The base commander replied that he could remedy that. So, even now, during the collapse of the German army, Papa and other Luftwaffe men were sent away to Oranienburg to take a course on the topic. "To a man," Papa said, they had the same low opinion of the political ideology they were supposed to absorb in order to save the fatherland, but as military men they had to do as they were told. At the end of the course each man was told to give a speech. Papa's was on "The Snoring Bird." He said it was well received.

In Schwerin, I remember feeling afraid. There were daily bombings and I remember the loud wailing of sirens that announced air raids. Whenever the chilling sound began, we ran into a dark bomb cellar, and there we sat like trapped mice trembling on the benches in the darkness. After a while Mamusha and Ulla no longer went into the shelters, but instead, on hearing the sirens, they ran out into the open fields to try to get away from the town and the base, carrying and pulling Marianne and me along. I recall being out there and seeing smoke trailing from a falling plane. For about a decade or more after that, the sight and sound of a plane approaching filled me with apprehension.

We were now getting close to the "Amis" (Americans), who had already come within a few kilometers of Schwerin. The Russians were still advancing behind us. Finally we knew that one or the other would arrive at any hour. On the evening of May 2, Mamusha was in the mess hall at the base when she heard a special announcement: Hitler was dead! The officers were jubilant. It was the end of the war for them (although the official armistice was not until May 7). They were now free. They and their families would no longer be shot for desertion.

Papa had foreseen the end of the war for some time, and he had prepared for it. He had decided he would not defend the air base, because doing so would be "sheer madness." Neither did he want to be taken prisoner, especially by the Red Army. When the enemy was getting close, Papa sent a patrol out to scout. They returned in a great hurry, having seen tanks. Papa immediately ordered the quartermaster to distribute civilian clothes and shoes to the soldiers and then persuaded them to disappear. One soldier, "who was a university student in civilian life, a very good-looking, very nice boy," replied, "I shall go on fighting in civilian clothes as a *Wehrwolf*." Papa let the men take what they wanted and advised them to try to get home as best they could without being taken prisoner. Then he jumped on to his bicycle, heading for a hiding place that had previously been prepared in the woods. Mamusha and Ulla had built him a little cave under a huge pile of pine branches at the edge of a clearing by some densely growing young trees. Once he was in his hiding place under the brush pile, Mamusha planned to keep in touch with him while we waited out events.

Ulla and Marianne and I had been in the nearby village of Suelzdorf, living in a farmer's haymow along with many other refugees. Mamusha, who was still with Papa at the base, set out to join us. Along the route she had to hide for a while under bushes while a battle was under way. She made it to our barn, and we all sat and waited expectantly for whatever army would arrive. The shooting subsided. Then it stopped. Then we heard only rumbling.

At around six o'clock in the evening we heard the rumbling stop in front of us. Ulla sneaked a look out to the street. What she saw took her breath away: a huge tank painted with star insignia. They looked like the Russian star, but they were white, not red. Then they heard voices in English: "Come out, you bastards!" *Bastards!* One of the two English words we had heard before. No words could have sounded sweeter to us. Amis! Along with the other refugees, we rushed out to greet them joyously: "Welcome to Suelzdorf!" The soldiers were just as happy as we were to see the end of the war, and to receive such a friendly reception. Stepping out of their tanks, they passed out chocolates and cigarettes.

A few hours later, well after dark, Mamusha left us and went to Papa's hiding place. As she came near, she imitated the call of an owl, the signal they had agreed on. There was no answer. He was not there! She left some food and a note to tell him what had happened, then returned to our haymow in the village. Near eleven that night, a couple of hours after she came back, we heard "our" signal outside—a unique human signature whistle with a cadence of seven notes, much like a very short birdsong. It was used in our family whenever we needed to locate each other in a crowd (or in the woods while hunting ichneumons). It was Papa. He had come to the village and wanted Ulla and Mamusha to come out and inform him of events, in case it was not safe for him, a soldier, in the barn's haymow.

Papa's plan for escaping capture had worked flawlessly, but not without a close call and also some unexpected good fortune. He had first gone to where he had hidden a set of shabby but precious old civilian clothes—such as a vagabond might wear. He had changed clothes and hidden his uniform and his bicycle. He had then sneaked cautiously through the woods to the hiding place under the pine boughs. He slid under them and eventually fell asleep, but was awakened by the frightening sound of crushing trees: a tank approached and passed immediately behind him.

A little later a truck loaded with men and girls in uniform appeared in his clearing. "I could see these were SS—Hitler's most dangerous fanatics." They collected sticks and started a fire to cook a meal. One of them came to his woodpile. Papa said he was tempted to grab the man's foot—but he kept still, listening carefully as they were talking while changing into their civilian clothes. The war was not officially over, and Papa feared they would not have wanted to leave a live witness to their desertion. They dragged their uniforms and a lot of baggage to the side of the clearing, then left.

Papa was at the right place at the right time. The SS had left pots, many pounds of butter, potatoes, a box full of Iron Crosses (military decorations that became useful barter with the Americans, who took them home as souvenirs), and a typewriter. The first three were like manna from heaven, and the typewriter would be used for decades by Mamusha to type Papa's manuscripts on ichneumon wasps.

THE AREA AROUND THE VILLAGE BECAME A HUGE camp for prisoners of war, and soldiers and civilians were camped in all

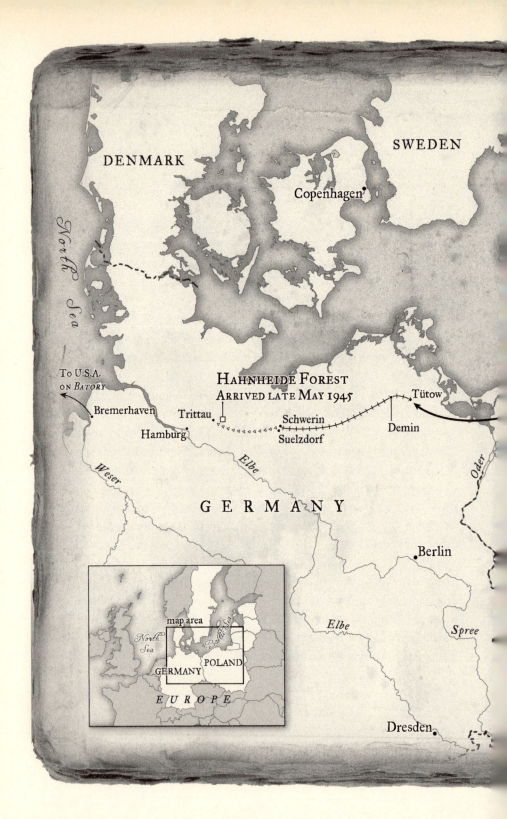

The path of our flight from Borowke to the Hahnheide.

Modes of Transportation

◁◁◁◁◁ Horse, wagon, or sleigh
ooooooo Tank
- - - - - Car or truck
+++++ Train
———— Plane

Baltic Sea

Kaliningrad
(Königsberg)

RUSSIA

Strebelsdorf

Vitzka-Strand

Slupsk
(Stolp)

Gdansk
(Danzig)

Kolberg

Braust

Chojnice

Obkaz

Tuchola

BOROWKE
DEPARTED 23 JANUARY 1945

Pita
(Schneidemühl)

Bydgoszcz
(Bromberg)

Vistula

Poznan
(Posen)

Wojarta

Warszawa
(Warsaw)

POLAND

LATVIA

LITHUANIA

0 Miles   25   50
0 Kilometers 50

the houses and in all the fields. Papa, disguised as a poor farmer, worked for a local farmer as a laborer. Shoveling manure was not, perhaps, what he was born to; and as a peasant farmer he had little ability to provide us, his clandestine family, with food and water—anyway, both were extremely scarce. He was, however, fortunate not to have become a prisoner, and had managed once again to avoid an awful fate. The Russians captured and retained German soldiers as prisoners for cheap labor right through 1956, up to eleven years after the war had ended.

For most Germans, the aftermath of the war would continue long after the official hostilities ceased in May 1945. This was the beginning of hunger. With so many men dead or displaced, the women alone bore the responsibility of providing food and caring for the kids. Their husbands existed only in pictures. I was lucky to have my father (who had brought along his thirty-two rattraps) as we entered the years of starvation.

We had escaped the Red Army and felt saved. Nevertheless, for the time being, we were in effect prisoners. We were not allowed to leave our village. It was not an uncomfortable place, at least for me. I remember a haymow crammed with people. Little more. Ulla went riding on stray horses that had been abandoned by the army. She gave some of the GIs riding lessons and befriended an American officer who said, "Take your family and go west!" He knew something, but he would not tell us what it was. Papa was uneasy. Somehow, he felt we should now try to escape our confinement. Then we received news that made us wonder whether we might still be delivered into Russian hands after all.

Mamusha met two young Polish men who had been freed from a concentration camp. To prove it they pulled up their sleeves and showed the numbers tattooed on their wrists. As fellow Poles (Mamusha and Ulla were both fluent in Polish, as was Papa) we became friends, and they told us the disturbing news that the Americans planned to withdraw from this territory and leave it for the communists. Two independent pieces of gossip meant something; Papa decided we needed to leave, and soon.

It would not be easy, because the Americans had already closed all the roads toward the west. They told us we could go back home, to the *east*, from whence we had all come. "Why run from the Russians?" they wanted to know, and added: "They are good democrats. They are such gentle people." Papa may have agreed. He had admired the Cossack horsemen and felt affection for the Russian peasants, but Joseph Stalin had killed 20 million of his own people.

Befriending the two young Poles proved to be yet another serendipitous stroke of luck. Unlike the rest of us, as former concentration camp prisoners they were granted special privileges and permitted to leave the village. We hatched a plan. We would pretend to the Amis that we were all one family who had been reunited and were going west together. Papa, then forty-nine, would not shave for a few days and would wear his shabby clothes and play the "grandfather." Mamusha would be the wife of one of the handsome boys, and Ulla would be the wife of the other.

A prisoner-of-war camp for German soldiers had already been established by the Americans, and Papa went there and introduced himself to the commanding officer, by showing him his book *Der Vogel Schnarch*. The commander was entertained and told Papa that we could have two of the canvas-covered Conestoga wagons that were standing around. They were, basically, houses on wheels and looked just like the famous wagons of the American west. We could also have any horses we could catch.

Ulla was a practiced horse wrangler and rounded up four loose German army horses in short order. They would pull the wagons. Furthermore, Ulla's friend, the American officer, came to our hay barn and offered her a car that had been found along the roadside along with several cans of gasoline. He said once again, "Leave! And take your family."

"Why?"

"Trust me, it will be better for you."

"Why," Ulla wanted to know, "do you care about us?"

"Because," he answered, "you remind me of my girl back home. I wish you well."

Among all of us, only Papa knew how to drive. We started out to the west, toward the Elbe-Trave canal, with the two young Poles in the leading wagon, Mamusha and Ulla and Marianne and me in the second, and Papa in the car bringing up the rear.

We had not gone far before we were stopped by a sentry at a crossroad. What proceeded seems trivial, but it wasn't.

As we came to a stop, the sentry, a British soldier with a riding whip in his hand and a thin mustache, asked Papa, "Whose car is this?"

"Mine," Papa answered.

"Where did you get it?"

"I found it." So far, so good.

Then: "You have no right to have it. This is a military vehicle." (The car was painted in brown and green patterns for military camouflage.)

Papa retorted, "No, it's not a military car."

The sentry then sent for his superior officer, who took one look at the car and said, "It is a military car."

Papa, always needing to have the last word, said, "No, it is not."

The officer then fairly roared, "That is a military car; leave it here."

It didn't matter; we had no spare gasoline; nor could we get any. But this incident marked, I think, the first time that Papa's authority had ever been challenged. The encounter seemed hilarious to both Mamusha and Ulla, perhaps because it illustrated Papa's stubbornness in the face of overwhelming odds, as well as the stark contrast between his presumed authority and his actual power. I think it was also the end of his old life as a man in control, and the dawning of a new life in which his wishes were no longer commands and his authority did not go unquestioned.

We anticipated both real and imagined difficulties. Since we were likely to be questioned at checkpoints, it was important to maintain an identity as a family of Poles, so Marianne and I received a crash course in Polish. We were taught how to answer simple questions like *Jak się nazywasz* ("What is your name?") and *Ile masz lat* ("How old are you?"), to which I was supposed to say *Pięc lat* ("Five years"). Unfortunately, I never got it straight—I was as likely to answer *Pięc lat* when asked *Jak się nazywasz?*

We lost the car, but they let us through, and after that our travel went smoothly. As we camped along the way and cooked over open fires, we feasted on the butter and potatoes that the SS had left at Papa's hiding place in Suelzdorf.

We were not stopped again until we reached the checkpoint at the bridge at the Elbe-Trave canal, the boundary where the American zone ended and the British zone began. Here, no civilians were allowed to pass.

One of the British sentries was an ethnic Pole. Like one of Mamusha's brothers, Jan, he had also joined the British army after the Germans overran Poland. We now learned the dreaded news that these "British" were guarding the bridge in order to prevent refugees from leaving the zone left for the Russians.

About three months previously, Churchill, Stalin, and Roosevelt had had their little tête-à-tête, the Yalta Conference. They had agreed that after the war they would divide Germany into occupation zones, and the Russians wanted a huge sector of eastern Gemany, Poland, and the Baltic countries, and with their sector, the people in it, as spoils of war. Presumably Papa had an inkling of what was to come.

Not knowing what to do, we camped on the shore of the canal, gathered wood, made a fire, and sang Polish songs under the night sky.

The next morning our two young Polish men, the former prisoners, visited the bridge guards and explained our situation to the ethnic Pole, a captain, and to the lieutenant, a Scot. After that these two officers came often to visit us at our campsite by the canal.

We stayed for days, but it was not a dreary time. On the contrary, everyone was still celebrating the end of the war, Ulla and Mamusha were pretty, and the two soldiers guarding the bridge came over with food and drink. Diplomatic relations between the bridge defenders and the two women were established, and the party was on, though it was not necessarily all that entertaining to "Grandfather." I later asked Mamusha if they had revealed their identity as a mock family. "Oh, yes," she said, "we told them the whole situation."

One day the bridge guards decided to allow us to cross the bridge. And so after more than three months of flight from the east, we finally crossed into freedom. We were saved from a communist future, and given the chance for an American one.

We later learned that shortly after we crossed the bridge, refugees started to arrive en masse, "thousands upon thousands"—those who learned too late the news of the Americans' leaving. Most had no canvas-covered wagons for shelter and no food, and none were allowed to cross the bridge. They ended up in the territory that became communist East Germany.

Once we were safely in the west, Papa got tired of playing the role of grandfather. He grew jealous of the two handsome boys, who continued to be friendly with their mock wives, and he asked them to leave. Decades later I asked Mamusha if she had ever asked them why they had been put in a concentration camp. She said, "No, never. We never asked. We were so happy to be free—all of us. We just wanted to be out of there. We were thinking only of the present day." But she added, "They took whole families, simply because of nationality. Young boys had a capacity to survive."

With the benefit of hindsight, one of the impressions I have is that the atrocity was perceived as being the result of a force of nature, an angry God, something out of control. It was all too overwhelming to contemplate, and existence was reduced to the immediate. But it had all been set in motion owing to ignorance and revelations that stood in for facts. For years afterward Mamusha would often exclaim, "We survived!" Every time she said it, I couldn't help thinking of the millions who were not around to make such a claim. Blind chance had played a role at almost every turn. Those who were lucky may have thought they had been favored by the gods, or thought they had been chosen. I did. But I was wrong. Statistics don't lie.

Though those were incredibly dangerous and unpredictable times, Mamusha and Ulla were, however, also resourceful and brave. Papa had the good sense to always be looking ahead and planning for the worst possibilities, which sadly were usually exceeded. I have wondered if his earlier experiences leading expeditions into the forests and swamps of Celebes and Burma and other places may have helped him and all of us get through the war. He knew how to rely on himself, how to have faith in his own wits and abilities.

On March 17, 1945, while still in Schwerin, Papa wrote to Erwin Stresemann, with whom he had remained in contact through most of the war:

I have a month of the most dangerous, exciting, worrying, and fast-paced experiences behind me. That I am at the end here, luckily landed in Schwerin, and accompanied by Ulla, Frau Bury, and the small children, is the work of good fortune. The capture of the Snoring Bird seems silly in comparison. . . . Maybe fate will bring us together again, so that I can describe it all to you. Once more, a bit of Life was granted. Around us there is spring, and peace, like a dream. The landscape is wonderful, with its lakes and birdlife. The forest extends its hand and delivers a happy medley of voices. There sing the black and the spotted thrushes, the robin, winter wren, brown creeper (!), and the titmice—the music of the laughing gull, grebe, moorhen, the woodpecker's drum, and jackdaw, crow, and starlings are active in the old trees—How wonderful the world could be for us!

Papa in uniform, 1915,
wearing his mother's ring.

Papa's mother, Margarethe von Tepper-Fergoson, with palette, around 1903.

Papa with his father.

ABOVE, RIGHT: Papa's sister, Charlotte, at Borowke, shortly before she died, 1909.

RIGHT: Papa and Liselotte Machatcheck in his study, around 1936.

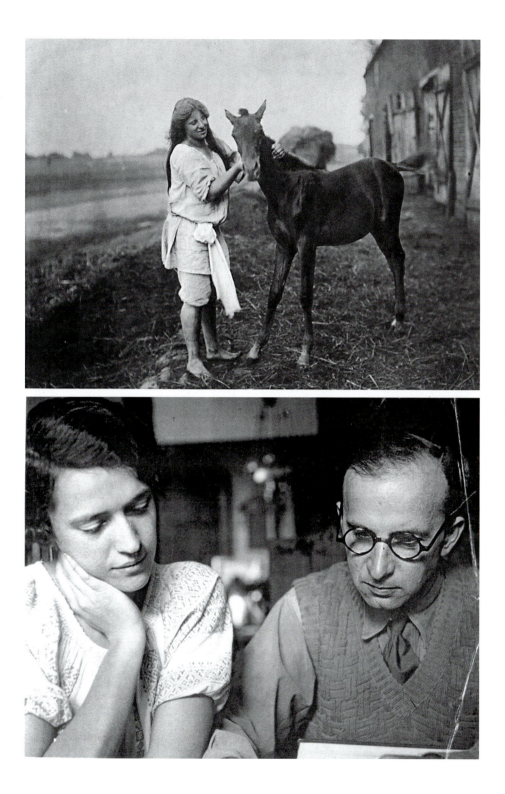

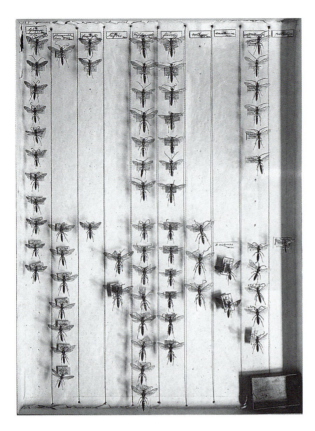

One box of Papa's ichneumon wasp
collection (now in the Warsaw Museum).

An ichneumon wasp.

ABOVE: Papa's reputation as a "rat-catcher" was well earned in Celebes, 1931.

LEFT: Papa with two young *Uhus*, 1935.

BELOW: Papa sleeping with "Dickie" on camp bed on Mount Victoria, 1931.

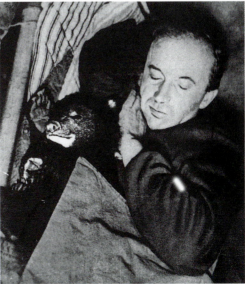

ABOVE: Papa with Dr. Lutz Heck, visiting his Persian panther, Peter, at the Berlin Zoo, around 1939.

RIGHT: Papa, the "rail king," in the Halmahera swamps, triumphant with Wallace's rail, 1932.

BELOW: The snoring rail.

ABOVE: With Anneliese at Borowke, in 1940.

BELOW, LEFT: My sister Ulla (Ursula) with
two lion cubs at the Berlin Zoo, around 1939.

BELOW, RIGHT: Me, 1943.

ABOVE: With Ulla and sister Marianne in the Hahnheide, around 1947.

RIGHT: Me with Marianne, around 1948.

BELOW: My sister Marianne and me shortly before we fled from Borowke, in January 1945.

ten   In the Enchanted Forest

The days of our future stand before us
like a row of little lighted candles—
golden, warm, and lively candles.
The days gone by remain behind us,
a mournful line of burnt-out candles;
the nearest ones still smoking.

—"CANDLES," CONSTANTINE P. CAVAFY

AFTER WE GOT ACROSS THE ELBE CANAL BRIDGE,
we headed for a country estate, the Sonnenhof, in the town of Trittau,
about 30 miles from the city of Hamburg. This spacious home was owned
by Baron von Stoltzenberg, a stepson of Papa's ornithological friend Herr
Hans von Nordheim, a judge in Berlin. Early in the war, old Judge Nordheim had told Papa that should we ever need shelter, we should try to
reach the Stoltzenbergs, who would have lodging for us.

Von Nordheim had disappeared during the war, but he miraculously
reappeared a year after it ended, having spent the time as a prisoner of
war in Russia. He had been put to work cutting peat from a bog where he
saw hundreds of breeding birds and watched the dances of the courting
black grouse. Von Nordheim said that the other prisoners didn't seem to
notice these marvelous birds at all, and that Papa had made him appreciate these things, which had "made life bearable."

We got to the Stoltzenberg estate a day too late to get the promised apartment. By the time we pulled in with our two wagons, the town was already inundated with refugees, and there was no more room. The baron had waited for us for a long time, and assuming that we had met the same fate as so many others, he had finally given the quarters to much more needy refugees, Frau von Gordon and her two surviving children. Her husband and two of her sons were *gefallen* in the war. *Gefallen* means what it sounds like. German soldiers were not killed in action—that would have been defeatist. They just "fell." Euphemisms and double talk were invented to disguise the gruesome reality of millions of German men dead.

When the war ended on May 5, 1945, most of the houses in Trittau still stood intact. However, even the noncombatants here had experienced the war. When Hamburg was bombed in July and August of 1943, the sky over the town had turned eerily black and red, and ash rained down to cover every house and tree. The "Tommies" (English troops) arrived on May 1, 1945, and there was little or no resistance and very little upheaval—perhaps even less than when we, as *Flüchtlinge* (refugees), arrived. Almost all the people in town had taken these strangers from the east into their homes, and there was no accommodation to be had anywhere, so we had to camp. The baron led us to a field with an animal shed along one edge of it. The shed had a roof and dirt floor and was open at the front. I recall the monocled baron squinting as he helped us spread hay for bedding. We parked the wagons next to the shed, and unhitched the horses to graze in the meadow. We built a campfire, hung our pot over it, and cooked a soup from some veal and peas the baron offered. This was meant to last several days, but I remember that while we were out scouting around the area after our first meal, the calves, with whom we shared our meadow, came over while we were not looking and ate the soup and knocked over the kettle.

Some days later Mamusha and Papa went into the nearby forest of tall pines and spruces and tall beech trees with smooth gray trunks. The forest was a nature reserve called the Hahnheide. In it were clearings where the ground was covered in patches of purple blooming heather and where blueberries, raspberries, and blackberries grew. They wandered around for hours and later told us they had seen deer and boar tracks, and had found a heron colony and a raven's nest nearby, up in the top of a great beech tree. As dusk approached and it was time to come back "home" to us at our shelter in the meadow, they had gotten lost and were forced to seek shelter in a thicket of young spruce where they spent the night in a cold drizzly rain.

The next morning they found a small footpath and, following it, unexpectedly came on a small cabin standing upon a bank above a crystal-clear brook, running over a pebbled bottom of black rocks. The tiny cabin was dwarfed by tall pines, spruces, and beech trees, and it blended into the dark surrounding forest. Its two windows were closed with green shutters, although there were fresh tracks to the door under a jutting roof. Papa knocked, perhaps half expecting a wicked witch from one of the Grimms' fairy tales to emerge. Instead, a man who had been a German soldier, and who had lived in this cabin for only a few days, answered. He said that he was about to leave the cabin, which was administered by the forestry department headquartered in Trittau, and he showed them the way out of the woods.

Papa and Mamusha quickly returned to Trittau in high hopes of getting permission to use the cabin. At the Forstampt they met Rolf

The sketch I made of our Hahnheide hut while we were there. Note my pet, Jacob, on the roof.

Grützmann, the forester of the Hahnheide, who, they were pleased to learn, was an avid amateur naturalist with his own collections of moths and also mosses. There was a great variety of both organisms in this forest, which promised many ichneumons as well. Grützmann collected caterpillars and raised moths from them in screen cages, and from those caterpillars he often got ichneumon wasps, which he didn't want but Papa was keen to get. Permission to move into the cabin was granted!

We hitched the horses up to the wagons, rolled out onto the dirt road by our cowshed, and turned onto a side road through a field where blue cornflowers and wild red poppies would soon bloom amid the newly planted grain. After crossing the field, we continued on a surface paved in smooth rounded stones that had been polished by a century or more of traffic by horse-drawn carts. Gnarled oaks lined the road, and on the side opposite the field we had just crossed we saw a clearing overgrown with bushes and raspberry vines. We clattered on over the cobbled road, passing through patches of young spruce forest, and after about a mile we made a sharp right-hand turn off this main road onto an unobtrusive path that led into a spruce thicket.

The loud, rhythmic clip-clop of the horses' feet and the rattling of the iron-rimmed wheels on the cobbles became muffled as we continued over yellow sand. The road curved around a tall sandbank and then crossed a small moss-draped bog with patches of stunted birch and heather. A mile or so later we reached the shallow brook, quietly murmuring over its bed of black pebbles. It meandered through a stand of thick alder trees, where patches of bright green nettles grew in the shade and a few stunted spruces were nourished by the light that filtered through the foliage. Soon we found a long-overgrown but faintly visible footpath leading up and along the steep bank to the cabin. The horses barely managed to pull the wagons up the steep slope and through the trees to our new home. Papa and Ulla unhitched the horses, and the wagons never again moved from the spot for all the five years that we lived there.

Grützmann and others would come to think of me and Marianne as Hansel and Gretel, as we would walk daily to and fro through the woods to the village school in Hamfelde, and later to the school in Trittau. Over four years of schooling, about 800 school days, we must have walked that route 1,600 times. That is why even now I remember almost every footstep of it. Only once or twice did we see a vehicle in our forest, and we probably saw fewer than a dozen persons. I remember only three. One time we came around a bend and saw a man sitting above us on the bank alongside the road. He smiled and seemed friendly, but we were still frightened because his penis was hanging out of his pants. He told us to come back to the same place the next day, but instead we took an alternative path. Once we saw a man wearing only shorts running down the road. He seemed curiously lost in his own world. We later learned he was a *Waldlaüfer* (woods runner, or jogger). Another time we spied a man at dusk on the road ahead of us. He stopped and stood very still as though surprised to see us. We bolted into a spruce thicket and ran with thumping hearts to take a detour through the woods that we knew well, back to the cabin.

It was during our years in the Hahnheide that my senses were awakened to the natural world. Everything was new. Nothing was anticipated, so nothing was colored by prejudices or expectations. Nothing would ever equal the sensations of discovery—except possibly in my research years later, in which I studied and experienced nature at other than the surface levels. The Hahnheide was my childhood paradise; I had all the time in the world to explore it; and there were few distractions of any kind, and certainly no bombardment of the fantastic such as one might get from television or books, so that the subtle could still reach me with power and immediacy. Half a century later, I have vivid memories of even small details of the forest and its inhabitants. I recall the feel of the warm, loose sand on our bare feet in early summer, as bright green metallic tiger beetles flushed off the road and flew a few yards ahead of us. In May we saw the brown june bugs (*Maikäfer*) with triangular white markings on each side of their bellies and their antennae that folded up into many tiny parallel plates. I loved to hold them in my hands, and I spent hours doing so. Each spring we heard the cooing of the wood doves, the harsh screaming of jays, and the sweet songs of the wood warblers and chaffinches. There were big red ants that built mound nests in the small clearings where they were warmed by the sun, and that gave off an acrid smell when we poked them. The agitated ants snatched the white pupae that we exposed and that I tried to grab as snacks for my pet crow. But the ants were faster; they scurried off with most of the grubs slung beneath their bodies. Papa showed me where digger wasps excavated tunnels in the sand

and stocked them with spiders that were comatose from a venomous sting, as red-brown cuckoo wasps hovered near the entrances trying to get in. I saw bright red snout beetles rolling soft fresh leaves, and discovered a little yellow egg inside each one when I unrolled it.

In late summer and fall the woods sprouted mushrooms. There were yellow ones, brown ones, and others with red, blue, and green caps. Some mushrooms had white flecks and others were smooth. Gradually the golden birch leaves fell down and the larches lost their yellow needles. And then one day the first flakes of snow drifted down. The new snow smelled clean and fresh and it transformed the forest. Barefoot until then, we were now wearing shoes that we received in CARE packages from Papa's ornithological colleagues in America. (The packages from CARE—the Cooperative for American Remittances to Europe—also provided food. Occasionally they included such luxuries as cigarettes and coffee; but Papa personally used neither and did not allow the women to consume them—he kept them for bribes.) Every step in the snow made a crunching sound and left its mark. The same was true for the animals— the deer; the wapiti, or elk (Hirsch); and the mice. We left tracks in these woods, but no permanent mark. None. We did not even think to carve our initials in a tree.

The door to our Hütte was tucked under a low roof overhang. From a bench under that overhang we looked down a steep bank to the brook where we went daily to fetch our water. Inside was a small brick fireplace with a metal plate as a top. A bench with its back to this fireplace formed a wall to the room on the other side. There, a window looked out on a vista of scattered pines where mouse buzzards nested, and someone had put up a platform to encourage peregrine falcons to nest. We made a small garden patch and tried to grow some potatoes. The women also grew tobacco, with its beautiful pink flowers.

Our bedroom, with bunk beds along the wall, was to the right of the main room and separated from it. This room was illuminated by a tiny window that looked out on a spreading beech tree, more woods of pine and spruce, and the Conestoga wagons in which Marianne and I played. A pair of pied flycatchers nested every spring in a box attached just above the window. I loved their soothing calls and recall the exquisite happiness and comfort I felt while watching them. The pair stayed busy flying

in and out of their snug home, first carrying grass and later carrying insects for their young.

The dense forest down to the brook kept the ground cool and damp,
so moss and mushrooms thrived here. I especially liked the bright red-
and orange-capped amanitas with their white flaky spots. I remembered
what George on the Borowke farm had told me about the magical world
of the forest gnomes. Marianne and I built little huts out of spruce and
pine sticks and covered them with glowing green moss. We built them
next to the red-capped amanitas, imagining that they might attract and
provide shelter for the gnomes. I doubt that I actually believed in gnomes,
but I liked to pretend. Anyway, the more I learned about the real creatures
that ran around in the forest, the less I was entertained by make-believe.

For many survivors, the hardest years were these right after the war.
There were few options for finding food, and what could be purchased
was strictly rationed with food stamps. We could buy only one loaf of
bread per week per family. Food was our main preoccupation, and although we could always find a few nuts, berries, and mushrooms in
the forest, feeding five mouths was a challenge. So Papa went into the
nearby village of Hamfelde and traded our horses with a farmer, Herr
Burmeister, for grain, potatoes, a weekly allotment of skim milk (Marianne and I would bring home half a liter on our way from school), and a
wagonload of sugar beets.

Even we Heinrichi were not so self-sufficient that we could live independent of the modern cash economy. The adults needed money, and
selling alcohol—which one can make from sugar beets—seemed like an
idea whose time had come. Papa had met a chemist, a fellow refugee from
the east, who knew about distillation. They set up a still using a few
twenty-liter metal milk cans for fermenting the beet mash, copper tubing
for distilling (secured by the chemist), and some bottles for storing the
product. The bottles were begged from Rolf Grützmann's wife "to store
berry juice"—only half a lie. The beets, which we'd buried in a pit next to
the cabin, were unearthed and scrubbed in the brook by the women. Then
they cut them up into small pieces, and boiled them in the milk cans.
Marianne's and my job was to keep bringing the wood to stoke the fire.
The boiled mash was then set aside to ferment in a warm place next to the
stove in the already crowded cabin.

Grützmann for a long time knew nothing about what was going on. He called our establishment *die höfliche Hütte* (the "polite hut"). He was deferential not only because Marianne and I were well-brought-up children who always curtsied and bowed when greeting him and other adults but also because when he came for a visit and knocked on the door, one of the women often announced, "We're bathing just now. Would you please be so kind as to come back later?"

Once we gained his confidence, we amused him by revealing that the politeness was meant to mislead him only so as not to burden his conscience. Had he come in at the wrong time he would have seen the milk cans with fermented beet mash cooking on the stove, connected to a long copper tube leading into a bottle, with cold water bathing not a body but the coil coming from the can. The vaporized alcohol condensed in the copper tube and dripped into the bottle. After that came the less volatile water. To keep from filling the bottle with water after the alcohol had been extracted, Papa used his taste buds. He had an uncanny ability to detect alcohol, and customers in the city of Hamburg would later declare his product "excellent." Papa, of course, never drank the alcohol. I never once saw him take even one sip, even years later at the farm when Mamusha would routinely make her "home brew" and keep a selection of various alcoholic beverages in her little liquor closet.

When Papa took his product (usually two bottles at a time) to market in Hamburg, he was prepared to be intercepted at any moment. There were sentries at the train station eager to intercept incoming black market goods, for their own use. One time, Papa was followed by a suspicious cop who nabbed him just as he was about to board a streetcar. "*Was haben sie da?*" ("What do you have in there?) he asked, pointing to Papa's knapsack.

"You'll laugh," Papa said. "I have flies." He opened the knapsack, and the cop took a brief look and did laugh! There were indeed flies—a box of pinned ichneumons. Papa had maintained his honor by telling the truth (unless one faults him for saying "flies," when he knew they were really wasps), and the officer had done his duty.

Successful as this operation was, it was no way to feed seven people, and Papa continued to try to find a job. After a six-month pause, he resumed correspondence with Erwin Stresemann. In a letter dated October 28, 1945, he wrote:

Finally I can try, after such a long eventful pause, to resume a connection with you. For a long time I have worried about you and your fate, until finally from your wife, whose address I had by luck noted, I got the good news that you were well, your house on Kamillen Street was not destroyed, nor the collection in the museum. You were busy retrieving it from the cellar. These messages from you were the biggest joys the entire last summer had to offer. All together and healthy, we eventually landed exactly where we had intended: in Trittau. It was the hardest expedition that I have ever led, and all experiences came to bear, else we would never have reached our goal. Herr Nordheim took me under his roof. . . . Since July we have lived in a tiny hunting hut in the middle of the most beautiful forest, in the midst of the deepest solitude. All around hammer and call the great black and green woodpeckers in abundance. We hear the croaks of ravens (!!) and in September we can hear the elks yell through the window of the hut, so that we can barely sleep. It would be wonderful, were it not for two big worries. The biggest and most pressing is about Anneliese, my dear partner of all victorious research expeditions, of whom I have no news. The second, about subsistence. We live temporarily from the proceeds of our sold trek horses. It will be sufficient for a few months more. What then? I would like now, after fate has freed me from all responsibility for a large estate, to dedicate the rest of my short, unencumbered existence to my great love, zoology. To that end I would need a position that pays a salary, since I have become a have-not, who has nothing left to offer. But all efforts to get a position in Germany were totally without success. I then contacted von Jordan [at Lord Rothschild's Tring Museum], who would gladly have hired me but who also could not accept anyone. Now I have applied in London, and will also apply in New York. But I don't have great hopes. Naturally, I would most like to work by your side. But that thought is too wonderful to even give me a quiet hope. Poland makes no sense, because there is little room for zoologists. Moscow would be more likely. If you hear of something, where an ichneumonologist (and a good ornithologist and bat collector) is needed, think of me. In the meantime, I have not been entirely idle and have systematically collected the ichneumons of this region, which probably has not been done before.

Papa had indeed not been "entirely idle" in looking for a job. In early April 1945, he had already sent a letter to Dr. Sanford at the American Museum of Natural History in New York (courtesy of a soldier returning home to the States; there was then still no mail service in Germany). He wrote to "beg" Sanford's help: "Thirteen years ago I brought you *Arami-dopsis plateni* from Celebes, and you had written me that you never will forget the thrill. . . . I am nearly fifty years old and strong enough for every expedition you would like to send out!" Not receiving any answer, Papa wrote to Ernst Mayr (on December 15, 1945) also asking for a job, as well as inquiring about payment for mammals he had collected for the museum before the war. He also asked for help "to pay the passage to New York." Mayr answered three months later saying that he was "not hopeful in the matter of employment at one of the museums." Ernesto Marcus, a professor Papa knew from Berlin who had emigrated to São Paulo, Brazil, also replied to an inquiry, saying that the job possibilities in Brazil were "problematical." Papa then wrote to Sweden, to René Malaise (who had invented the Malaise trap that Papa had used to catch most of his Burmese ichneumons); he too confirmed that there was "no possibil-ity" for a job there, but sent thanks for the sawflies Papa had collected for him from Burma, and said that he had named two new species after Papa.

It looked as though prospects were brighter in Germany. Strese-mann had given Papa an offer in East Berlin—a chance to work beside him—and had helped to arrange for two other offers. Papa refused all. In May 1946 he had already written to Stresemann, "I can be quite pleased that I am at this time *Torfstechmeister* [chief of the peat diggers] and not there [in Berlin] as assistant." Six months later he wrote to Stresemann, "I have been unable to decide to crawl behind the curtain," meaning, I think, the iron curtain. Papa's old mentor then helped him get a "serious offer" from a "Professor Krieg in Munich," but again Papa declined, tell-ing Stresemann, "Berlin would be preferable to me, although for the mo-ment there is time, because I still tolerate being a 'freeman of zoology.' "

Stresemann then helped him get a job offer in the "wild Harz moun-tains, for the British." (Again guessing, I think he meant the British Mu-seum.) Papa thanked him profusely but soon backed away from the job: "You forget that as a successful collector I function as a 'diplozoan' [an animal consisting of two kinds in symbiotic relationship] in which the male is only the untiring hunter and tracker, who is of little help as a taxidermist. . . . The employer would need to buy the total diplozoan."

A year later, on November 28, 1948, in a letter to Dr. van Tyne of the Museum of Zoology at the University of Michigan in Ann Arbor, Papa recounts how he has received $1.50 per bird from the museum in Stockholm and asks what he could collect for Michigan; but he excuses himself, saying that he would not be able to collect "systematically" because "I can't get permission for a shotgun, so I have to catch birds by different, difficult, means." He adds, "The coming summer probably will be my last in Europe, because I started an application for immigration to the United States." I do not know what Papa was thinking he would do in the United States, especially as he had just received a letter from the ornithologist Frances Hamerstrom in Wisconsin, who said, "So you won't be disappointed, I write you that there is very little paid for bird skins in the United States, usually nothing." But the cold war was starting to heat up, and Papa once told me (many years later, when I was in college), "In the next war, I want to be on the winning side."

While Papa looked for a job, we had in effect become hunter-gatherers. In the summer there were always raspberries and blueberries to pick, and we ranged far and wide in search of them. In late fall we gathered beechnuts and acorns. Papa could catch a trout with his bare hands, and he taught me how to do it as well. We would lie on our bellies along the brook, reaching out and gently probing underwater with our hands along undercuts from the banks. When we felt a trout with the tips of our fingers, we would then carefully work the hand near the fish's head, and then suddenly clamp down on it just behind the gills. The thrill of grasping a beautiful pink and blue spotted trout is not easily forgotten, especially when it becomes a delicious and precious meal.

I usually accompanied Papa on his rounds. I don't remember if we did any talking, and I now remember only snippets from these excursions—the fresh morning smells of a foggy dawn as we walked down the path to the brook, where long green weeds flowed like waving green hair hiding sleek trout in the current. Papa showed me an undercut where the water had receded to leave a dark passageway. "This is where water shrews live," he told me, and he got down on hands and knees to show me how to set one of his weathered rattraps. He carefully pulled grass from the top of the bank to hide the trap. We checked the traps twice a day, and I went along, eager to see what we would catch. When searching for caterpillars with Grützmann, we would spread a sheet under a small tree and shake. Every tree rained down some exciting new discovery. I became fascinated by caterpillars and eventually learned to find them by scanning the foliage for partially eaten leaves. Feeding-damage, I found, is a clue, and following clues made the caterpillar chase fun.

On December 5, 1945, Papa again wrote to Stresemann, expressing disappointment that in a recent letter he had "not given one drop of hope regarding a position at the Berlin Museum. . . . If you, the old companion and impresario, can't offer me a job, then where can I hope to find any?" He then mentions that Anneliese, "*die schöne Helene*," has arrived unexpectedly and safely—"a most beautiful Christmas present. . . . So, there is again a 'highly qualified staff' capable of great efforts! Please try for a shooting permit. And where shall we go? I think an expedition to the Black Forest, the Bavarian forest. . . . Do you have special requests for birds from this or other areas?"

For ten months, ever since we had left Borowke, we had no idea what had happened to Anneliese, Liselotte, and Erna—our trusted maid in Borowke. The last we saw of them was when the Russians were coming, and they were staying. I have never heard any details of what happened, except that Anneliese told me they made themselves look as ugly and shabby as they could, and posed as old "workers" and as Poles, and were accepted as such. After the war, transportation resumed and possibly they checked—through the Red Cross, where we had left word of our intended destination in Trittau—and were as surprised to find a trace of us as we were to be traced. Anneliese then took a train west, and Erna accompanied her. Liselotte stayed and never left Poland. Anneliese then wrote Stresemann on March 3, 1946: "The ex-beautiful Helen must herself write you—because her friend and patron reputedly has no time—the godlike one splits wood—how low have the Greeks sunk!" She asks for thread and labels for specimens and says, "The worst for us is the awfully small space—it grates on the nerves." With her and Erna's arrival, we were now seven persons crammed into a space measuring four by six meters. Even for Papa, being with four women in such a small space was too many. But tensions were highest between Anneliese and Mamusha.

Apparently things improved on the economic front. On June 27, 1946, Papa wrote to inform Stresemann that he had a small order for mammals from the American Museum of Natural History in New York.

By February 1947, things started to look absolutely bright. Papa wrote to Stresemann:

> Dear Impresario! The future is promised by a small ray of hope—at least for us—in a CARE package from Sanford [at the AMNH]. The whole hutful of people are *aus dem Häuschen* [literally: "out of the house" but meaning "beside themselves"] in expectation of the possibility of for once eating their fill. That is, in fact, a big wish of us all. How much we would now love the *Frass* [gross amounts of food] of the Latimodjong: *Nasi puti sama ikan kring* [Bahasa Indonesian for "white rice with dried fish"]—at least twenty to thirty sparrows in the soup! Here I had to feed on in this winter: two foxes, one house cat, and one rabbit, and it was luck that, with a lot of work, I got these.

There were few occasions of plenty in these lean times, and so food was memorable. I recall one day when we had consumed all our bread except for one small piece. Papa and I were going into the woods to search for food and wanted to take the remaining piece of bread with us. Mamusha objected, but Papa persisted and she eventually relented. We wandered about all morning and I clearly recall that it was a glorious, sunny early spring day. Although the beeches had not yet unfurled their delicate leaves from their long brown buds, there were yellow flowers coming through the winter-matted brown leaves on the forest floor. And in a pool of shallow water I saw a mottled green frog. The brilliant green excited me, and that frog and everything associated with it became even more memorable when we sat down and leaned up against the large trunk of a beech tree, to eat our crust of bread. Everything was quiet. Suddenly we heard a dog barking in the distance. Then also a different call. Papa sat up straight, then jumped to his feet and ran in the direction of the barking dog, with me at his heels.

We found it panting, with blood on its muzzle. As Papa had hoped, nearby lay a dead roe deer, whose dying call he had heard. Roe deer are not large like Maine whitetails—Papa managed to stuff the deer into his knapsack, although he could not hide the long legs or the head. He could not be caught carrying it, or the carcass would immediately have been confiscated by the British (who now owned the forest and everything in it). Papa sent me home alone, and then he waited till darkness and stalked home through the thickets, avoiding the road.

The piece of bread invested in our hunt had been worth it. Other memorable finds would follow, and all would be documented in Papa's ledger, much as he would keep track of his pennies in later years. He listed everything he found, including "one egg," "one chicken without head, killed by a fox" (which turned into chicken skin fried with lots of fat and became a recurring fantasy for me), "deer killed by the dog," "one bottle," "one partially rotted boar, "and "one elk."

The boar and the elk were found by Marianne and me. One winter day, as we were walking to school, we heard ravens in a thicket. A light dusting of snow covered the ground and I went into the woods to investigate the commotion. I saw something big and black—a partially eaten boar. There wasn't any meat left, but the hide still had fat on it. We ran home and told Papa. He secured what he could from it, and we made delicious lard and crispy crumpets.

The dead elk, arriving when our diet frequently consisted of fried mice, provided an unexpected and memorable *Schmaus* (feast). Marianne and I had the daily chore of collecting firewood, and on one of our forays, Marianne spied the huge brown animal lying, seemingly dead, on its side, near the brook. We rushed back to the cabin to report to Papa that we had found a dead *Hirsch*. At first he didn't believe us, but when we showed it to him, he immediately covered our treasure under brush. Under cover of dark, all seven of the hut's residents returned to the carcass, working by candlelight in a wild scene I'll never forget. First we dragged the elk to a spot closer to the hut, and then began the process of skinning and butchering it—a process that went on until dawn. Some of the meat was smoked (and hidden) by suspending it inside our chimney. Papa had learned in Persia how to preserve meat by boiling it with suet. The fat rises, then hardens and forms a seal at the top. We saved even the parasites that had set into the skin and bones—Papa thought he might eventually be able to sell them. He did, along with a large number of small mammal specimens—mice, shrews, and birds (collected with rattraps, snares, and slingshot, as I will describe later)—to the American Museum of Natural History in New York, where he had and was still accumulating credit for future payment.

One of Papa's rattraps that survived all of his expeditions.

The rattraps and insect nets were more important than anything else we owned; of all the treasures Papa could have brought from Borowke, he had been lucky or full of foresight (or both) to bring them. For him they were basic living equipment—the way a toothbrush might be for others. He used them now in the Hahnheide to run trap lines for small mammals. The specimen skins could be stuffed and sold to the museum in New York, the bodies fried and eaten by us. I now own one of these traps—it must be almost a century old. I still use it occasionally, and it is one of my prized mementoes.

Stresemann, meanwhile, was stuck in what was now East Berlin, and not faring as well as we were. On August 1, 1948, he wrote:

> Now we have reached the point where you can not even afford a letter stamp. I can no longer afford a western newspaper, because for that you need West money but I get paid in East money, which the

owners of things of real value, like fruit, vegetables and potatoes, look at the way *Kauri* [cowrie] shells would be viewed on Wall Street. . . . I say this to comfort you: You at least have the freedom of the western democracy.

He ends his letter: "When will we be together again on soft soles stalking snipes? When I go into the woods these days, with Verla [his wife], it is only to gather cow flops for the vegetable garden! In friendship, your Erwin Stresemann."

About a month later Stresemann wrote again, commiserating with Papa because he had not yet been offered a job. "Your letter is a document of this terrible era for intellectuals," Stresemann wrote. "What does the world care for nature-researchers when it is interested in only the most banal urges. Anything that doesn't have a direct use for consumption is, for most people, laughable."

Papa's ledger indicates that he performed hard physical labor when the need for cash became pressing. He dug out pine stumps. People needed heating fuel, but they were not allowed to cut trees or gather firewood in the forest. Papa proposed digging pine stumps out of the ground of a landowner, Dr. Döhn, from Trittau, and converting them to firewood to sell in Trittau. He got permission, and a contract was drawn up (signed May 31, 1946) for 500 stumps. I suspect, however, that he also dug out the stumps for a second reason. Pine trees have huge deep taproots. Some of the pits that resulted from digging them out he was able to modify as pitfall traps for catching mice and shrews, to supplement the catch from the rattraps.

Papa took me along to check his pits as well as his rattraps. It was fun, because we never knew what we'd find. Along with a vole, a mouse, and sometimes a tiny shrew, there were often carabid beetles. *Carabus* is the genus of the large, flightless running or ground beetles, and the largest are as big a small shrew. Whenever Papa lifted the green moss cushion he had put on the bottom of a pit (to lull whatever fell in to seek refuge under it, rather than have it spend much effort trying to escape) I expected to see one of these beetles. Papa would grab the exposed beetle and hold it up for me to see. The jointed legs would flail. A pair of short antennae sprouted from near the mouth, equipped with sharp pincers, and the beetle would discharge a gob of brown fluid from its mouth.

Childhood drawing of a long-tailed tit, whose nest I found in an alder tree near our cabin.

I was entranced by these creatures and learned all the species' names—there was the big black *Carabus coreaceus*, the delicately grooved and studded copper-colored *C. canellatus*, the black *C. intricatus* with violet on the thorax and edges of its elytra. Sometimes we found the shimmering gold and green *C. auratus*, with its orange legs; or the black *C. clathratus*, with its red spots. Papa gave me a field guide to beetles, Hans Wagner's *Taschenbuch der Käfer* (*Pocketbook of the Beetles*) and I dreamed of finding those I had not yet encountered.

Unlike us civilians, the forester, Grützmann, was allowed by the British administration to have a shotgun, and he and Papa occasionally went hunting for songbirds together. I accompanied them, feeling for the first time what it must have been like to stalk strange birds on an expedition. My passion for carabid beetles waned in comparison with my new love, birds. Holding a bird in my hands, feeling the softness of its feathers, and seeing its incredible colors and patterns from up close were truly something. The boys in the village regularly raided the nests of crows and wood pigeons, plucking the young out to eat just before they

The *Zaunkönig* [winter wren] in the Hahnheide that so much intrigued me because of its energy and secretive habits.

were ready to leave the nest. No amount of hunger could induce me to kill a baby bird. It was one thing to shoot a bird from a distance while hunting, quite another to kill a helpless nestling. Baby birds inspired in me the urge to nurture and protect, not kill and eat.

I especially identified with the winter wren, which I often saw by the brook where we got our water. I wondered where this bird lived, and Papa told me to look amid the soil and the roots of a windthrown tree. He was right. In a cavity behind roots I found a roofed-over nest, like that of the long-tailed tit, except that this one was made of bright green moss and fine spruce twigs. I stuck my finger into the entrance hole of the tiny nest, and curled it down to feel the tiny eggs. This tiny wren lived close to the ground in the dark, damp forest places where moss grew. It made me think of the forest gnomes, and I was always transported to a magical place in my imagination whenever I saw this tiny brown gnome-bird, with its short tail straight up in the air and as it sang its exuberant, tinkling refrain.

Papa (and Mamusha and Anneliese) encouraged me to adopt nestlings, and once grown the fledged birds roamed freely in the woods outside our cabin. This was fun for all of us and provided an opportunity to get to know the species intimately. Taking care of Jacob, my baby crow, was a constant chore, one that continually kept me out in the woods to provide for him. After I had fed him perhaps a thousand ant pupae, a dozen grasshoppers, ten beetles, five bird's eggs, and one mouse in one day, a hawk swooped down and had him for a snack. I think such experiences forced me to accept and even appreciate the cycle of life from an intellectual and objective rather than just a sentimental perspective.

On the way back and forth to the village school, Marianne and I went by a patch of low, scraggly pines which were easy to climb, and where once I found a wood pigeon's nest. It was a flimsy see-through platform of a few dry twigs, home to two pink squabs with kinky yellowish feathers. I decided to take one. Then I found a jay's nest and could not resist taking one of the pinfeathered young jays home too. I bonded with these babies at once. They grew up and were free to roam near the cabin, but both were also all too soon caught by a Sperber (like the sharp-shinned hawk in America).

The Sperber, Papa taught me, nests in thickets of spruce. There were such thickets not far from the cabin, where I had found a song thrush nest earlier. That nest was memorable because it was decorated with the tinfoil confetti that had been dropped from the sky to confuse radar during the bombing of Hamburg. But there was no sign of the Sperber here. In another spruce thicket, I found a pile of feathers from a recently plucked bird. Soon I saw feather piles everywhere. Then I heard a high-pitched rapidly repeated "keek, keek, keek . . ." I was near the nest! And suddenly there it was—a platform of sticks tucked under the canopy of green spruce branches. There were feathers adhering to the twigs, and white feces were splattered all over them. These birds made no attempt to hide their nest, whereas in every other nest I had found previously, there was never a speck of feces that might give it away.

I started to climb the tree, and immediately the hawks came zipping like arrows through the branches from behind, skimming over my head. My heart beat faster and I felt respect for the power of these birds. The fact that they had killed and eaten my pets had little bearing on my emotions. They were quite simply spectacular.

In retrospect it seems we were living an almost primeval existence, although to me at the time it seemed totally natural and I did not give it a thought. I suppose at a young age one accepts whatever is experienced, and that can be almost anything. For the most part almost all of my experiences every day were indeed in the forest, and the exotic ones were outside it; and even to this world of the past and of the future it was Papa who provided the introductions.

Once, I accompanied Papa to a lecture he gave about the Snoring Bird at the nearby University of Kiel. I sat near the front, among hundreds of students, and afterward he took me to the local museum, where the curator pulled out the collection of carabid beetles for me to see. She was surprised when I started spouting the Latin names for the different species.

Although Papa's lecture did not make a particular impression on me, it did on two young men, Rolf Grantsau and Werner Bachmann, who afterward visited us frequently in the Hahnheide. Rolf drew ichneumon wasps for Papa's manuscript in progress on the Oriental ichneumons. He also made a watercolor sketch for me depicting some boars along the brook where we sometimes saw them. He later drew a fanciful picture of Mamusha celebrating the New Year at the cabin in the Hahnheide. This watercolor depicts Mamusha in vibrant colors, wearing a blue dress and seated at a table in the cabin. A half-full glass of red wine is in front of her, and a tipped-over empty bottle is at her feet. Two human-size brown wood mice (the European equivalent of the common American deer mice) join her in drinking wine at the table. Smaller-size mice are scurrying all over the floor. Red and blue confetti falls from the ceiling. On the back of the card is written, "Happy New Year's Wishes, Rolf Grantsau and Werner Bachmann."

An even more fanciful sketch Rolf made parodies our luck. It depicts Papa trying to catch a rail in front of the cabin in the woods—one of the most unlikely places in the world to find a rail. Papa trips on a four-leaf clover as he tries to sneak up on the rail, and a big, beautiful black-and-yellow ichneumon lands on his butt. Mamusha, who is behind Papa, sees the ichneumon wasp and strikes at Papa to try to get the wasp. The handle of her net snaps in half as it hits his butt, and both the rail and the ichneumon escape.

Rolf lent me his watercolors, and under his guidance, I did my very first sketch. I painted a stag beetle for Papa as a birthday card. On the back I wrote that I now had 447 beetles of 135 species in my collection, and I wished him a happy birthday. I was nine. (The collection did not survive. It was in cigar boxes and was destroyed by dermestid beetles, but I found this little card in the packet of Papa's remembrance letters in the barn in Maine.)

Papa may not have been permitted to have a shotgun, but both Rolf and Werner were experts with a slingshot, and they taught me. We made slingshots using strips of rubber cut from discarded automobile inner tubes. We used either black or red rubber, but the red rubber was best. I can't begin to describe the excitement I felt when searching for just the right branch for the body of the slingshot, and then making and using the final product.

On one of my very first hunts with Rolf, we got a kinglet, a gorgeous blue titmouse, and a long-tailed tit. I was shaking from the excitement of seeing these birds, which I had seen only from a long distance, almost magically transferred into my hands. In birds, I had found my own ichneumons.

I remember our years in the Hahnheide as some of the best of my life. We had a real-life Robinson Crusoe existence. I had no responsibility for ensuring that we did not starve, although I helped out in finding food, and I was free to explore the forest. I was pretty much oblivious of the tensions created by overcrowding that made life difficult for the adults.

Papa had been delighted when Anneliese arrived (with Erna, who quickly got a job working for the mayor in town); he had feared that she had perished at the end of the war. But Mamusha, who had been the mistress of the hut for a year, complained bitterly that she could not escape from Anneliese, who now acted as though she was the boss, trying to revive, in a way, the hierarchy that had existed at Borowke. To help resolve the issue of overcrowding, Papa first told Ulla, by now twenty or twenty-one years old, that she had to leave and find a job or a husband or both. Furious and hurt—and supercompetent as always—she did find both. She got a job as a translator and secretary with UNRA—an international organization that helped soldiers of the occupation army—and there she met Leon (Loschek) Wartowski, a Pole who was in the British army. Not long afterward she gave birth to a son, Thomas, and then she and her family immigrated to Chicago. Ulla never forgave Papa for evicting her from the cabin. "It was not that he did it, but how he did it," she would write to me many years later. Then, when Erna's boss in Trittau, the mayor, "got involved with some shady dealings, and had to leave town immediately," Mamusha recalled, "she then had the first shot at getting his apartment. Everyone had to be a little bit crooked to survive."

But Papa could not get rid of either Anneliese, who was still legally his wife, or Mamusha, who was effectively his wife, even if he had wanted to. Temporary separations of the two women, then, were the only solution. These were partially accomplished when Papa took Mamusha to southern Germany on collecting trips. They went to the Bavarian forest and the Austrian and Bavarian Alps, staying in lodgings and castles arranged through Papa's zoologist friends. Meanwhile, Anneliese took care of Marianne and me back in the Hahnheide. As at Borowke, she was once again my adopted mother, and now she was Marianne's as well.

I remember Anneliese as a warm and loving person, and I have very few memories of my mother from that time. Anneliese took it on herself to try and civilize us. At least she is the only one I remember ever trying to do so. She tried to educate Marianne and me so that we would become presentable to the high society to which she had been hostess at Borowke for decades, even though we now lived in a hut in the woods.

I recall hearing the words *gut erzogen* ("well trained"). We were quiet and *gehorsam* ("obedient") even by German standards. We learned to defer to authority almost reflexively. Anneliese taught us to say "thank you" and "please," but we were never allowed to speak unless spoken to. We were taught the proper way to greet an adult. As a boy, I had to hold the right hand of a man firmly, then bow. When greeting a lady I was required to kiss the back of her extended hand, as a proper gentleman should. Marianne had to curtsy, by making a quick bend in her knees as though her leg muscles had momentarily gone weak. My underwear always had to be clean. "Why?" I wanted to know. Anneliese's explanation: "Suppose you had to go to a doctor."

And table manners—they were the most difficult to master, but even in the cabin nobody lifted a spoon at mealtime until everyone was seated and the lady of the house had taken her first bite. It was: sit up, don't slouch, elbows to the sides, no talking, no reaching, mouth closed while chewing, empty your plate completely, and don't ask to be excused until everyone is done. I learned that in polite society there would be several spoons, knives, and forks, each at a special place with respect to the table setting, and that these had to be used in a certain sequence. We didn't have the proper utensils in our hut, and I had trouble catching on to the idea, wondering why one or two utensils couldn't do the trick, and what difference it made which hand they were held in or how they were used. I still don't know the answers, and I daresay Anneliese would not approve of the casual manners I have adopted as an adult.

My posture was another of Anneliese's concerns. She stood me up against the wall to point out that my back was crooked; she found a space between the wall and my lower back. I didn't master getting my back "straight," either, and I think she eventually gave up on that project.

Papa was even less forgiving of any defect. He was concerned with my "nerves." He thought I had "bad nerves," which came from "over-stimulation," something he spoke of as if it were (along with food coloring) the most hazardous of all threats to health. Even later in his life he went to great lengths to avoid this danger, never ever drinking any coffee or alcohol, never eating more than one square of chocolate at a time, never turning on the car radio. To demonstrate his own nerves of steel, he would stretch his hand out toward me and hold it as still as a statue's. Maybe I could not hold mine still as long, but I was absolutely quiet as we hunted for hours at a time for ichneumon wasps. This did not impress him.

One day while we were chopping open rotten stumps, looking for hibernating ichneumons, I spied in the loose soil at the base of the stump a *Carabus intricatus*. There was little chance of its running away from me, but in my eagerness to have this rare beetle in my growing collection, I was beside myself with joy and nearly pounced on it. Clearly, I was "over-stimulated," and perhaps because of my eagerness, Papa decided to teach me a lesson: He took the beetle from me declaring that I'd been too hasty, declaring it was his beetle.

Several years ago, when I was trying to revive my memories of our time in the forest, I began to wonder how Mamusha had felt about having Anneliese around. Mamusha told me that it was stressful for her that Anneliese, from whom she thought she had finally escaped during "the flight," had unexpectedly caught up to her in the Hahnheide. But then I wondered: if Mamusha left us in Anneliese's care for months at a time when she and Papa went collecting, wasn't she using Anneliese?

"Was it hard to leave us when you went away to the Alps in the summers with Papa?" I asked.

"Oh, no—there was never any problem. Erna and Anneliese took

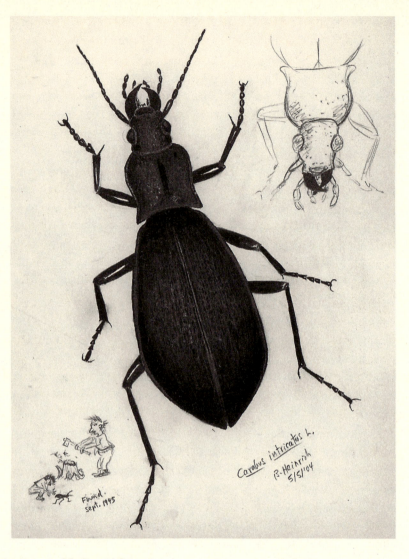

Carabus intricatus, and finding it hibernating
in a stump.

care of you." Ulla had gotten a job—at Papa's instruction—and left the
forest. "You liked to be in the Hahnheide. You were more attached to
place than to parents."

"Didn't you feel strange using Anneliese to take care of us, when you
wanted to get rid of her?"

"She would have loved to go with Papa, but Papa wanted to go with me, not her. She was still his wife. But he was in love with me."

"But weren't you depending on her?"

"No, never. Papa let her stay. You didn't need to be taken care of. You were always outside. You only needed to be fed."

THOSE YEARS WE SPENT IN THE HAHNHEIDE were the formative years of my life. I was five when we arrived and eleven when we left. I think children have a proclivity to be interested in the natural world, but often this is thwarted by a lack of exposure. Other things fill the void as their eager minds seek to learn about the world. I consider myself lucky. I had, by force of circumstances, only the natural world as entertainment. I learned to navigate my way through the forest, and to notice details, which assumed great importance in my child's mind, and my adult mind too.

One of the vivid memories that I have carried with me is of an ichneumon I found one day in the early fall when Marianne and I were walking home from school. I stopped on the shady bank by the brook to peel a bit of moss off the ground and look for hibernating insects. To my utter surprise, I struck gold immediately—there, right in front of me, lay the most gorgeous ichneumon wasp female I had ever seen (only females hibernate; males die off in the fall). She was shiny blue-black with bright yellow markings and a light spot on the thorax. I had never seen one like her.

I had by then seen and netted or found hundreds of ichneumon wasps, but none made such a singular and long-lasting impression. Even though she was not objectively any more beautiful than another, I knew she was a keeper—one that Papa would want.

The image of that spot, and that ichneumon wasp, was sharp in my mind for many years. I wondered if it was "real" or just something I had imagined. Memory can be faulty, but I could see no moral, political, social, or other reason why my mind might have altered the colors to red and purple or changed the place where I found it to some other location.

Many of the images and memories I held from the Hahnheide— images of beetles, wasps, caterpillars, kinglets, bumblebees, ant lions, green grass, ravens, jays, titmice, butterflies, birds' nests—would turn out to be seeds that would germinate and grow into questions and ideas that I later investigated with scientific method. I wrote hundreds of papers on research subjects such as how birds use signs of leaf damage to locate caterpillars, how ravens find and share food, how bumblebees choose which flowers to visit, how winter moths keep warm, how bees shiver, how wasps catch their prey, why birds' eggs are colored in diverse ways, how kinglets keep warm in winter. I made sketches and watercolors and eventually wrote books about all these phenomena that had first come to my attention in the Hahnheide. But the image of the blue-black ichneumon continued to lie quietly, as though in waiting, for more than half a century.

The ichneumon finally came back to me, with vivid and surprising clarity, when I found among Papa's old rat traps and the remembrance letters that I rescued from Mamusha's barn in Maine, an epic poem worthy of the Greeks, written by a dentist friend of Papa's from Trittau, Dr. Gerhard Walz. This clever composition is not directly translatable from German to English, but the story is as follows. Walz first establishes that Papa is obsessed with ichneumons. He points out that Papa is so passionate about his wasps that he is quite willing to leave everything aside to chase after these insects, even if it means coming home late and risking the derision of "his women." One day when Papa arrives at his "quiet little hermitage in the woods," I greet him excitedly and declare: "I have a beautiful one, a rare wasp female, one that I know for sure you have never seen. Oh, how beautiful is this wasp!" Walz then paints a picture of Papa's face paling when he sees the wasp in the jar and he dreamily whispers to himself that he finally, after twenty years, has this wasp he has so desired. He mutters that "this crazy little kid has found it." Papa then turns to Pise (meaning me; my Borowke nickname "Pise" was from *Protichneumon pisorious*, a species of ichneumon) and says, "I'll give you the blue *Carabus* in trade and as a reward." The beetle, Walz then points out, had recently been the subject of disagreement: some days earlier Papa had hacked open a stump looking for hibernating ichneumons and a carabus beetle fell out. Pise, eager to grab the beetle, had exclaimed, "It's mine!" But Papa had thundered, "Oh, no!" and taken the beetle for himself. Now, on seeing this magnificent ichneumon, Papa recants: "Pise, listen to me, my son! Give me your hand. I want to give you this beetle." And as Papa takes the wasp from Pise, his mind is drunk with thoughts of traveling to

Burma, Persia, and other faraway lands, and he thinks to himself, "What in the world I missed to see, the Hahnheide forest has given to me."

I was stunned when I read Walz's poem for the first time, in 2003, because it occurred to me that Papa had apparently *hidden* it in the barn among his remembrance letters. Why had he never shown it to me when he was alive, if only to explain my nickname that Anneliese used in the Hahnheide? Mostly I was stunned because I felt that this was the very wasp I seemed to remember so clearly. If so, it would be in Papa's postwar collection, which was now in the Bavarian State Museum in Munich. He had sold the collection to that museum shortly before his death in 1984.

I knew Erich Diller, the curator of that museum. So I e-mailed him, describing the wasp. He e-mailed me back to say that he had no trouble

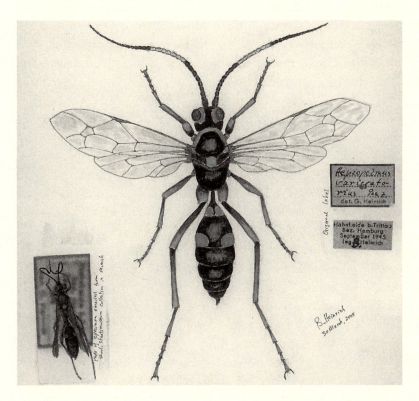

The ichneumon *Hepiopelmus variegatorius*, mentioned in Walz's poem, that I remembered finding in hibernation under moss.

locating the wasp species I described. He attached digital images of it, and I opened them eagerly.

The first image was a close-up of the label attached to the bottom of the pin. It read *Hepiopelmus variegatorius*, the same species mentioned in Walz's poem. The second picture was the data label, which read: "Hahn-heide b. Trittau. Bez. Hamburg, September 1945, leg. B Heinrich." It was one of the many labels which all originally had G. Heinrich preprinted on them. Papa had crossed out the "G" and written in the "B." The next two pictures were of the wasp itself.

Fifty-nine years of my memory now suddenly met the real thing head-on. She was indeed as I remembered her, deep blue-black with yellow markings. I had been five when I found her. Reading the label caused me to choke up with emotion. Papa had put my initial on the label: he had given a five-year-old credit for his scientific discovery, a gesture that could have meaning *only* to me. I pulled out my pencils and watercolors and began to sketch the wasp to fix her even more indelibly in my mind.

In my drawing I spread the wings, the legs, and the antennae, just as Papa would have liked it to be done.

eleven Leave-Taking

Adieu, adieu! My native shore
Fades o'er the waters blue;
The night-winds sigh, the breakers roar,
And shrieks the wild sea-mew.
You sun that sets upon the sea
We follow in his flight:
Farewell awhile to him and thee,
My native Land, Good Night!

    —"MY NATIVE LAND, GOOD NIGHT," LORD BYRON

Freedom's just another word
For nothing left to lose.

    —"ME AND BOBBY MCGEE," KRIS KRISTOFFERSON

DESPITE HIS ATTEMPTS TO GET JOBS, PAPA DID not have a bad life in the Hahnheide and was not in desperate straits. Although adept at spreading ichneumon wasps, he didn't have the manual skill to operate a typewriter or even to skin and stuff a mouse, but as always throughout his whole life, he had a wife who doubled as his secretary and tripled as his technician. He had no mortgage payments, no car loan, no taxes to pay. He did not belong to any club or organization that asked for membership fees. For him, if work was only for the money, it meant being a "wage slave." He, however, was free, and he often lived in the distant lands of his imagination. In a letter of November 13, 1948, he wrote to Erwin Stresemann, who was still director of the ornithology division at Berlin's Museum of Naturkunde:

> Shall we leave today for the kingdom of memories and fly on the pinions of the blue bird of romantic imagination to the Latimodjong, He-He, and Mengkoka . . . ? Yesterday evening sixty industrious coolies hacked a clearing for the first camping place. Then they got up and left. The night was quiet, and still I lay awake in anticipation until first dawn. Before me now is the long-awaited and unknown world of the primeval forest of the mountains! Quietly I step outside the tent, shotgun in hand—almost in the same moment a ground bird rushes in long thrush-like jumps out of the thicket. . . . The breast is rust-red, the back gray, the bill is dark. All that I recognize in an instant—and then the bird once again vanished in the thicket. Was it *Turdus celebensis*? But the bill was surely not yellow. What then could it be?

Papa then thanks Stresemann for an earlier suggestion that he earn money by writing popular articles: "Have not tried yet. I had prepared myself for lectures, but got no requests. Not a one! But I have no specific troubles. For this winter I am all set. For later, I am preparing for emigration to the United States. My ichneumon colleague there has already applied to bring me over. Maybe he will succeed."

"Preparing for emigration" left a lot unsaid. His American sponsor agreed to pay the way over for a traditional family; but it would not do to bring two women. Papa had planned to take only his wife, Anneliese, and me. Mamusha then went berserk—if he wanted to take me, then he had to take her too. So Papa divorced Anneliese and married Mamusha. Probably, Anneliese would, as always, have deferred to his plans and perhaps planned to follow him to America later. She could instead tag along with her daughter, Ulla, who would get in through her Polish husband, Leon Wartowski, who had served in the British army.

A month after writing to Stresemann about preparing to emigrate to America, Papa wrote to him again: "I have received news that my number on the waiting list of émigrés to the United States will come up shortly. So in 1950 I will be able to leave the old continent, in the hope of finding some meager bread, which in Germany unfortunately was not possible—a reality which for me brought with it a certain bitterness."

Stresemann answered Papa, writing that he was surprised to learn of Papa's wish to leave "the old continent" and of his "bitterness." Stresemann pointed out that he had attempted to get Papa positions at the museums in Berlin and Munich, but that at the time Papa had preferred to remain "free as a mouse catcher." I believe that is literally true. Papa was never happier anywhere than out in the field, and to have some justification for being there, to him, meant hunting or trapping for the rare, the small, and the exotic.

Stresemann expressed doubt that finding work in America would be easier, adding that he would be happy to be proved wrong.

A number of letters were exchanged over the next year between the two old friends, mostly concerning the taxonomy of wood mice (*Apodemus sylvaticus*) that Papa and Mamusha had collected in the Alps on their leaves from the Hahnheide, and that Papa wanted to publish. Then, on March 2, 1951, Papa wrote to his "Dear Master!" to say that he has returned the proofs for the publication of his work on wood mice and also to say farewell as he embarks on "my newest expedition—to America. If the United States does not please me then I can wander on, since the world is from there open. A homeland I have, in any case, no more."

WE BOARDED THE POLISH OCEAN LINER STEPHAN *Batory* in Bremerhaven and left the European shores in early April 1951, bound for New York City in the land of freedom.

Our ship was what I imagined a city might be like: it had big smoke-stacks, a restaurant, a theater, a swimming pool, and shops. We had a tiny room with a small round window, down many flights of iron stairs. We slept on double bunk beds. I can't remember much except that I felt sick to my stomach and spent a great deal of time on deck leaning against and over the railing and looking down and into the ocean. I listened to the deep rumbling roar of the giant engines far belowdecks, and watched the frothy propeller-churned water recede into the distance. Giant swells rose and fell in a slow irregular beat. I watched a white albatross with a black back accompanying us for hours. It skimmed the waves on stationary outstretched wings. Silver flying fishes occasionally leaped up out of the waves to glide along on their winglike fins, only to be reclaimed by the dark spooky waters.

Papa sometimes joined me on deck and told me of the wonders we were about to see. He talked about the *Kolibri* (hummingbirds) that we would find in America. He said they were the tiniest birds in the world. Some of them were no bigger that a bumblebee, and all of them shone and twinkled like polished jewels. He told me they had long thin beaks and would hover like sphinx moths and suck nectar from flowers. In America there were also *Klapperschlangen*, poisonous snakes with rattles on their tails. He told me about *Waschbäre*, little bears with ringed tails who washed their food; and some kind of weasel-like animal that sprayed with a powerful stink anyone who tried to touch it. America would be a land of great wildernesses of forests and deserts, mountains and prairies, all harboring amazing birds and animals.

One evening during our passage a movie was shown in the ship's theater. It was only the second movie I had ever seen (the first was Walt Disney's *Snow White and the Seven Dwarfs*, which my teacher in Trittau had taken our class to), and it was about Indians, the people of America. I was spellbound, and I can still see images from that movie in my mind. I recall a blazing campfire in a dark forest; partly naked warriors clutching bows and wearing knives at their sides, crouched around the fire; and a robed chief wearing a fur headdress, giving a speech. My heart raced when I saw a savage in a loincloth, clenching a knife in his mouth, sneaking on hands and knees through a thicket. In another scene I saw a mounted rider with a feather headdress up on a hill, watching like an eagle from a crag. Down below them were Conestoga wagons—just like the one we had left standing in the Hahnheide!—crawling along in a sea of grass. A band of warriors started whooping and raced down a grassy slope, and the people in the wagons got out and lay on their bellies and fired muskets. The Indian riders slid to the sides of their running steeds, shooting arrows over their backs as they raced by. America, I knew, would be exciting.

part TWO

The New World

twelve   The Beginning

WHEN I CAME ON DECK ON THE MORNING OF
our twelfth day at sea, the water was suddenly calm. Rising ahead above
the water was a monstrous green-robed female figure holding aloft an
equally monstrous green torch. I saw the tall, angular structures in the
distance behind her that Papa called *Wolkenkratzers* (literally, "cloud-
scratchers"). The tiny cars crawling along the shoreline in front of them
reminded me of my carabid running beetles.

Tugboats came alongside the *Stephan Batory* and nudged us to a pier,
and we walked down a long gangplank, and into a sea of people, more
than I had ever seen at any one time. There were two among them who
recognized Papa and stepped out of the throng to welcome us.

Many years later I would learn that one of them was Dr. H. M. van
Deusen, of the mammal division at the American Natural History Museum. He had a wonderful surprise for us: $1,000. This was payment for
the mammals Papa had collected in Burma and sent directly to the museum before the turmoil of the war. The museum would give Papa an additional $2,000 for all the mammal and bird skins he and Mamusha had
collected since 1945 in the Hahnheide and on their trips from there to
southern Germany. It had been very wise of him to let the money accumulate in America, because he would have lost it all in the *Währungsreform* of
1948, when everyone started anew with his forty deutschmarks (about
$400).

Van Deusen was with Gretel Mayr, the wife of Ernst Mayr (who then
worked in the bird division of the museum, but on the day of our arrival

was out of town). Mayr was an academic "relative" of Papa's. He had studied ornithology with Stresemann, who sent him to New Guinea and the Solomon Islands on a two-year bird-collecting expedition, which solidified his career and brought him to America. Later, he became a professor at Harvard University.

Papa's mouse catching, bird catching, and bootlegging may have been successful by the standards of Hahnheide, but they were unlikely to be more profitable in America than they had been in Germany. Expeditions to collect for museum specimens—the one thing Papa really knew how to and loved to do—were now out of fashion. An ornithologist friend in Michigan had warned Papa before we left Germany: "In America they don't pay much for mouse skins, if anything," echoing what an ornithologist friend from Wisconsin also had told him of bird skins.

But Papa was undeterred. He had come to America mainly because his colleague in the study of ichneumons, Dr. Henry Townes at the University of Michigan, had suggested that in the United States he could finally publish his manu-script on the Oriental Ichneumoninae. Townes, being an academic, was not rich, but he had managed to secure us a financial sponsor, someone who might provide a safety net in case we needed it while we were getting settled. This man, a rich businessman—I'll call him "Dr. C."—was eager to "do something in zoology," we were told, but, alas, had had "no time to finish" his doctoral work. (Still, he referred to himself as "Doctor" and we were also encouraged to refer to him that way.)

Townes had convinced Dr. C. that we were worthy of support, presumably by telling him that Papa was a potato farmer who'd been driven out of a remote corner of Poland, where he'd been an entomologist of unusual talent. Dr. C. had a farm in Maine that he wanted to get up and running and growing potatoes. He also promised to build Papa a laboratory on the farm. From the sound of it, we would be stepping into an American Borowke. Papa must have thought this a promising start in the New World. It seemed his good luck had followed us across the ocean.

Gretel Mayr took us to her house in New Jersey, where Dr. C. was to pick us up in a couple of days. Before our scheduled departure, we had a little time to explore New York. There was one thing that Papa wanted to see above all: the Museum of Natural History. Van Deusen and Gretel took us there. I still remember my fevered anticipation of what might be inside as we ascended the marble stairs and walked past the museum's tall columns to its grand entrance. I felt like a midget alongside the carved stone panoramas depicting buffalo, elk, mountain sheep, ibex, rhino, and wildebeest—and a great statue of Theodore Roosevelt riding on a horse with an African on one side of him and an Indian on the other. The animals were those that Roosevelt had hunted in America and Africa, first as governor of New York and later as president of the United States, from 1901 to 1909. The words carved in stone (as I would later read) were "rancher, scholar, explorer, scientist, conservationist, naturalist, author, historian, humanitarian, soldier, and patriot." I did not know a word of English at the time, but from the grandness of it all, it was plain to see that the museum was built by and for heroes.

When we finally entered the museum itself, we saw displays that captured some of the views of nature that must have captivated Roosevelt and people like him. They were people who (in his words) seek "adventure in the wide, waste spaces of the earth, in the marshes, and among the vast mountain masses, in the northern forests, amid the steaming jungles of the tropics, or on desert sands or on snow." Such people, Roosevelt said, "must long greatly for the lonely winds that blow across the wilderness and for sunrise and sunset over the rim of the empty world . . ."

His words described Papa's passion, and the museum embodied it. This was the land of freedom, and freedom was wilderness. Papa was a hero as well, one who worshipped wilderness, and to me freedom implied wilderness above all else.

Of the many displays that we viewed, the one I remember most vividly was a pair of grizzly bears on the tundra. I was awestruck standing before a bear that was rearing up on its hind feet, towering far above me. Nearby was a dazzling display of birds in the Sanford Hall of Birds, named after Leonard Sanford, the very same ornithologist whom Papa had met in Stresemann's house in Berlin and who had commissioned Papa to find the Snoring Bird in Celebes.

Over the next couple of days, the live American birds at Ernst and Gretel Mayr's bird feeder made an even greater impression on me than did those in the museum. I watched a jay unlike any jay I had ever seen before—it was vivid blue all over. The jay of the Hahnheide had only a patch of blue on its wing. There were also fantastic finches—purple, brilliant vermilion red, and lemon yellow ones with ink-black wings. What more would we see *outside* the city? I looked forward to our trip to the farm in Maine.

I was within days of my eleventh birthday, small and skinny for my age, when Mamusha, Papa, Marianne and I sank down into the seats of Dr. C.'s car. On this trip to our new home I was hoping to see birds—but I kept an eye out as well for mounted warriors with their bows and quivers full of arrows. As we headed north, we went through small towns with what we were told were called filling stations (where people got something called "gas"). I had, of course, never known of any fuel, except the twigs Marianne and I gathered daily in the woods to burn in our stove. Also totally alien to me were the sawmills and loudly clanging factories that we passed. Every town had several white churches, each with a sharp steeple. We were told that on Sunday "everyone" went to church. I had never even been inside a church. We knew nothing about Dr. C. except that it was rumored that his business was connected with "drugs," and so we suspected it was a prosperous business, because every town had a drugstore.

After hours and hours of this, we turned off the highway. But it was "mud season" and soon we could go no farther; our wheels were mired in Maine mud. We doubled back a few miles and stayed as guests at a dairy farm nearby. A couple of days later the road was better and we tried again. We saw thick gray-barked trees with little spouts on their trunks, out of which something dripped into buckets attached to the trunks. This something was called "sap," and we were informed that it was boiled and eaten with pancakes.

Eventually we crept up a steep, rutted hill, where we dodged mud holes and rocks. At the top, we finally drove up to a house with a gigantic old barn next to it. This was Dr. C.'s summer place, where we would stay. There were woods on one side and fields on the other. No other house was in sight, but we were told that we were close to the town of Wilton, Maine, of about 5,000 people, most of whom worked in the "woolen mill" and the "Bass shoe factory" and in the surrounding woods and apple orchards. As soon as we got out of the car, I knew that we had come to an alien and exciting world. One moment of listening: What were those strange noises? Were they birds, squirrels, frogs, or insects? In one look around, I saw endless walls made of gray stones piled together. These stone walls surrounded entire fields. I saw big trees around the house and strange rusted machinery lying by the large barn. Even the smells were different, new, and exotic. There were few signs of farming. No cows, no sheep, no people. But the barn was impressive.

Like most other barns (as I would learn later), this barn was about forty-five feet wide by seventy feet long and fifty-six feet high. All the beams, even the roof supports rising at an angle of forty-five degrees to the ridgepole, were square and had ax marks showing that they were hand-hewn (some nine inches per side). How had those huge timbers been maneuvered so high into the air? The whole structure of beams and braces was put together without nails, using neatly cut joints held together by wooden pegs.

I entered this place of mysteries through a twelve-by-twelve-foot opening guarded by two doors that were suspended from above and could be rolled on small wheels to open and close sideways. Stepping over a portal of two twelve-foot-long square-cut granite blocks, I encountered the central hub, where there still stood a half-loaded hay wagon that had once been pulled by horses. A huge four-tined hayfork hung above the wagon by a thick rope and pulleys suspended from the roof. The hay was lifted by the fork to near the ridge of the roof; then the fork's grip was released by another rope pulled from below, so that the hay fell onto the center loft. From there it could be distributed onto the two lower lateral lofts, where it was stored. It was later pitched down to the barn floor to be fed to the livestock kept in pens behind walls on each side of the main floor. One room after another on the left had been used to house chickens, pigs, and horses. Each had a door that opened on the main floor. The whole right side of the barn was one long cow stall with twenty cow hitches. Each had a trough with a flap door that could be supplied with hay from the front.

It was not hard to imagine the whole place teeming with livestock, but the only remaining sign of domesticated life was a large heap of decomposing manure overgrown with weeds at one end of the barn. The whole structure now served almost entirely as a giant birdhouse. A starling was singing and its mate was flying in and out of a hole in the eaves. Barn swallows twittered merrily while perched on the ropes that ran along the ceiling. Half a dozen of their mud nests were plastered onto the log beams of the ceiling, and the tips of white chicken feathers extended over the nest rims, along with the inquisitive heads of swallows busily incubating. A pair of little falcons (kestrels) entered a cavity in the old stacked hay on one corner of the mow through a crack from a loose board in the wall. They chittered pleasantly, flew back out, and hovered over the nearby overgrown field to hunt for mice. Among the pile of rusted machinery, in a tangle of dead growth from the previous year, I saw a brown bird (a song sparrow) that occasionally ventured out of hiding to a dead burdock stalk, where it would sing unlike any bird I had ever heard before. My spirits soared as I listened to its cheery song. Best of all were the soft soothing notes of a small bird with a sky-blue back and russet breast (a bluebird) that perched on fenceposts along the idle fields of matted yellow grass. I suspect Papa's enthusiasm for Dr. C.'s farm waned as quickly as mine grew. The promised shining "laboratory" turned out to be some low walls of cinder blocks with empty window frames out in a wet cow pasture overgrown with scraggly cedars and juniper bushes. Apparently Dr. C. had at one time started to build his wife an art studio, a project long since abandoned.

The chances of growing potatoes on this farm seemed slim. The fields were overgrown with weeds and brush. I suspect Papa told Dr. C. that there was little possibility of farming under the present circumstances. Dr. C. told us that in the meantime we could busy ourselves by extracting beans from a great mound of dried vines that had been stored in one of the haymows many years earlier. Then he got back into his car and drove away, back to his villa on the Gulf coast of Florida. Neither Papa nor Mamusha shelled any beans, but Marianne and I got to work.

The spring birds were just now returning, each more wondrous than the last. Soon the new green shoots poked through the dried matted grass from the year before and cheery black and white birds fluttered close over the fields while singing a melodious, sparkling song. The naturalist John Burroughs had described the bobolink's song in 1877 in *Birds and Poets*: "Sometimes he begins with the word *gegue, gegue*. Then again, more fully, *be true to me, Clarsy, be true to me, Clarsy, Clarsy*, thence full tilt into his inimitable song, interspersed in which the words *kick your slipper, kick your slipper*, and *temperance, temperance* (the last with a peculiar nasal resonance) are plainly heard."

I didn't hear any of this plainly, maybe because I didn't yet speak English, but I was nonetheless captivated by the tune. One "tune" that did help me learn English was the chant "Behold the busy baker, bread, pie, and cookie maker," which Marianne and I learned in the daily lessons that Papa gave us from a children's book that a neighbor had donated.

On the other side of the hayfield over a small hill lived our nearest neighbor, Mr. Cunningham, whose house and barn were surrounded by ancient maple trees and an apple orchard. Mr. Cunningham was probably as old as the maples themselves, yet he tended his herd of about ten Guernsey cows by himself. Every day he tottered across the bobolink's singing grounds to bring "some milk for the kids." He grinned from ear to ear, and almost every day he told us what a beautiful day it was. He was always right.

As May progressed and ever more birds arrived, it became increasingly onerous for Marianne and me to thrash beans in the barn. Mercifully, Papa allowed us to quit. I was free to explore down a long gentle forested slope leading to the beautiful, clear Pease Pond. Great trees leaned over its shores, and lily pads hid bass, pickerel, catfish, and sunfish. All were for me exciting new creations.

Our other neighbors, about half a mile farther down our dirt road, were the Floyd Adams family. One day Floyd came by to see if he could offer us any help. Papa, in his passable English, must have told him about our surprise at encountering the song sparrows, bluebirds, bobolinks, and kestrels—because in his slow soft drawl Floyd told us about the other fantastic birds we would find. He spoke of a wood snipe that circled in the sky in the evening and whistled like a skylark, but by using its wings. There was a woodpecker (the yellow-bellied sapsucker), which, he said quite seriously, "drank sap out of the trees," and another woodpecker (the yellow-shafted flicker) that hopped on the ground like a thrush and fed on ants. He continued with tales of a meadow bird (the meadowlark) with a bright yellow breast crossed by a black bib that although related to orioles flew like a grouse; and a real grouse, the ruffed grouse, living in the woods that drummed on logs. Hummingbirds? He had them right next to his house on his blueberry bushes! Everything he told us seemed fanciful, but would gradually be revealed as true.

I went over to see the hummingbirds, and soon we all became friends with Floyd's family. His wife, Leona, was slight, pretty, and blond. She had married at age seventeen and was now in her mid-twenties. Floyd had dark hair and a thin mustache. He was several years older than Leona. He walked with a limp due to wounds he'd suffered during the recent war. He had been a Marine and served as a sergeant in the Pacific (he still wore his *Semper Fidelis* ring). The couple had three boys: Jimmy, Vernon (known as "Butchy"), and Billy. Jimmy and Billy were close to my age, but Butchy was still a toddler. In their farmhouse also lived a boy of high school age named Robert, who was related to Floyd, and an elderly woman related to Leona. I quickly became the latest adoptee into this family, because after the boys and I got acquainted, I spent most of my time there.

Jimmy, Billy, and I didn't immediately communicate in language, but we had common interests. One of our first projects was to nail old barn boards together to build a raft that we floated on the farm pond. We spent countless hours on it, watching schools of baby catfish and white-bellied dragonflies, finding water bugs, and catching frogs. Jimmy and Billy showed me these wonders and taught me how to catch fish, and incidentally improved my English so that I was ready for school by fall without falling a single grade behind. I, in turn, showed them how to shoot birds.

I made a slingshot out of a discarded rubber inner tube, and having just turned eleven, I was at the apogee of my skill in slingshot use and maintenance. Billy and Jimmy were impressed, especially after I got one of those hummingbirds. I excitedly ran into the house to show Leona. But she wasn't pleased, and her reaction took me by surprise. After all, we boys had speared frogs and fished for hornpout down at the pond, and the men had hunted rabbits and coons and met no such disapprobation. Why was catching a hummingbird different from catching a butterfly or a fish? Didn't Leona known how frustrating it was to see those iridescent green ruby-throated jewels flitting about in her highbush blueberry patch and yet so far out of reach? It was an unsolved enigma that Henry David Thoreau had pondered a century earlier (in his Journal *about the Fur Trade*, 1859): "We will fine Abner if he shoots a singing bird, but encourage an army of Abners that comprise the Hudson's Bay Company.") There were plenty of hummingbirds. I doubted that one would be missed.

One morning the apple tree next to the shed where Jimmy and Billy gathered the bantam hen's eggs from the hay-lined nests was flushed in delicate pink. The fragrant blossoms were abuzz with honeybees and bumblebees and even more hummingbirds. We heard their excited high-pitched chirps above the low hum of their wings as they hovered, dipping into blossoms then out again in jerking motions. The throat-patches of the males looked velvety black, but sometimes when the angle to the sun was just right, they flashed a brilliant ruby-red. I had satisfied my need to hold one of these jewels in my hand and I no longer wanted to shoot one. Maybe something else.

"Can you kill squirrels with that slingshot?" Jimmy wanted to know. I heard one chattering in a maple tree by the stone wall and said, "You wait, I bring you one." As I came close, it stamped its hind feet on the branch it was perched on and scolded even louder. I pulled the leather holding a big round pebble back past my ear, aimed, and let go. The squirrel tumbled down and I jumped after it. As in the Hahnheide in summer, I was barefoot, bare-chested, and in shorts. The squirrel disappeared from sight and I rummaged for ten minutes or more in the green growth where it had fallen. Leona easily identified the plant species without having been there. It was poison ivy. She knew, because the skin all over my body had turned into blistery and itchy pustules. They were so bad that I had to spend a week in bed. Doctor's orders. Leona had called the town doctor, Herbert Zikel, who drove out to attend me, as he would do for almost anyone, day or night and all year long. It turned out he could speak German, although, like us, he now seldom did. He had left his homeland almost twenty years before us. We became good friends. As far as I know he never charged us, and on weekends he often took us to his "camp" at Lake Webb in nearby Weld, Franklin County.

In late summer, the apple tree by the henhouse ripened a crop of fragile apples with pale yellow skin and sweet white meat. But this tree was special to me not only for its fruit. The ground under it was strewn with bits of rusted iron from old farm machinery, including a still functional whetstone in a frame. We turned the stone with a crank to sharpen scythes and our jackknives. Billy and I ate the bruised apples that fell onto the hen-scratched earth and we reclined on the branches of the tree, under its thick layer of green leaves. The chickens scratched below us, dusted themselves, and brought their broods of downy chicks. I was absolutely content. But Mamusha and Papa had to attend to serious matters.

Within a month, Dr. C. came back to see us.

"Are you a *doctor*?" he asked Papa.

"No," Papa replied.

"Well, this is *America*. In America you can be whatever you want to be. Call yourself Dr. Heinrich and you will *be* Dr. Heinrich."

Papa didn't realize it at the time, but Dr. C. had had a plan. Apparently, as he footed our grocery bill, which included a lot of ice cream for Papa, he decided he wasn't getting his money's worth out of Papa as a potato farmer.

"Gerd," he told him, "I want you to write me a doctoral thesis." The topic was to be mites. I assume Dr. C. chose mites as his dissertation topic because they were even more obscure than ichneumons. There were other problems that he may not have been aware of. Only select qualified centers of higher education, with certified PhD programs, can grant a PhD degree. A candidate at one of these schools must satisfy a whole program of expertise, and even then only a committee of experts in the thesis topic is empowered to judge one of the requirements, the dissertation. "Dr." C. apparently thought that *any* expert could, at his or her own whim, grant a PhD. For a variety of reasons, Papa was obviously not qualified to serve as an expert on a new topic, or to grant degrees.

I cannot understand how Dr. C. thought he could get away with such a brazen hoax. I suspect his thinking was similar to that described by E. B. White in his famous book *Charlotte's Web*. In this book, Wilbur the pig, the runt of the litter who is slated to be killed and turned into bacon, becomes everyone's darling, thanks to a campaign devised and executed by Charlotte, a lowly but smart spider. Charlotte rescues Wilbur from his terrible fate by weaving words like "some pig," "terrific," and "humble" into her web at the edge of the pig's sty. Finding these surprising words inscribed in a web directly above the pig's pen, the farmer and a multitude of onlookers who come to see this "miracle" credit Wilbur with being "humble," "terrific,"—in short, "some pig." The farmer's wife is more skeptical. She says, "It seems to *me* we have an extraordinary *spider*" But her husband replies: "No, *see*—it's right there—just an *ordinary* gray garden spider." Similarly, Dr. C. assumed that everyone would credit *him* with the miracle of a PhD thesis on mites, never suspecting that anyone would think "an ordinary potato farmer" from some obscure corner of Poland could have performed the miracle.

No miracles happened, and Dr. C. returned to Florida without any assurance that Papa was his ticket to an imminent PhD. Two months later (by the end of June), Papa received a telegram informing him that Dr. C. was selling the farm. We would have to leave. But to go where? Dr. C. made an unexpected and attractive offer: "Come to us in Saint Petersburg, Florida; take it easy for a while; and collect ichneumon wasps."

Going to Florida was not exactly on Papa's agenda. He had just received a superb job offer—to be an entomologist in the Wisconsin forestry department at a salary of $350 per month. Papa had been ready to take us with him to Wisconsin, but Dr. C. strongly objected: "Oh, no, you can't afford to live on that in America! I can offer you much better. If you come to Florida, I'll build you a museum. You can then call the shots, because you'll be the director."

To start out in a new country, as the boss! Finally, a dream job after all. Only in America! Papa immediately declined the job offer in Wisconsin and he, Mamusha, and Marianne took the bus to Florida. I asked and was allowed to stay back in Maine with the Adams family. Floyd, Leona, and the boys and I had become, as they taught me to say, "thick as thieves," and it was decided that I would join my family later, when the Adamses traveled south to visit relatives.

Papa was pleased with America. In a letter he wrote to his "home" newspaper, Der Westpreusse (published February 1952) he described Maine as "so beautiful that one could lose one's homesickness" and added: "We know now where we will settle." He also talked of a "never-before-experienced feeling of personal freedom and genuine friendship from many people in all professions and all classes with whom we have come in contact—irrespective of whether you are of German extraction, French, English, or whether you are Republican or Democrat." His only complaints were "sponge" bread that could be "compressed to the size of a fist," and medicine that "has been reduced from a profession to a business."

The summer passed splendidly for me. I scarcely thought of my family in Florida. Finally we piled into Floyd's long, low maroon Pontiac for the 1,500-mile trip to Florida. Again I saw no Indians, but I made the acquaintance of another dangerous American.

We had stopped along the way because I needed to "go" and there was no bathroom in sight. I rushed into the bushes. When in new territory, I usually kept a keen eye out for zoological and botanical novelties.

This time I saw the strangest plant I'd ever seen. I yelled for everybody to come quickly to look at it. But the others were in a hurry to drive on. Leona hollered back: "Pull it up and bring it along!" By this time I already spoke passable English. I understood her words, and being obedient, I reached down and pulled the plant up. For the rest of the way to Saint Petersburg I paid the penalty—busily extracting the very tiny but potent opuntia cactus spines from my hands.

In the meantime, Papa, Mamusha, and Marianne had not been as happy in Saint Petersburg as I had been in Maine. As soon as they stepped off the bus, Dr. C. met them and dropped another bombshell: he had decided not to build the museum after all. Apparently he found out that he had to disclose his funding sources to the IRS before he could use it as a tax write-off, and he was not willing to go that far. He now offered to help Papa catch ichneumon wasps, and have him write a dissertation on these insects. Almost anything was acceptable to Papa, so long as he could collect his precious wasps.

Dr. C. kindly rented Papa and Mamusha a cottage. Mamusha was to be the nanny for their baby, and Papa was to earn his keep by being their gardener and taking care of the lawns. In all of his fifty years Papa had never mowed a lawn or had one mowed. He had cut wheat with a scythe, but that had a purpose. He asked Dr. C. *why* the lawn had to be mowed. Why fertilize and water the lawn to make the grass grow, only to cut it all off? I don't know what answer he got, but Dr. C. would later say that Papa was "very arrogant."

The dispute worsened, especially after Papa started trapping gophers, which Mamusha stuffed as museum specimens. Dr. C. did not appreciate their work and informed Mamusha that "people in America don't skin rats."

Dr. C.'s wife then had a sobering thought that canceled the ichneumon "collaboration" her husband had planned. She, like the wife in *Charlotte's Web*, thought it might attract attention if her Wilbur started spinning a fine web just when a world-renowned spider, who was known for spinning exactly *that* kind of web, showed up in his pen. Papa then wrote to the Adamses that life had become "unbearable" for them and that he and Mamusha missed them and Maine.

When I arrived with the Adamses at the cottage in Saint Petersburg, it was a tearful reunion, not so much between me and my parents as between Mamusha and Leona. But the Adamses' stay was short, and we all cried again when they prepared to go back to Maine. As they were leaving, Leona said to Mamusha: "Come back to Maine and stay with us until you find a job." This was a sincere offer and we knew it.

We remained in Florida for a while longer, and I remember being fascinated by the strange sea creatures at a beach that was a short walk from our cottage next to a giant fig tree. I became fascinated by sea creatures. I had never before seen starfish, sand dollars, blue crabs, shrimp, seahorses, dolphins, and schools of tarpon. When I was not in the water hunting for blue crabs and shrimp to eat, and collecting natural curiosities to keep (they would get a blue ribbon in a hobby show that fall in school), I was out in the palmetto scrub with Papa collecting ichneumon wasps. I also went to a nearby cattail swamp that fascinated me, because it reminded me of the romantic swamps of the Dobruja that Papa had told me about. It was alive with noisy red-winged blackbirds, and I came face to face with the tiniest heron I had ever seen, a least bittern. It stood absolutely still on its nest with its white-downed young. Its bill pointed straight up so that it blended into the cattails.

Meanwhile the Adamses drove north to their farm in Maine. Shortly after getting back, Floyd wrote to us, on July 16, 1951. It seemed that Dr. C. had been up to Maine to sell the farm and had had a conversation with Floyd about our family:

Dear Hilda and Gert (sic):

While here, [Dr. C.] tried hard to ensure that you would not return to this part of the country. It would seem that he is very anxious to keep you under his control. We were told that we should have nothing to do with you for a number of reasons. One being that you had smuggled large sums of money—some hundreds of thousands—into this country when you came. That we could get into all manner of trouble by aiding you. That you were to be sent back to Germany, etc. I mention this only to give you an idea as to the kind of "friend" you have. I certainly have more faith in you than him, and feel I will never be sorry for it. We miss you folks very much and *please* feel sure that we would be glad to have you come this way anytime. . . . Our garden is very good now. We have potatoes, peas, lettuce, radishes, beets, carrots, and other such from it. It has been

hard to get the hay this year. It rains often. The place seems to be missing something without Ben [as I was called by them]. Will hear from you again soon I hope, and I will again write in a few days—
    Always, Floyd

Three weeks later, on August 6, he wrote us again:

Dear Gert:
    Received your letter today. Was very glad to hear from you. [Floyd discussed several options that we might take, and then continued.] We would like very much for you to stay with us until you find a job and housing that you like. If you would like to come to Maine that would be best. We would very much like you to come. We will do everything possible to help you. We will be more than glad to have you stay with us. If you decide to come, I can meet you in Portland or Lewiston any evening to save you money on the bus. I hope to hear from you soon—also to hear that you will soon be with us. . . . Haying is almost over now. Finish this week.

We took the next bus home to Maine, and never regretted it.

    The writer Louise Dickinson Rich, who lived near our town in western Maine (and who later moved to Prospect Harbor on the coast), writes in her book *The Coast of Maine*, "I like the way they treat me, as though I were a reasonably sane human being." We had possibly less claim than Mrs. Rich to be considered sane, but we were nevertheless treated as though we were.

thirteen   The Adamses

Maine is a unique place. It seems to breed and
attract unique people. To most, it is an end-of-the-
line kind of place, somewhere you have to go to
because it's not on the way to anywhere else. As
such, Maine stands apart. It succumbs to change
in fits and starts only as it has to. Outside influ-
ences are farther away from Maine than in most
places. If there is anything that seems closer to
Maine than other places, it's the old ways. They
seem to hang on longer here.

—*MAINE: A PEOPLED LANDSCAPE*, R. TODD HOFFMAN

I WAS HAPPY TO COME BACK TO THE ADAMSES'
farm, to the barn pulsing like a beehive with activity—a grunting sow and
her squealing pink piglets, white ducks waddling down to the little pond
where we frolicked among the garrumphing bullfrogs and the white-
bellied dragonflies sunning themselves on the rocks. Nesting starlings
made rhythmic churring noises in the eaves of the barn. Every evening
Robert or Floyd hollered, "Come boss—come boss—come," and the little tan Guernsey cows would come trudging up the path from the pasture, each heading straight to its own individual hitch, where it would be
fed its daily grain ration while being milked by Floyd, Robert, or sometimes Leona, who would sit on a wooden stool under the cow and pull the

teats by hand. I liked the sound of the alternating jets of milk splashing into the bucket. After the cows were fed and milked they were unhitched and let back out to pasture, where the Hereford beef cattle ran free all the time.

In the morning we woke up to the sound of roosters crowing. They were bantams of the traditional "wild" type—brown-mottled hens and golden russet roosters with long curled green and purple tail plumes. The chickens ran loose all around the house and the barns and made dusty molds in the dirt under the pippin apple tree and the giant elm that dominated the barnyard. The chicken house was a separate gray clapboard structure with a row of straw-lined cubbyholes. Every morning several of these nests sheltered a hen hunkered down to lay an egg. Jimmy, about eight years old at that time, peeked into each of the nests in turn and reached in and under the only mildly protesting hens to take the eggs and put them into his basket. In the summer, some of the hens made their own hidden nests somewhere near the barn, and we would eventually see them strutting slowly about the yard with a dozen or so little peeping fuzzballs in tow. Individually the chickens, ducks, pigs, cows, maple sugaring, and garden were never a big production on this farm, but all together they brought the place to life.

Aside from the farm animals there were various pets and other hangers-on. Jack, the big shaggy red dog, spent summer days lying in the dirt in the shade under the front porch. Jimmy picked up a baby skunk one spring. "Skunky" was never "de-scented," but it never smelled or became otherwise objectionable. Jimmy sometimes kept it in his bedroom, and in the winter when Skunky was full-size and fat, it hibernated in the cellar behind the shelves of preserves that Leona had put up during the summer.

The Adams place was much like many of the small farms in western Maine at the time. A casual visitor then or a time traveler now would see a muddy farm pond surrounded by rocks and weeds; an old house with a decrepit porch and a sprawling lilac bush in front of it; rusty farm machinery lying about; and cows, cats, chickens, and kids wandering here and there. There would be pigs, a garden plot, and a barn with falling clapboards. The visitor might smell poverty. But to me, this tangle of people, animals, machinery, and junk had a throbbing, captivating vitality. If I thought the Hahnheide was enchanted, then this was paradise.

The Adamses' homestead.

Leona served pearly white Wonder Bread that was available at the store, but it was less than wonderful to Mamusha and Papa. They, who in the preceding years had not infrequently fed on fried mice, declared it "inedible." Maybe they didn't appreciate the wonder in something like foam that could occupy the maximum amount of space with the minimum amount of nutrition and taste. Perhaps Papa didn't realize that it was meant to serve not as nourishment but rather as a vehicle for a piece of cheese, or for peanut butter and jelly. Then again, he wouldn't eat peanut butter, either.

It was summer and early fall when we lived with the Adamses, and the boys and I stayed outdoors most of the time. After haying or doing other evening chores and having a quick supper, we would grab our fishing poles and a can full of freshly dug worms and head out. We hurried past the chicken shed and slid under the fence through the gap in the stone wall by the big elm tree, into the pasture. The honeybees were making their last excursions on the sweet-scented milkweed. A monarch but-

A singing katydid.

terfly or two might fly by, and the big green katydids with their long slender antennae had started their crisp, incessant chirping. Their metronomic stridulations suffused the balmy air of each late summer evening.

As we hurried along the worn cow paths past the apple orchard and down into the cool shady woods, our chatter mingled with the evening song of the white-throated sparrow. Finally we reached the alder thickets along the shores of Pease Pond, where our rowboat lay hidden. After bailing out the water, we took our seats, and with glimpses of a monster fish or two gliding underneath us (once in a while we even caught a sunfish, yellow perch, pickerel, catfish, or bass), we pushed through floating lily pads and pickerel weeds out into the pond. When the fish rose out of the water to devour the insects hovering on its surface, they left concentric circles on the glassy pond. Another late summer day would fade into night as the whip-poor-will sounded from the hill and bat silhouettes fluttered erratically against a darkening sky. Some of the bats skimmed low over the water, which was riffled here and there by the V of a speeding water beetle. As it got darker, the lightning bugs (lantern beetles) shone their cool white lights along the shore. Each would make a rapid series of flashes, then pause for a second or two, and then repeat the flashes a few yards farther on. The white perch started to bite around this time, and the tugs we felt on our lines had an intoxicating effect on us. We stayed till our worms were gone, and then stumbled back up through the woods. When we tumbled into bed, we gave only brief thought to the kettle of fish chowder made with our own milk, butter, onions, and corn that Leona would probably serve for supper the next day.

On some nights we went into the swamp with a neighbor who had hounds. "What are we going to hunt at night?" I asked Floyd.

"Coons!"

"What's 'coons'?"

"It's like a big cat with a ringed tail and a pointed nose." These were the "little bears" that Papa had told me about on the boat over to America. I sure wanted to see one of those! Two men and three boys charged into the woods, followed by as many baying hounds. I don't remember catching any coons, but tripping over logs in the dark and sloshing through the mud dampened my curiosity only slightly. Coming back one time through the apple orchard, we saw something big and black sitting in an apple tree. "What's that?" I wanted to know. "That, my boy," said Floyd, "is a porcupine, and don't go up there and grab it or you'll be *awfully* sorry!" He chuckled loudly, perhaps recalling my run-in with the cactus. The porcupine was rattling its teeth when we got close, so I decided not to mess with it. Floyd then told me that I should not mess with any skunks, either. "If one ever gets you," he warned, "we might not take you coon hunting with us for maybe a whole month."

Floyd and the boys told me about beavers and the "painted" turtles in the swamp at the outlet of Pease Pond. "And come fall we'll show you how to find bee trees and get buckets and buckets full of honey." I couldn't wait. Floyd kept a beehive at an attic window of the house. It provided some honey, but not nearly enough.

By September the purple and blue New England asters had started to bloom, and the goldenrod shone bright yellow under the apple trees in the old orchard. I remember the cool morning air sharp on my face and the scent of ripening apples in the air. By noon the air hummed with bees working to top off their honey stores from the last flowers of the year. It was then that all five of us—Floyd, Robert, Billy, Jimmy, and I—went "bee lining," an activity that was not yet extinct in Maine.

Bee lining in the eastern American forest has a long history. In 1782 Hector St. John de Crèvecoeur, a farmer in Orange County, New York, described it thus:

> After I have done sowing, by way of recreation, I prepare for a week's jaunt in the woods, not to hunt either the deer or the bear, as my neighbors do, but to catch the more harmless bees. . . . I proceed to such woods as one at a considerable distance from any settlements. I carefully examine whether they abound with large trees, if so, I make a small fire on some flat stones, in a convenient place; on the fire I put some wax; close by this fire, on another stove, I drop honey in distinct drops, which I surround with small quantities of vermilion, laid on the stones; and then I retire carefully to watch whether any bees appear. If there are any in the neighborhood, I rest assured that the smell of burnt wax will unavoidably attract them; they will find the honey, for they are fond of preying on that which is not their own; and in their approach they will necessarily tinge themselves with some particles of vermilion, which will adhere long to their bodies. I next fix my compass, to find out their course . . . and, by the assistance of my watch, I observe how long those are returning which are marked with vermilion. Thus possessed of the course, and, in some measure, of the distance, which I can easily guess at, I follow the first, and seldom fail of coming to the tree where those republics are lodged. I then mark it.

Our basic tools in the early 1950s were also the compass, a watch, white flour (instead of vermilion), and a "bee box," an improvement on Crèvecoeur's technique. We made our bee boxes by nailing together thin boards, about three inches wide by five inches long, and then lining the bottom with a piece of honeycomb. Beneath that we'd build a screen window into the box, to admit light through the honeycomb. After we captured a bee from a flower by trapping it in the box and slapping the lid shut, the bee would see only a faint light shining through the wax combs on the bottom. The bee would try to escape toward the light, and in doing so would practically divebomb into the bonanza of sugar syrup that we had poured into the comb.

We knew the bee had started lapping up sugar syrup when it stopped buzzing, and we could then safely remove the box cover without disturbing her. Carefully, we set the bee box on a post and waited until the bee was done, whereupon it crawled up to the edge of the box, wiped its antennae with its forelegs, then lifted off. It flew back and forth to investigate the source of the newfound bonanza, and then circled in wider and higher arcs until finally making a "beeline" toward its nest. We hoped this would be a hollow tree somewhere in the forest. We would sit down near the box and wait for the bee to return with recruits.

We could then determine the direction to the nest by averaging the departure directions of several bees. We guessed the approximate distance to the nest by dipping the end of a grass stem into flour paste and daubing a white fleck onto the abdomen of an individual bee. We could recognize this bee, and time how long it took to travel to and fro. This was scientific research, though it occasionally felt like magic. We had a question—where is the bee tree?—and we did the simplest and most direct experiments to try to get an answer. What could be more pleasurable than watching and manipulating bees in a field of goldenrod in order to decipher the location of a hive full of honey? I could not imagine. But not even lining bees could distract Papa for a few minutes from his ichneumons.

After we knew the direction to the tree, we established a "cross" line by putting the cover on our bee box when many bees were feeding in it at the same time, and then releasing them all in a field or clearing close to where we thought the bee tree might be. Most of the bees, when released at this unexpected place, would fly off in the wrong direction (they apparently did not know they had been moved and went off in the same direction as they had gone when departing from the other site). However, if we were close to the bee tree, some of the bees apparently recognized landmarks and reoriented themselves sufficiently to make it back home. These then returned to the *new* bait location, bringing more recruits. When this happened, we had successfully established a cross line and we had a fix on the location of the hive. It usually took us several days, and sometimes a week or more, of catching bees, lying in the grass, marking, timing, and more lying in the grass before we deduced the approximate location of the tree. We then waited for a warm sunny day when there would be a lot of bee traffic to help us locate the tree.

Walking through the woods, we searched for black specks of flying bees silhouetted against the sky above the tree canopy. Coming across a possible bee tree, we would press our ears closely against the trunk in order to listen for a low hum within. What a thrill to find a humming tree! As we all gathered around it, Floyd would pull out a jackknife and carve "F.A." in the bark. By common convention going back to pioneer days, initials signify ownership. It did not matter whose land the tree was on. The bees were ours. We only needed to perform the courtesy of asking the landowner's permission to cut the tree. But this was just a formality. It would not cross anyone's mind to refuse bee hunters their prize.

We harvested the honey long after the haying was done and before the apples were all picked. One morning, after finishing our chores more eagerly than usual, our procession walked single file, carrying a crosscut saw, an ax, a couple of iron wedges, bee veils, gloves, a smoker, a knife, lots of buckets and pans, and an empty hive. Wild bees are protected by law; we could take their honey, but we could not leave them to die without food and a hive for the coming winter.

Our excitement reached fever pitch when we arrived at the tree and Floyd and Robert started the sawing. "Swish—swoosh—swish—swoosh" went the rhythm of the crosscut saw. It ripped into the bottom of the tree as spouts of white shavings flew to the sides. As soon as the tree fell, the air turned black as thousands of bewildered bees milled around the gap in the sky near where the entrance to their hive had been. Floyd and Robert made two more cuts, each halfway through the tree about two feet above and below the nest entrance. They then pounded in the iron wedges and split off the uppermost section of log to reveal the golden combs of the nest within. While they were busy with the tree, we boys kept working the smoker, making sure it was kept half-filled with dry punky wood and pumping out smoke to distract the bees. Under the cover of a thick smoke screen, we closed in on the honey.

We chewed on the newest, whitest combs. The darker, older combs with brood (eggs, larvae, and pupae) were trimmed with the knife and tied into the frames of the new hive, along with a portion of the honey. This new hive was left for a week or so alongside the opened tree to allow the bees to move over fully into their new home. They would wax in their combs solidly, chew up and remove the string, and clean up the spilled honey from the tree cavity and carry it into their new hive.

We took the honey and broken combs we collected to the farmhouse, and in one of Leona's big kettles on the kitchen stove we heated and strained the mixture by wringing it through cheesecloth. Every pound of honey wrung out of these combs was a pure distillate of adventure and fun. I loved bee lining as much as I loved honey. Later on I maintained some hives of my own (I do so even now) and did research on bees that led to some surprising discoveries about their behavior and physiology. Bees are of the same order as ichneumon wasps—Hymenoptera, insects with two pairs of wings that are hitched together to work as one pair—yet they are very different from each other.

My interest in bees kept me not too far from Papa's interests, though my approach was radically different. I would become an experimentalist, concentrating on trying to find and solve new problems, while he would forever try to find and describe new species.

IN EARLY OCTOBER THE AIR BECAME DRIER, THE skies became bluer, and the first frosts dappled the morning grass with fine white crystals. This, I was told, was "Indian summer," and the maples around the house turned such a brilliant gold that they glowed in the sunshine; even the air beneath them was luminescent. The red maples turned red indeed: brilliant vermilion and deep ruby-red. Whole hillsides that were green one day became the next a patchwork of contrasting shades of green, yellow, orange, purple, and red. It was a breathtaking and fabulous sight that would have been impossible to imagine and took my family by surprise. There was nothing even remotely like this in Germany or Poland. Often a single leaf (of red maple) would contain many of these colors, as though someone had painstakingly painted it.

This was a good time of year to find ichneumons, just before they went into hibernation, and Papa and Mamusha found a few ichneumon hot spots along the edge of the Adamses' apple orchard—basswood leaves where the wasps sunned themselves and a stand of young pine trees sprouting in a field, where the wasps came to feed on the sugary secretions of aphids, which had established colonies in the trees. I was by now becoming less amenable to hunting ichneumons, as I became increasingly more occupied with my social life and activities with the Adamses and other neighbors; we also liked to climb up into the mountains of Weld, about fifteen miles away, where we picked berries—we would come home with backpacks full of blueberries and wild mountain cranberries. And in addition to the bee lining and fishing, there was hunting.

By fall, the grown-ups went to work picking McIntosh apples in Mr. Shardlow's orchard in nearby East Wilton. Mamusha and Papa went there with Floyd, who could drive them in his Pontiac on afternoons and weekends when he wasn't working at the wool mill in town. Both Mamusha and Papa said that they enjoyed the work. They were in the open air, setting up a ladder against low trees, climbing up with a basket and then filling it and emptying it into wooden crates. They could work at their own speed and were paid by the crateful. The alternative was the loudly clanging, regimented mill. Marianne and I were now busy with school, but we helped pick apples on weekends. We liked the work less, but we did not consider or experience the unanticipated largess that Papa encountered while moving his ladder from one tree to the next. Once, a man walked by and without stopping pressed a ten-dollar bill into his hand. On another occasion, Papa and Mamusha were invited after apple-picking to stop in at someone's home near the orchards. Again, Papa was flabbergasted when this complete stranger pulled a twenty-dollar bill out of a pocket and pushed it into his hand, without saying a word. In Maine, the old ways seemed to hang on. And so would we.

 *fourteen* | Home

If your days and nights are like sweet scented herbs,
that is your success! All Nature is your congratulations!

—HENRY DAVID THOREAU

OUR SOON-TO-BE HOME WAS CALLED THE "OLD
Dennyson place." It was a homestead close to Pease Pond and only a mile
through the woods from the Adamses, as the crow flies. The last Dennyson to live there, an old man, had died three years earlier; before him, the
farm had raised sheep and cattle for over 150 years. More recently, young
aspens, pines, and alders were springing up over the abandoned fields,
and junipers were taking over the old sheep pastures. It once had a giant
barn, like those at Dr. C.'s place and the Adamses' place, but that barn
was partially collapsed and a small, more recent barn (made of sawed
rather than hand-hewn beams) had been built as a replacement. At one
point along the long driveway to the house stood a carriage shed, which
still sheltered an old horse-drawn sleigh and a wagon.

The original house—the main house—was of the usual plain square
saltbox construction. Like all other farmhouses thereabouts, it stood on
an unmortared foundation of fieldstone that enclosed an earthen floor
cellar, and the foundation was bulging inward. The roof leaked. The plaster was coming off the walls, and the ceiling in the living room was sag-

233

Our old Dennyson place.

ging. The wallpaper in most of the rooms was water-stained and peeling. The windows, each with four to six windowpanes, were drafty because many of the panes were loose, set now in only remnants of their original putty. This house of course didn't have electricity or running water, except for the rain coming through the roof in the summer and seeping up from the foundations into the cellar in the spring.

Despite this panoply of woes, the place had its charms. It had a three-seater outhouse attached to a roomy woodshed. As at the Adams place, a huge spreading elm tree stood like the hub of a wheel in the center of the yard, and the house was surrounded by six century-old sugar maple trees; large lilac bushes; a jasmine bush; and several viburnum, which bore white flowers in spring and red berries in winter. Of this place, Papa said, "I loved it at first sight." He was enamored mostly of the big maples—they reminded him of the chestnut trees at Borowke. The asking price was $3,500, and for $3,000 it was his.

The house was absolutely empty of furniture or anything else except dust, cobwebs, and mouse turds. Then a miracle happened. The neighbors organized a housewarming for us; they came in their pickups and

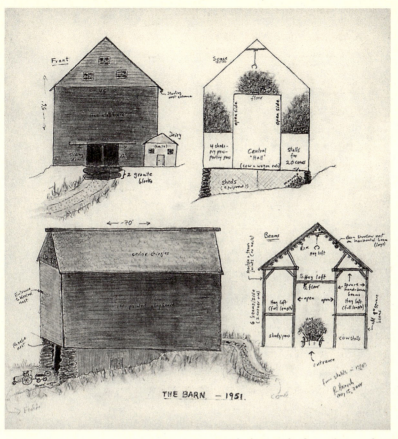

The barn at the farm when we first arrived in Maine.

cars and station wagons from miles around. In all about 100 people came, and they brought all the things we needed to make our house a home: tables, chairs, beds, a large cast-iron stove, dishes, pots and pans, kerosene lamps, clothes, wallpaper, paint, preserves, blankets, axes, wedges, a crowbar, a crosscut saw, and the free use of Susie, the horse belonging to our neighbor Erland Adams (no relation to Floyd). That evening one of the neighbor families, the Earl Ellriches, hosted a party, in the American pioneer tradition of welcoming newcomers. And somebody came by almost daily to see how we were doing and to help out. The hospitality, Papa said, was "overwhelming." In a letter to Stresemann, dated November 30, 1951, he wrote: "I cannot say that I'm sorry about my jump over the ocean. From everyone in the circle of my neighbors here (small farmers and people who have a farm but work in the nearest factory) I have experienced so much substantial help and warmhearted friendliness, as I never had even in my dreams."

I loved staying with the Adamses; but at the old Dennyson place, which Papa would later sometimes call "New Borowke," I felt that it was safe, finally, to put down roots in America, to become part of it. This was not someplace to pass through on the way to some other place. I took it for granted that I'd be here as long as I lived. If I were forced to be away, I would always come back. I dug young trees up out of our farm's own forest and planted them along the driveway. They included basswoods for their aromatic flowers in the summer, to attract bees; American ash, whose seeds would feed grosbeaks that came south from the Arctic to spend the winter; oaks for acorns to feed squirrels and blue jays; and red and sugar maples for colorful foliage in the fall and sugar sap in the spring. Papa planted two weeping willows on the opposite side of the driveway to be reminded of those near the Borowke house. Much later—in the 1970s, when I was in California—he wrote to me: "You made a nicely diverse selection of the trees of the woods! I shall not see them as beautiful old trees, but I may still be pleased to see them grown from childhood into adolescence." In another letter several years later he wrote:

Time and again during the last week when I went up the avenue [of trees I had planted along our driveway] from the highway to the house, I was fascinated and filled with joy, by the sight of the gigantic, marvelous maple trees shadowing our little, old house. An old, tall, powerful tree represents for me one of the deepest expressions of beauty nature has to offer. I hope that these trees, for which once I bought this new home, will still stay a long time, after I am gone, to give their beauty and strength to all of you, who may understand it.

When we moved into the house in the fall of 1951, the closest bit of wilderness was an alder bog within fifty feet of the house. This bog, unlike the one at Borowke, did not have cranes residing in it, but the wood snipe (called woodcock, as well as timberdoodle) made its home there, and for me this bird was a nice substitute.

In April, as soon as the sun had melted the snow off the matted grass, the woodcock came out of the alder bog at dusk and flew over the stone wall and into the adjacent field. This field became his display ground, and he put on an unearthly performance each dusk and dawn, before again retiring into the alders. I camped out here alone at night; I would, for years on end, never tire of hearing and watching him. His dancing arena became my favorite field, and eventually it became my bumblebee meadow, where I studied bees' foraging behavior in the summer.

The queen bumblebees came out of hibernation shortly after the woodcock returned, and they started their colonies in the mouse nests in the dry grass. They fed on the pussy willows that were flowering in late April in and at the edge of the field and in the bog. In the summer, the young trees shooting up on the woodcock-bumblebee meadow were my prime caterpillar hunting ground, and it was here that I found the only io (pronounced "eye-oh"; it is a saturniid) moth caterpillar that I have ever seen. Jewelweed, goldenrod, and asters bloomed here in that sequence in late summer and into fall. Honeybees came, and—lining them as Floyd had taught me—I found their colony in one of the 100-year-old hollow hemlock trees in our woods. Over the years, young white pines sprang up in the meadow, and as at the Adamses' place, there Papa hunted ichneumons that fed on the honeydew produced by the aphids that fed on the pinelings' sap.

Our woods began near the meadow. To get to them you passed a small apple orchard that was frequented by grouse and porcupines. Stepping over a shallow hemlock-shaded brook that was big enough for small brook trout, you continued on a rough path with forest to the right and sheep pasture to the left. The pasture had been grazed bare and then left unattended, so that by the time we took possession of it in the 1950s it was overgrown with spreading juniper bushes. These provided an ideal habitat for snowshoe hares. Young larch trees, balsam firs, pines, and ash were starting to replace the juniper bushes, starting the process of reforestation that had already been in process in Maine for almost 100 years, after the sheep industry failed and farmers moved west.

The woods that already existed were a mixture of young and old sugar maples, American ash, elms, basswood, and sugar and red maples that had not been disturbed for decades. This mature forest held a scattering of the old hemlock trees that housed honeybees, porcupines, and raccoons. Years later, whenever I came home from California, I would—almost before doing anything else—go into these woods to reconnect with home. I went in rainstorms and snowstorms, on bright sunny days, in all seasons, and I even "sampled" them once on a windless midnight in July. I recall hearing the occasional cheeps of sleeping birds, and the barely audible but steady patter of caterpillar feces falling on leaves.

These woods sloped gradually up to the top of what was called Picker Hill. Here, near the top of the hill, was a second long-abandoned apple orchard. Like the lower one, it was surrounded by stone walls, indicating that people had at one time spent a great deal of time and effort to clear the ground of rocks. Most of the apple trees were centenarians or even older, and they were just barely hanging on as they were being overtaken by ash, pines, and maples that had seeded in.

The field and forest were all the more appealing because of being neglected—unlike our house. There was much work to be done in the house when we first came. The rainwater that had been entering for years was decaying the hand-hewn beams in the roof, and unless attended to quickly the house would have suffered the same fate as the old barn and the carriage shed, both of which collapsed in heaps within two years of our arrival. Neighbors helped to set up scaffolding on the roof, and I climbed up (I was now twelve years old) to rip the old shingles off the roof and to nail new ones on. As I did, chimney swifts made repeated passes overhead before dropping down into the cavernous black hole of the house's chimney, presumably to settle on their white eggs in nests of twigs glued together with their saliva and attached to the blackened brick walls.

The chimney rose barely two feet above the ridgepole of the house. Up there nothing was extraordinary, but below it everything was. On the first floor, the chimney flared out into three fireplaces: in the living room, a bedroom, and the former kitchen. In the last, a brick facade covered nearly half a wall, and it incorporated not only a fireplace but also ovens for baking, and other nooks and crannies of now obscure function. There were black iron hooks and levers in the fireplace for hanging and retrieving pots. We saw no similar brickwork remaining in any other house and were informed ours was "old-fashioned." Within several years Papa had the whole thing ripped out so that he could use the big pile of bricks to build a fireproof addition to the house for the accumulating glass-covered boxes of his ichneumon collection.

But not even three fireplaces could, in an old house without any insulation, prevent the pipes from freezing in the winter. Of course we had no plumbing, so there were no pipes to worry about, and there was no need to heat the whole house. We pumped our water up from the well with a creaky hand pump that was set directly over the well, so that the water spout emptied directly into our black cast-iron sink. To start the water flow, you needed to prime the pump by moving the handle up and down a few times while pouring a little water from a pitcher into the pump at the top. Only then would the leather seal of the plunger be airtight enough for the pumping action to produce the suction that would draw up water. In the evening we would raise the pump handle high enough to allow air to get past the pump's plunger, and gravity would pull the water back down into the well where it never froze.

When we moved in, Mamusha heated and cooked on the cast-iron wood-burning stove we'd been given. She put the wood in the top and took the ashes out the bottom to fertilize the garden. The stove had a tank on the side for heating water; four lids, to put pots on; and an oven, for baking. Papa cooked his oatmeal here every morning. All year the fire was lit and kept going, but it would always burn out at night. We tended to live in the kitchen around the wood stove. On winter evenings, before we blew out the kerosene lamp, Marianne and I each heated a brick on top of the stove or in its oven, wrapped the brick in a towel, and brought it upstairs to put under the blankets at the foot of our beds.

The outhouse was strategically located, attached to the end of the house near the garden. Admittedly it was chilly there in the winter, but you could get to it without having to take one step out into the snow. It was connected to the main house through the woodshed that, in winter, was a great cold storage room in which to hang the snowshoe hares that Papa and I shot in our juniper pasture. And whenever you went through the woodshed, you brought back a couple of sticks of wood for the kitchen stove.

To me, however, the main advantage of the outhouse over an indoor facility was aesthetics. Every spring a pair of phoebes built their nest on a beam near the three seats. At one's morning visit the bird would flush off her nest, and one could admire her four or five immaculate white eggs in the nest of green moss held together with dried mud that attached the nest to the beam. A phoebe's nest differed from the swallows' nests in the barn in being lined with fur rather than feathers. The swallows' nest cup, on the other hand, was made only of mud and incorporated no moss. The phoebes in the outhouse, the swifts in the chimney, and the swallows in the barn had all adapted to their new conditions—they almost ceased to exist in the wild where they formerly nested on cliffs and in giant hollow trees.

Marianne and I "nested" in our own little rooms. They were upstairs, one on either side of the chimney, and each had a small window. My view was directly into the branches of one of the great maple trees surrounding the house. I built birdhouses out of the plentiful boards from the collapsed barn and installed them so that I could hear the birds sing and watch their activities as soon as I woke up in the morning. Starlings sang directly outside my window, but then a pair of kestrels evicted the starlings from their nest, to take the young to feed their own young. I eventually took one of the kestrels' young for a pet, first raising it in my room until it could fly and then giving it the freedom of the fields. It followed me to beg for food and kept me busy catching insects and mice.

I loved my cozy little bedroom. It was nearly bare, but I had a terrarium at the foot of my bed where I kept a variety of snakes that I caught near the house and the barns. I also had a pile of *Nature* magazines on a table, along with a book of American birds. Both were presents from Mary Gilmore, a neighbor, and I still own them. My huge milk adder, which like all the others in Maine was nonpoisonous, soon escaped; it probably pushed open the cover, and it was last seen slithering down the stairs. The rest followed suit.

Like the wild birds adapting to nest in and around the farm, I adopted the American culture of living close to the great outdoors and being in touch with the idea of "wilderness." Wilderness is thought of as a uniquely American value. Americans were the first people to enshrine specific areas and designate them as wildernesses. Many people have connected the birth of our nation to the wilderness, and see it as a source of the country's strength and character. In the early 1950s in Maine I still felt a romantic sense of frontier and the wilderness. (For Papa, though, the love of wilderness had nothing to do with frontier and everything to do with the roots of our biological heritage.) In school we sang songs about the prairies and the Oregon Trail. We also sang "O! Susannah," about going to Louisiana "with a banjo on my knee." We heard all about Daniel Boone, Davy Crocket, the mountain men, the trappers, and the fierce swift Sioux and Arapaho. The Bass shoe company in Wilton specialized in moccasins, and I wore mine proudly. They were thick rawhide without any soles, to keep them authentic. I did not have the means to wear a coonskin hat with a tail hanging down my back, or a buckskin jacket with fringe on the sleeves. But I would have liked to, if it would not have seemed fake and presumptuous to do so, since I had done nothing to deserve the honor of wearing that uniform. Our teacher told us how the Indians used to run down deer, and how they could silently become one with the forest and see everything in it. Here was something worth aspiring to. This athletic and utilitarian view of nature and those who lived in it, coupled with Papa's abstract naturalist passions, melded me into a hybrid—a boy who watched caterpillars and lined bees, but who also hunted rabbits and fished for trout, and kept wild pets that kept me connected to the wilderness.

On one of my excursions down to Pease Pond I heard the begging long caws of young crows. I found the nest near the top of a balsam fir tree, at the pond inlet by old Frank Currier's pasture. I climbed up and brought one of the pinfeathered babies home with me. Like the kestrel, it stayed for a while in a substitute nest near my bed. As in the Hahnheide, I was kept busy hunting for caterpillars, mice, and frogs to feed it. After hopping out of the nest Jacob II, like Jacob I, was permanently free, outside. The kestrel had left by the end of August, but Jacob II was still my constant companion. He would intercept any caterpillar or field mouse as soon as I found it.

But by fall Jacob II had also started to become independent. He was finding some of his own grubs. He would wander as far as three miles away, into town, and if I met him on one of his excursions, he would come to me and land on my shoulder, making soft cooing sounds in my ear. Jacob and I were similar; we were both young and driven, if not controlled, by passions and emotions. The more intense the emotions, the less the power of reason, and the stronger the power of imprinting. Jacob was unnaturally, unconditionally bound to me; I was irrationally, irrefutably in love with a chubby little dark-haired girl in school. I was bonding with my new home and becoming anchored. As with Papa, who had been anchored with a trance-like fascination to his Borowke and his ichneumon wasps, I became imprinted on everything in, of, and around the old Dennyson place, and my routines and memories began to braid me into the landscape and its cycles of the seasons.

fifteen   New Jobs, Old Life

PAPA EXPERIENCED THREE MAJOR DISAPPOINT-
ments related to jobs, in as many months after arriving in America, and
these presumably dampened his initial expectations. And economic
prospects in rural Maine were not bright, as the farm economy was then
dwindling.

In earlier days there had been sheep farms, like our own Dennyson
place, in the nearby countryside. Sheep farming had then moved to New
Zealand. By the time we arrived in Maine in 1951, the wool business had
already collapsed. The sheep farms were finished, and the woolen mills
were closing.

Our neighbors who were living and working on farms along nearby
country roads were making a living through diverse economies. They
kept a few cattle, plowed a few acres, cut a little pulpwood for the paper
mill in Rumford, and worked at the mill whenever it was up and running.
Farming had been in decline since the end of the Civil War, when people
left the hilly, rock-strewn soil for the wide-open prairies "out west." Dairy
farming hung on a while longer, maybe because cows, like sheep, could
coexist with rocks. But the bureaucracy learned that the raw milk from
the small dairy farms contained germs. Germs had been around all along,
and they were killed by pasteurization at the dairy, but soon that wasn't
good enough. By the 1960s, the farmers were required to pipe the milk
directly from teat to cooler, a process which involved buying expensive
equipment that no small-time farmer could afford. The message was
clear: get big or get out. In western Maine few got big.

Papa and Mamusha, however, had one fallback that their neighbors did not: collecting specimens. Papa had a hand net and his thirty-two rat-traps. In America, however, he also needed a vehicle, and he needed at least a little cash to get started.

On October 27, 1951, just six months after we arrived, Papa bought a used 1948 Ford half-ton pickup truck. He got it from the Main Street Garage in Wilton. According to the purchase statement, which he saved (and which I found, coated in pigeon guano, in our barn), the truck cost $825. He got it for a down payment of $275, with the remainder payable in twenty-four installments of $31.76 each. I can't imagine how the Wilton Bank agreed to finance his purchase. As collateral he listed "none." He indicated his age as "fifty-five"; salary, "zero"; occupation, "farmer"; number of dependents, "four." (I assume he included Anneliese, who had emigrated to Chicago, where Ulla and her family lived, shortly after we'd come to Maine.) His address is left blank (we were then staying with the Adamses), and for former residence, he put "Europe."

The truck now allowed Papa to go with Mamusha to pick apples in Shardlow's orchard independently of Floyd. After the apple harvest, they worked in a canning factory. But the canning of locally grown beans ended immediately after the harvest. What then?

Next they tried logging. Papa and Mamusha had a crosscut saw, a cant dog (or peavey, a logger's tool for gripping and handling tree stems), an ax, and a toboggan (all gifts from the housewarming party). Our neighbor Erland Adams had offered them the free use of Susie, his big brown mare, as well as her harness and a sledge for hauling the logs out of the woods. These were priceless gifts, because winter was coming and trees had to be cut and firewood prepared, independent of any wood for sale. Mamusha said later (in an interview with a friend, Alice Calaprice): "There was one person who contacted us—he wanted to buy some wood for the factory. So that's what we decided—we would cut our trees and our wood. We had beautiful, beautiful old trees. You ask Bernd, he was crying when we had to cut them. All the big, big trees." Actually, it was not all of them, at least not just then.

That "one person" was Mr. Leon Ogilvie at the Wilton lumber company. He had a little factory shop across the street where he manufactured kite sticks and wooden dowels. When Mr. Ogilvie heard about Papa and Mamusha working in the woods he told them, "In America women don't work in the woods." He offered Mamusha a job in his woodworking shop, and she immediately took it. The problem was that it takes two people to operate a crosscut saw: one to pull it one way, the other to pull it the other way. Papa was forced to quit working in the woodlot, and Ogilvie offered him a job as well. Papa had to lug boxes and crates around. He was not the speedy type, and his female supervisor told him to get the lead out. Mamusha counted the kite sticks into bundles of 100. "After counting to thirty-seven I got distracted and lost track—Oh, God, it was an *awful* job," she said. They soon quit, and both returned to cutting logs in the woods.

Our first winter in America was one of the snowiest on record. Once when I came out to the woods after school I climbed up into a maple tree, and jumped—and landed in snow up to my neck. They had to dig me out, as they had to dig out every tree they felled, in order to saw it into pieces. They earned $200 that winter, after which they both decided that women indeed should not be working in the woods in America. And neither should men, unless they could bring in a supplemental income. As usual, it was by collecting.

"Chalk" Allen, another friend from town, made that possible. He lent Papa a shotgun, and when Papa went to work in the woods he took the gun with him to shoot birds. He shot a grouse, and Mamusha prepared a live-mount of the skin. It was sold and displayed in the window of Cram's jewelry store on Main Street. One day the game warden came to our house; Papa had just gone into the woods and Mamusha was still busy in the kitchen.

The warden asked Mamusha, "Did you prepare the grouse in the window at Cram's jewelry?" We often received visitors who'd seen and admired Papa's collections.

"Yes, I did." she answered. "Do you like it?"

"It's beautiful," he replied, and then explained that he noticed the bird because it had its snowshoe fringes on its toes, which meant that it was shot in the winter—which was "out of season."

"Season?"

"Yes, hunting season for grouse is in the fall," and only grouse, ducks, and woodcock had such a hunting season.

He then complimented Mamusha on the beautiful job she'd done stuffing the grouse, and asked her if she had any other birds.

"Thank you, yes, I'll show you." Mamusha then proudly showed off more stuffed birds—chickadees, woodpeckers, finches . . . none of which had a hunting season. The game warden then wanted to see a "license."

"License? What for?" Mamusha wanted to know. She thought it was a free country. "But apparently," she told me, "it was not as free as we were used to" in Europe. She knew you needed a license to drive a car, but to stuff a bird? She told the officer, "This taxidermy is all I know how to do, and we need a few dollars to make a living." But her logic fell on deaf ears.

"Where did you get these birds?" the game warden wanted to know.

"Well, my husband shot them."

"Where is he?"

"He is cutting trees in the woods." Was that illegal too? They were our woods! Surely it was not illegal to take a robin or a chickadee that nobody else wanted. Our friends hunted lots of grouse, geese, ducks of various kinds, woodcocks, and doves. Papa didn't have a license for anything, and Mamusha, who was getting worried that he had done something wrong, wanted me run out and warn him to hide. But the game warden snagged me first, and had me lead him to where Papa was heaving logs onto a toboggan.

Chalk's shotgun was leaning against a tree next to Papa. The game warden immediately confiscated it.

Chalk and Floyd both later went to the game warden's house and explained that we were just immigrants and didn't know any better. That helped. Chalk got his gun back, and nobody went to jail.

AFTER WE MOVED INTO THE OLD DENNYSON place I left the tight orbit of the Adamses and got caught up with our new neighbors, Phil and Myrtle Potter. When we first arrived, Mamusha and I had once walked by their house, about a quarter mile up the road. I was barefoot, and we were both carrying insect nets.

"You were quite a sight," Phil said to me later. He had an angular face and thick, graying hair that was trimmed into a crew cut. He seemed always on the verge of exploding into guffaws—he would crack a ribald remark and double over, bellowing with laughter at his own joke, slapping his knees for emphasis. Myrtle was of Penobscot Indian descent and came from the nearby village of Norridgewock, where there had been an Indian settlement. Together they hunted deer in the fall, and they later took me with them on trout fishing trips up-country.

Phil and Myrtle had always wanted a son, and they became for me almost surrogate parents. I do not recall any tension with my real parents. Instead, I think I was pulled to the new and the exotic, and others offered it more than my parents, who could only try to be at home in the new environment while the others already were. These dynamics would later make it easier for me when Papa and Mamusha went away and left us at a boarding school. Also, entering my teens, I was already seeking independence in any case. I sensed Papa to be "an outsider" where I wanted to belong as an insider. I did not yet appreciate uniqueness, so I reacted almost with irritation, rather than taking it as a compliment, when I was at times considered a "chip off the old block." I thought of Papa as an intellectual, which, to me, meant ineffectual.

The woodsmen I admired, like Phil Potter and Floyd Adams, were everything Papa was not. The Potters' two daughters, Phyllis and Joan, had just graduated from Wilton Academy, a high school. They were more interested in reading romance novels than in carrying a fly rod or a deer rifle. I, on the other hand, wanted to do nothing more than carry these things. In my preteens, I was at the social evolution stage of the hunter, learning to provide for the clan. Like a caterpillar who must go through more molts before it can become a butterfly, I also had many more stages to look forward to, and sometimes to endure.

To me, Phil was the embodiment of the Maine outdoorsman. He had been a logger up-country in the northern part of the state when he was a teenager. He had once been a trapper, and he knew the skills of deer hunting, trout fishing, bee lining, and fixing looms at the woolen mill where he now worked part-time. The Potters also kept a few cows and had a large garden.

Phil taught me the proper way to paddle a canoe and how to hunt deer, cast a trout fly, drive a car, and plink cans and bottles thrown into the air with my new Remington bolt-action single-shot .22. I recalled leafing bug-eyed through the Sears Roebuck catalog that Phil owned, and it was from this that I ordered, with what I had earned doing his chores, both the rifle and a Kodak box camera. I "captured" a deer, a porcupine, and a hawk's nest with my camera and lent it to Mamusha and Papa for their trip to Mexico, and they brought the trip back to me in pictures.

Whenever I came to visit the Potters, it took Phil only a few minutes to think up some often dangerous deviltry, such as throwing firecrackers into a steel pipe stuck into the ground and topping them off with golf balls that would then be shot into the stratosphere, or sticking a stuffed deer head in the bushes where hunters were apt to pass. He had many stories to tell of the jackasses who emptied their rifles into the dummy, even out of deer season.

Sometimes we went trout fishing up-country, leaving on weekends before daylight. Phil and I drove for hours and hours on lonely roads, where we seldom saw a house. Then we turned off onto gravel dirt roads, where we continued for yet another hour or two and where we saw no cars, no humans, just the occasional moose or deer. Eventually we'd stop at some gravel pit, unload his Grumman aluminum canoe from the top of his car, and head off on an overgrown logging trail before finally coming through the confining trees and onto the shores of a pond or stream. We'd set down, pull out a whiskey bottle, and take a nip or two of "fly dope," and then we'd get out our fly rods and push off into the pristine water. We braved the swarms of blackflies and copperheads, and in the evening we'd fry our fish and warm our baked beans over a smoldering fire in some gravel pit, sometimes in a drizzling rain. I tried never to complain—I wanted to become a man. Phil took me to many distant places, and the names still ring romantically in my ears—Wytopitlock, Enchanted Pond, the West Branch (of the Penobscot River), the Dead River, Umsaskis, Chesuncook, Mooselookmeguntic Lake . . . We'd get home at dusk, in time to call in his two cows, milk them, and centrifuge the milk. This is how we got the cream to make our butter in the hand-turned churn, and our ice cream, which was flavored with pints of fresh strawberries from the garden, and honey from the honeybees that I was then keeping.

Sometimes I stayed at the Potters' house late into the evening, relaxing in front of their television, watching *Tombstone Territory* or some other western. I think Papa may have been disdainful or disappointed, but he never said anything about my fascination with the bygone days of the frontier. Maybe he lacked empathy for hunting and fishing because he often heard the deer hunters and fishermen talk about the size of the antlers of their deer, or the length of the fish they caught, as though they killed for bragging rights. As I've mentioned, at first Papa and I did occasionally hunt rabbits for the pot, so he was not against hunting or fishing on principle. But he believed "real" hunters were woodsmen who stalked and shot not for sport, but only when they had a sure shot, and who used all the meat themselves.

Of course, Papa was a hunter too, of zoological specimens, and those were almost always to be found elsewhere. After catching some mice and birds near our town, Papa soon set his sights on Mount Katahdin some 200 miles to the north.

In the spring, after their first hard winter working in the woods, Papa told Mamusha: "Let's just go on an expedition!" And so in May 1952 he loaded up his Ford pickup with oatmeal, dried fruit, and dried milk for food, plus of course his thirty-two rattraps and his net for catching ichneumons and, of course, Mamusha (his cook, taxidermist, and supplemental ichneumon catcher), and they drove north to the mountain. Marianne and I happily went to the Adamses.

The small mammal collection that Papa and Mamusha brought back from Katahdin included some rarities, in particular a pygmy shrew and a yellow-nosed vole. The collection was purchased by Professor Raymond Hall, the famous mammalogist at the University of Kansas. Dr. Hall was apparently very favorably impressed with the collection, and he offered a contract of $3,000, all expenses paid, if Papa and Mamusha would leave that same summer and collect small mammals, particularly gophers, in Mexico. Papa jumped at the chance. He wrote to Erwin Stresemann from Lawrence, Kansas (while already on the way to Mexico, on October 17, 1952), that he had taken me and Marianne to a boarding school (more on this in the next chapter), that he had stopped to see "the beautiful Helen" (Anneliese) in Chicago, and that he was now happy "in the late afternoon of my life to finally be allowed to do what gives me the most enjoyment and zeal."

The next letters, however, told of disappointments. He complained that he was supposed produce at least 3,650 mammal skins per year. To do that he would have to hurry from one locality to the next and set perhaps 200 traps—which would involve having to rush and would leave little time for biological observations or for preparing the skins properly. Papa was, if anything, always frugal and leisurely. Thirty-two rattraps had always been sufficient, and maybe the boundary between enjoyable activity and work, and he usually did as he pleased.

On this expedition Papa had a list of assigned missions. One was to get specific gophers from specific localities that had been wilderness when these forms were described but were now urban settlements. He was to find out if the two forms of similar gophers were species, varieties, or subspecies. He finally located an area (a small valley) where the ranges of the two forms not only met but overlapped. One form lived only on the bottom of the valley and fed on grass roots, while the other form lived on the slopes and fed only on the roots of yucca. As with the woodpeckers collected for Stresemann in Persia, such ecological separation is proof that the two gophers were indeed different species and not subspecies. Papa asked for his boss's blessing to publish his finding, but was told that he was paid to collect, and could not to publish scientific papers of the data he might collect, because he was only on salary as a "field collector." Papa was incensed. He could always carry on despite all difficulties, but he also recognized that he had no control over others; he had control only over himself. Feeling marginalized as "a salaried drudge," he immediately contacted the Chicago Natural History Museum, which offered him a job collecting birds and mammals in Angola, Africa.

Papa eagerly accepted the offer. After completing their yearlong expedition to Mexico, he and Mamusha briefly stopped at the farm in Maine, and then departed for Africa. However, before leaving Papa wrote to Stresemann on September 30, 1953, with some concerns as he reflected on collecting in Angola. He wrote about his own waning strength and his fear that a disease of the African jungle might now be fatal. He complained that collecting "is an expensive bread for me at my age, as it needs my extraordinary efforts. I look around for a rest to do my own research, but I fear it will never come." He concluded: "My new home is beautiful. I have a wonderful piece of wilderness, my own, rich in birds of all sorts, and ichneumon wasps. And still and in spite of all I have homesickness for my dear old Borowke."

Despite his fears, misgivings, and homesickness, Papa and Mamusha went on two almost back-to-back expeditions to Angola. Each lasted two years. In between, on October 10, 1955, Papa wrote to Stresemann that he was again, as four years ago, "living hand-to-mouth as an apple box carrier in the apple harvest at ninety cents an hour for ten hours a day." Stresemann replied, "In the United States there will be no laurels for you because there reigns pure capitalism, which lets people like you haul apple crates!"

Papa was at that time also starting to do some work for the Entomology Research Institute of the Canadian Department of Agriculture and wrote: "The salary is good, the money badly needed." His job was to identify the insects in the Canadian collection. However, not much identification was possible, because about half the ichneumons in that collection were undescribed species. Papa had to document, describe, and name them and write identification keys. He eventually (after five years of part-time work) wrote the seven-part series "Synopsis of Nearctic Ichneumoninae Stenopneusticae with Particular Reference to the Northeastern Region (Hymenoptera)," published in *Canadian Entomologist* in 1961–1962. This work required him to spend months away identifying Ichneumonidae in the collections of the National Museum of the U.S. Department of Agriculture in Beltsville, Maryland; and at Harvard University and the Smithsonian.

In the spring of 1956 Papa and Mamusha left for their second expedition to Angola; they returned from it in the summer of 1958. They had brought back several new species of birds, including a striking-looking thrush with a pure white head. Dr. Rand of the Chicago Field Museum (for whom they were working) named it *Cossipha heinrichi* and published its description.

Cossipha heinrichi, adult *(left)* and juvenile *(right)*, the new thrush Papa discovered in Angola.

The discovery of this new bird induced Dr. S. Dillon Ripley, then director of Yale University's small but excellent Peabody Museum (and later director of the Smithsonian Institution), to offer Papa and Mamusha a third expedition to Angola. Because of a bloody uprising in Angola, the expedition was diverted to Tanganyika—now Tanzania—instead. This expedition, the last for Papa and Mamusha, would not begin until 1961, and I would end up joining them, for an experience of a lifetime. But in the meantime precisely that which had given Papa and Mamusha "the most enjoyment and zeal" would turn out to be somewhat less pleasurable for Marianne and me. For us it meant many long years away from the home and friends we had bonded with, and from our parents—who for years we would know only through letters.

sixteen   Good Will Home

> *They hunt no more for the possum and the coon,*
> *On the meadow, the hill, and the shore;*
> *They sing no more by the glimmer of the moon,*
> *On the bench by the old cabin door.*

—"MY OLD KENTUCKY HOME,"
STEPHEN COLLINS FOSTER

ON THE MORNING OF AUGUST 30, 1952, MARI-
anne and I packed up our few clothes. We were being plucked from paradise and taken to a school-home-farm for disadvantaged kids. We knew
we'd be at the school for a long time, but we had no idea that it would
eventually stretch to six years for me and eight for Marianne. For much of
this time (except the last two years) the old Dennyson place stood empty,
to be used only during Papa and Mamusha's short breaks between expeditions. And even then the school forbade us to leave, because it would be
unfair to the other kids who had no home to return to.

Marianne, at the age of eleven, was left at Averill Cottage on the girls'
campus of Good Will Home, only a few minutes away from the old Indian
settlement of Norridgewock, and thirty-five miles as the crow flies from
our farm, with deep wild woods separating the two. I was dropped off at
Pike Cottage, a similar white house, a mile farther. Papa walked me up
the path across a trimmed lawn, and as I had been taught, I bowed and
kissed the hand of our housemother or "matron," who answered our

knock on the door. Lying that night in a bare room on an iron bed with a thin mattress, in line with a dozen other boys in identical beds, I knew that from now on I'd be on my own. I was twelve years old, entering the sixth grade.

Many of my classmates were from the less well-to-do parts of New York and other cities. Here they, and I, would learn about country living. We would learn that the beans we ate on Saturday night came from long rows in fields that had to be plowed and planted in spring rain, weeded in the summer sun, and picked on the sunny days followed by cold nights of Indian summer. On Fridays the beans were shelled and sorted on the kitchen table. Next they were soaked overnight, and then cooked all day in the oven of a cast-iron stove that we fed with wood or coal. The older boys tended a herd of more than 100 dairy cows, and all our milk and butter came from these cows. We would learn the source of a chicken drumstick. We found out that the warmth of the fire on a winter night came from the trees we felled with our axes in the woods and limbed and hauled out with horses, then sawed, split, and piled. We would learn many good and valuable things. But education was not the reason we had been dropped here. And we knew it.

Several days after we arrived, I wrote to Mamusha and Papa: "I am very sad that you are so far away. I hope you come back soon with lots of money and animals. I borrowed a moth book from the Library, and I saw illustrations for ten species of sphinx moths." I also told them I had not yet seen Marianne and that I hoped that they would be home soon. I thanked Papa for the caterpillar he sent me. "It feeds on poplar and willow and is beautiful."

Two and a half months later, Papa replied in a four-page letter from Tucson, Arizona, where they had stopped on their way to Mexico. He wrote that they had just arrived from Texas, and he had received my letters and told me that if he didn't answer right away I should not think he had forgotten me. It was simply that he was so busy in the field and exhausted in the evening.

In one of my letters I had apologized for not sending him a birthday present. Papa told me that my report card was for him a beautiful present—"Es ist gut," he said, and "keep on this way." He urged me to think about what he had told me: namely to always try my best, regardless of whether it was in school, work, or sports; and to keep my money, in-

GOOD WILL HIGH
Hinckley, Maine

Report of *Bernd Heinrich*

from *10/10/55* to *11/18/55*

English	*B-*	Physics	....
Latin *I*	*C*	Chemistry	....
French	....	P. Geometry	....
World History	*B-*	U. S. History	....
Algebra *I*	*B-*	Biology	....
Conduct	*a*	Absent	....
Effort	*B+*	Tardy	*3*

Promoted to Grade ....

Repeat Grade ....

F. E. Graffle
Principal

Industrial Arts, B+

My report card from high school.

stead of sending him something that he might not need anyway. He noted that I was not selfish and he knew that I would help him if he ever really needed it. "*Hab ich recht?*" ("Am I correct?") He said he was sad that I was unhappy because he had hoped I would make friends and like it there. He said:

> As they say in English, "make the best of it," and that applies to everyone all their lives, and means that we must try in every situation we find ourselves in to always find the good and the useful, to build up the soul and the body so that the time is not wasted. You are young, and healthy, and not dumb. That alone is something to be happy about.

In the second half of his letter, Papa described the fauna of Texas and Arizona, and talked about finding rattlesnakes and a coral snake. He closed with: "Enough for today, my dear son. Continue to do well in school and be happy about life in whatever it may bring. Better times will come at some time." And a PS: "Did you find any ichneumons?"

BEFORE LEAVING OUR HOME THAT SUMMER I HAD collected ichneumons for Papa and hawk (or sphinx) moths for myself. I had found the latter hovering at dusk on the sweet-scented milkweed blossoms near the barn. I was fascinated by these large and sometimes colorful moths because they resembled hummingbirds. I wondered what the caterpillars of each species would look like, and I hunted for and found some of these caterpillars on different kinds of bushes in our fields. It was a never-ending game to detect these marvels of camouflage, and every one that I found gave me a thrill.

I had smuggled some of my still-growing caterpillars, and the pupae derived from them, into Good Will, but I did not dare let anyone know about them. Who would understand my passion for creatures that many people loathed? I found a broken window on the ground floor of the school's L. C. Bates Museum of Natural History, slipped through, and left my charges in the cool basement, where they overwintered in safety.

Good Will was founded in 1889 by George W. Hinckley, an idealistic Methodist minister who took in homeless boys to give them a "family," a home, and a country environment. His ideals for the school were enshrined in the "roundel," which shows a picture of a barefoot boy sitting on a bench and reading. In a circle around him is written: "Home, Education, Discipline, Industry, Recreation, Religion." The school's mascot is the beaver, "who works when he works, and plays when he plays."

The school owned more than 3,000 acres of farmland and forests along the tranquil pine-bordered Kennebec River. "Adirondack" (Henry Harrison) Murray, the founder of the "outdoor movement," whom GW (as we came to refer to George Walter Hinckley, to differentiate him from two other Hinckleys there at the time, namely WP, Walter Palmer, GW's son, and NH, Norman Hinckley, who was of no relation to the other two) admired, had said: "You can't grow strong trees under a glass roof any more than you can grow boys into strong men by any indoor culture. They need freedom of the fields and streams." Of the latter I would soon feel deprived.

G. W. Hinckley was a visionary admirer of nature who believed that the way to God lay through God's creation, nature. It followed that nature study was allied with religion, in what had once been called "natural theology." To make sure that the kids would be exposed to the outdoors and see God's order in nature, he had cabins built in the woods and provided numerous groomed trails throughout the extensive property.

As the roundel made clear, discipline and industry were high ideals at this school; everybody had a job. Papa had never in his life done housework. Nor had I, up to this point. But on my first evening at Good Will, a generations-long Heinrich tradition was broken. I was last (third) in line at the kitchen sink, drying dishes with a towel. The next morning, and endless mornings thereafter, I dusted windowsills, and crawled on hands and knees scrubbing the floors in the dormitory rooms. Other boys did the same in the dining room and kitchen. We did this each morning right after breakfast. Afterward the matron ran her long probing index finger into every corner, to make sure we had missed no speck of dirt. Cleanliness, obedience, and daily Bible readings were at the top of her list of virtues, and those caught leaving dirt were punished with extra duties. I was terrified that she might find some evidence of my ungodliness. Although GW had dedicated his life to helping boys by letting them experience nature, the matron, who soon referred to me as a "little Hun," used any excuse to keep me locked up in the house and forbade me to write to my parents in German. I was again called "Ben" (maybe because almost nobody could pronounce the soft-sounding German "er" in Bernd). But I didn't object. To the contrary, I liked my new name because it sounded more American. And I had other consolations. On January 3, 1953, I wrote to Papa, then in Mexico, that I found out there are six species of rails in Maine and the smallest is the black rail. Only one nest had been found in Maine. I concluded, "I wish I could go sometime into the big marshes." A month later I wrote again: "Our matron has left for a few days and I can again write to you in German. . . . I do not like it here, and doubt that I ever will."

Two years later I was still there, and I still didn't like it. By now I knew the routines. As cook boy I got up before everyone else and tended the furnace. I shook the grates to loosen the ashes, poked the burning coals, shoveled new coal in on top, and took ashes out from below the grates. While everyone slept I lingered awhile, drowsily waiting for the coals to glow and blue flames to flare up. Steam would soon rattle the pipes and hiss from the radiators upstairs. I would go upstairs and check the kitchen stove, putting water on to cook the Quaker oatmeal, Cream of Wheat, or Ralston. While waiting for the water to come to a boil, I fetched the milk from the refrigerator, poured it into two pitchers, and set the two tables. I brought out the butter, warmed the doughnuts in the oven, sliced the bread, and made sure the sugar bowls were filled. When all was ready, I rang the bell at the foot of the stairs. The boys would come pouring down and pile into the bathroom to pee and wash up. Then they would loiter briefly in the study while waiting for the second bell, on which they would rush into the dining room, each standing behind his assigned chair, waiting for the matron to come in to stand at her very own small table by the window. She would call for one of the older boys to lead us in a prayer, and we bowed our heads as Paul (who later became a real preacher) mumbled the prescribed words. The matron sat down, and there followed a loud scraping of chairs as we followed suit. There was no talking and nobody got up until the matron had given permission.

I was cook-boy, and after breakfast on any one day I got out the bread mixer; measured out the flour and the warm water with yeast, salt, and lard; and started turning the crank. Eventually I produced dough and dumped it onto a layer of flour on my dough board to knead. I greased metal pans, cut the dough, and put it into the pans to rise. On other days I made doughnuts, cookies, or pies.

I had barely enough time to wash the cooking utensils; the dish crew was long done with the breakfast dishes. By now the linoleum floor was splattered and grimy. I was responsible for cleaning it. In the meantime, I'd kept the fire going and put away the food from the table that the dining room boy had brought back to the kitchen. On Saturdays, I needed to start preparing the beans for the evening meal. After study hour the night before I had sifted through the dried beans, one handful at a time, taking out all small clods of earth and the pebbles, and then soaked the beans overnight in a kettle. In the morning I added mustard, slabs of salt pork, and molasses; and after putting on the cover, let the beans simmer in the oven of the kitchen stove. Meanwhile, the matron had inspected each boy's work and retired to her room.

I could now join the other boys going to school or (if it was Saturday) report to Walter Price, to work at the barns till noon. In the summer we were then separated into different work details, such as weeding or haying. In the winter, Saturday mornings (after we had finished our house duties) we were sent to work in the woods. Felling gray birch and piling brush was hard work, but we got to build a bonfire in the snow, and Don Price, the overseer from the farm crew, usually let us take breaks. On cold days we were allowed to stand around the flames to warm up.

I liked the men on the farm crew who supervised us, and they considered me "a good worker." Years later Don told me that while we had been working cutting wood in the winter under his supervision, many of the boys had complained of the cold, but I hadn't. He remembered asking me why I wasn't cold, and my telling him that my solution was to work harder. This strategy worked for many problems, but not with my first matron. With her it backfired: the harder I worked, the more she found for me to do.

On April 9, 1956, I wrote Papa a letter which appears to be a significant regression from my already poor writing efforts of the past. By then I had almost forgotten German, but my English had improved only slightly, and I mixed the two languages. Translated into standard English, my letter reads: "I have again fixed my knife, and it has a good point. I can now throw it fairly well." (My friend Bob Nagy and I were in our Daniel Boone phase, trying to make a perfect throwing knife out of ground-down files pilfered from the barns.) "Thursday I went to a beehive and a bee stung me behind the head and one on the hand. Friday I didn't go to school, but into the woods. I fell into a stream that I tried to jump. I found a dry spot, took my clothes off and dried in the sun, and lay in the sun and almost went to sleep."

One forbidden pleasure I remember with fondness was sneaking off to swim in the hemlock-bordered Martin Stream. Running along the woods trail to get there while evading "capture" was half the fun. On one occasion I discovered a saw-whet owl, clothed in soft powdery feathers all the way to its feathered toes, in a hemlock tree along the bank. I had read about this tiny owl in my bird book but had never before seen one. It was peering down at me with its big yellow eyes. I wanted it.

Could I stun it with a soft clay ball? I picked some blue clay out of the bank of the stream, the same clay that we rolled in naked to make ourselves mud men at the swimming hole. I made a ball, put it in the sling-shot leather, and brought the owl down. A minute or two later, the bird revived. I had always handled one problem at a time. Now what would I do with it?

I could not let it go—that was clear. So I wrapped it in my shirt and started to walk back to school. I had a screened cage that I used for rearing caterpillars, so I put the owl into it and maneuvered it up into the dense branches near the top of a balsam fir tree in the woods by my cottage. There the owl would be in its element, and nobody would find it. The next problem was food. I set mousetraps and climbed up with a dead mouse to feed and socialize my pet. The owl just sat and looked at me. After two days I felt bad and set it free.

These antics earned me the title "nature boy," which I was not entirely happy about. I recognized that I needed to excel in one of the more socially acceptable domains of activity if I was to survive at Good Will. Church was one such domain. It was also one of the very few occasions when we got to see the girls. In church, two of the older boys, acting as ushers, were front and center. One, tall with angelic blond curly hair; and the other, short with dark straight hair slicked up over his forehead, marched side by side down the center aisle. Every Sunday they stepped in tune with "Pop" Hamlin's somber organ music. Pop sat on his bench right in back of the preacher facing the choir, which faced us.

The ushers would walk down the center aisle, past the congregation, past the boys sitting to the right of the aisle and the girls sitting to the left. When the pair came to the front of the congregation, they each picked up a wooden bowl lined with green felt. Then they turned around, and went from row to row, passing the bowl to collect money. We were supposed to put money into the bowl when it came near us. I felt that the matron, sitting close by, was watching, and I usually put in a nickel, which was a lot—it would buy almost two letter stamps or a whole candy bar at the store in Prescott, the administration building. When the two boys got to the last row at the back of the church, everyone would be in front of them and facing forward, and they could then finally reach in for their unofficial cut.

Most of the big (high school) kids had an even more awesome job than the two ushers. They stood up in church where everyone could see them and they sang at various times during the hour-long service. They wore long black robes with golden-orange drapes around their necks and over their chests, and sat high up on benches near the minister. All the older girls made it a point to get into the choir, and most of the big boys therefore tried to get in as well. I was too little, and it never even occurred to me to try out for something so lofty.

Book learning was another socially acceptable domain, though to me it was often a dreary one. "Study hour" started with a Bible reading and prayer in the evening at seven o'clock sharp. We cottage mates sat at our assigned seats along two oak tables with the matron at the head of one of them. If any of us finished our homework before eight o'clock, we still could not get up (or else most of us would have finished in fifteen minutes). My friend Bob, who, as I said, was modeling himself after Daniel Boone and who in his spare time practiced throwing his bowie knife (I was his promising apprentice), sat up against the wall with his eyes down, reading a well-thumbed comic book about Tarzan of the Apes, which he kept folded inside the Bible or a schoolbook. He soon ran away and got as far as New York City, where he jumped onto a freighter and got a job, before ultimately ending up in California and operating a sawmill. Forty years later I caught up with him through the telephone, starting the conversation with, "Do you still wear a hunting knife on your belt and throw it so it turns three times in the air before hitting a knothole at ten paces?" He answered: "Is this Ben?" Likely as not, Buggsie, one of the other big boys—our tractor driver, who later went into business in heavy equipment—would be deep in thought or busy composing a love letter to his girlfriend. "Misty" Moran's deep concentration and glazed eyes signified that he was once again reading Batman. Buddy was memorizing baseball statistics and the names of ballplayers. Most of the boys could reel off the names of the players in the major leagues as well as their batting averages. I didn't even know who the Red Sox were.

I spent my study time poring over the pages of my bird book, especially the pages about rails. Sometimes I made sketches from the illustrations, feeling myself drawn into the lives and the places of the birds. There was never enough space in my brain for the names of both ballplayers and birds, and I had no interest at all in the subtleties of fender shape that distinguished a 1951 Studebaker from a 1953 Oldsmobile.

On the other hand, what concerned nobody else—what I saw in the woods or read about in my bird books—seemed to stick in my brain without effort.

Our seventh-grade teacher, Miss Dunham, sometimes read out loud to us in class. She chose wonderful books. I particularly remember the story about Glenn Cunningham, the champion mile runner; *Uncle Tom's Cabin*, a book about slavery; a book about Abe Lincoln, who read by candlelight in his log cabin; and *The Adventures of Huckleberry Finn* and *Tom Sawyer*. The big river, the Kennebec, was right next to the school and it could as easily have been the Mississippi. What Miss Dunham read felt real. Somehow, she managed to steer me toward the library.

I was usually alone. The place was always absolutely silent. The aging librarian, Annabelle Jones, as I recall, sat at a desk just inside the entrance and asked if she could help me. I said I liked nature and adventure books, and she got up, showed me to the stacks, and pointed out the books that were just right. I read them, and found a new world. Osa and Martin Johnson's, Richard Halliburton's, and William Beebe's adventures in far-off Africa became to me especially exciting, maybe because I pictured Papa and Mamusha there in the jungle.

In my bird book I found fuzzy black-and-white photographs that left much to the imagination. Perhaps a Florida gallinule, *Gallinula galeata*, on a nest of cattails? I would think of Papa's stories about his trip to the swamps of the Dobruja. The black-and-white picture of the Virginia rail, *Rallus limicola*, looked as if it could be a stand-in for Papa's *Aramidopsis plateni*.

In the spring Miss Dunham posted a competition: listing spring flowers and birds. We were to report the first species seen, and each species won a point. As I was in the woods a lot, often taking a diversion through them on the way to and from school and on Sunday mornings and Saturday afternoons, I didn't mind telling her what I had seen. It turned out I found the first yellow violet, saw the first black-throated green warbler, etc., and was soon so far ahead of my classmates that it was embarrassing, because it solidified my reputation as "nature boy."

It seemed that everything that really interested me was impossible, not allowed, or considered sissified by the other kids. To top it off, I continued to be undersized, and I wondered if I had some kind of a hormone problem. I longed for some whiskers and bulging muscles. I thought I might escape from my "nature boy" image if I became a baseball player.

I had figured out that "baseball" is synonymous with "all-American," which is in turn synonymous with all that is good. I figured I had potential at the sport, as I already had much practice throwing my hunting knife and maple-pole spears. So I traded them for a baseball mitt and practiced throwing and catching a ball with my cottage mate, Buddy. We were preparing for the inter-cottage championships; and Buddy, the biggest and most athletic of us, was the natural captain of our Pike Cottage team. He chose himself as the pitcher, and because I was his friend I became the catcher.

We met Fogg Cottage on the baseball diamond in the field down by the "stockade," an enclosure of vertical logs where we had a campfire once a year and played Indians and roasted hot dogs and marshmallows. I was not exactly a star catcher, and we just barely survived the first elimination round. In our next game, against Golden Rule cottage, Buddy decided to be the catcher, and he made a wild gamble and had me pitch.

I was on top of the mound, and on top of the world, surveying the bases, ball cocked with my right arm ready to let fly. Yes, I *could* throw a sizzling fastball. I felt a smooth motion that rippled up from my toes to the tip of my fingers as the ball rolled off along the threads and ripped with a "thunk" into Buddy's mitt. "Strike one, strike two . . ."

We beat Golden Rule. Keyes cottage was next. Same lineup. When their star pitcher—whom I didn't like, because he slicked his long hair back with Vitalis (I had a short crew cut)—stepped up to the plate, I had my chance to show who was a nerd, and who wasn't. "Strike three— you're OUT." We, Pike Cottage, had won the championship!

ONCE I WAS IN HIGH SCHOOL I NEARLY QUALI-
fied as a "big" boy, and I would no longer be caught advertising that I'd seen a blue or a yellow violet. My outdoor activities then revolved around snowshoeing, skiing, building log cabins, and going into the woods to impress my buddies with more socially acceptable nature pursuits, like catching and roasting a porcupine or a squirrel.

Living on this land that still breathed the spirit of the Indians inspired some of us to try to follow their ways. We harbored dreams of staying away for days—or forever. Maybe we could build a log cabin like Abe Lincoln and the pioneers and live in the wilderness. We had little idea what lay in the endless woods beyond the boundaries of our campus world. The woods were so large that as far as we knew, there was no real end to them. We could walk as far as we cared or dared. I dreamed of homesteading in Alaska, living as a trapper in a log cabin in the wilds, with a tame crow and perhaps a pet skunk for companions.

My drives to seek a habitat and build a home had evolved since my fantasy of a moss hut in the Hahnheide. I experienced a warm feeling of satisfaction hunkering down in a hollowed-out space in a thicket. If I could, I would have hollowed out a giant tree to live in, like a woodpecker, because I felt an attraction to snug, enclosed spaces, especially those with a view. I found such a spot on a hill far in the woods. Here, next to some balsam firs, I escaped on some Saturday afternoons and started chopping down trees to construct a log cabin. I notched the logs at the corners and fit successive ones together, first two on one side, then two on the other. But after I had the four walls up, I could not locate a saw to cut out window and door openings.

That project was abandoned, and several of us then started building another cabin from two-by-fours and boards scavenged from our sawmill. We allowed for doors and windows, but this cabin was also never finished. Still, finishing the cabin to live in it was not really our goal in the first place. We were driven by the process of homemaking as if blindly possessed, like a bird building a nest once it has found a suitable habitat.

I was growing up, living through successive stages of being, starting at the least derived or "primitive" condition when we were hunter-gatherers, and then moving on to a more settled life in society. But we never really cast off our roots—we just grow more trunk and branches. For humans, the basic hunter stage meant running and tracking. As John Haines wrote, "There is a close connection between reading signs in the snow—the imprint of a bird's wing or the scattering of leaves and seeds over the surface—and reading words on a page; the same inherent order and process is at work in both of these."

I went from tracking weasels in the snow to tracking the habits of birds through the seasons, and I started to "keep" observations by recording them in a little spiral notebook, putting in entries by date.

May 12, 1957: "Partridge has eggs. Trees fully leafed out and choke-cherry blooming. Sparrow hawk just laid eggs."

May 22: "Tree and barn swallows building nests. Crows [baby] starting to get feathers. Flicker already made hole [nest cavity]. Sapsucker still making hole. Phoebe just laid eggs."

And the next year, May 19: "Young crows have feathers."

May 29: "Blue jay young starting to have pinfeathers."

June 27: "Great horned owl young expert flyers—two babies at Bog Stream on right bank just below where other stream enters. Black duck has down-covered babies. Chestnut-sided warbler fresh eggs. Veery [a thrush] half grown young. White-throated sparrows out of nest."

My interest in bees had continued since my bee-lining days with the Adamses, and I was now also reading about bees and keeping a hive. I wrote to Papa on September 2, 1958: "Saturday afternoon I went with some boys and we felled a bee tree. It had a lot of honey and bees and it was close to the water so that we could dive in when the bees got angry. The bees are now well established in my hive. Yesterday we felled a second bee tree."

After my fourth year at Good Will, Papa sent me a little book, Karl von Frisch's *Bees: Their Vision, Chemical Senses, and Language*, as a Christmas present. He inscribed it: *Bernd Heinrich, dem Imker, von seinem Vater zu Weinachten 1956*, "To Bernd Heinrich, the beekeeper, from his father for Christmas 1956." Two years earlier, my seventh- and eighth-grade teacher, Miss Dunham, had also made an inscription to me, in a little book with blank colored pages for collecting autographs as remembrances from my friends and teachers. In it she quoted from *The Trail Song*, by none other than Good Will's founding father, GW! It read: "There are cool, shady trails in the Good Will woods, where the fir and the hemlock grow, where the blue violet opens its petals to the sun. And the thrushes sing sweet and low. When the day's work is over, and the bright sun has set, in a sea of purple and gold, I will think of the trails and the quiet of the shades. And my spirit will never grow old." Then she added: "Bernd, I think you know best next to Dr. Hinckley himself, what these lines mean!" I was fourteen years old.

I was not at first consciously motivated to make a getaway into the wilds. But as in most things, the tail was soon wagging the dog. I got handed a big dose of punishment duty inside the house, and then even a breath of fresh air seemed exhilarating. If a little can feel so good, why not more? Finally, I suggested to two friends—who happened to be the church ushers—that we walk through the woods all the way to my family's farm, pick up my .22 rifle (which I had not been allowed to bring to Good Will), and then trek into the hills by Mount Bald, where Phil had taken me deer hunting and shown me where bears had scratched the beech trees. I loved those woods, and had long dreamed of going there with just a few essentials—a hunting knife and a bowstring—and maybe killing a bear and living for a full year by hunting and by gathering berries and mushrooms.

Looking back with almost half a century of hindsight, I find it incredible that three boys, sixteen to seventeen years old, would pack up their knapsacks, each with only a spare shirt, a pair of socks, a can of Spam, and one or two cans of beans, a couple of boxes of raisins, and a hunting knife strapped to the belt, and wander off to live in the woods. We had previously tried to make a fire without matches, but so far had been able to achieve only smoke. We had killed a porcupine and roasted it on a fire (made with matches), and as one of my accomplices would remind me later, eating it "made you sicker than a dog." A friend and I, desirous of acquiring a proper Indian hue, had disrobed and lain on a sand dune by the river, only to burn bright red, down to the soles of our feet. Being so obviously unqualified for wilderness living, what were we thinking? We probably weren't.

At that age, we were all still in the happy world where the consequences of one's actions are rarely or only faintly anticipated. We planned for the next step, but not the one beyond that. The future didn't exist. To look ahead more than a few steps, much less to be considering possible careers, was to us an abstraction and an impossibility, given our abilities up to that point. Papa knew the names of thousands of species of ichneumon wasps and could differentiate them, a feat that was far beyond my capability and even my dreams. I could never sit still at a microscope, not even for an hour. Nor could I envision myself working in the mill like Phil Potter, or teaching school, or standing behind a store counter. But living in the woods as the Indians had lived for thousands of years, tracking deer and bear? Yes! That seemed real.

We had on a couple of occasions crept out of the cottage while our matrons were asleep, to make test probes into the dark winter woods. We went to an abandoned shack in the woods and built a fire in the stove to fry some Spam, and then returned home to our warm beds. Finally, late one evening in March after the first thaw, followed by a freeze that had left a solid crust on the snow, we could wait no longer for the real thing. We picked up our knapsacks and walked into the woods, past our camp, on a dead-reckoned compass bearing toward Wilton and the farm. We made good progress, until late at night rain suddenly came pouring down in sheets. We had to cross Martin Stream, and to our surprise when we got there near dawn, the water was rushing over the ice in torrents. We could not cross. But the die had been cast. This time we could not return to our warm beds.

Drenched, cold, and miserable, we stumbled and wallowed through the soggy snow, following the stream up to its mouth at the Kennebec River, where there was a bridge. This happened to be right in the middle of some woods in the center of the Good Will campus, next to the president's house. This was the only bridge, and we had to cross it. But there was morning traffic on the road by then. Luckily we found an overturned rowboat that provided a hiding place and a shelter for the rest of the day. We could resume our trek in the evening, under the cover of darkness. Since the crust on the snow was softened and could not support us, we had to take to the road late at night when the highway was deserted. We walked the distance of about forty miles all night on what seemed like totally deserted roads and reached the old Dennyson place late the next morning.

Mamusha and Papa were by then back from one of their expeditions, but they could not afford to take us home, presumably because they might soon have to leave again and the school did not permit its students to leave and come back at will. Besides, they paid almost no money to send us there: $697.80 for our combined fourteen years as students, according to the records kept at the school. Papa was then staying in Ottawa working on the ichneumon collections of the Canadian Department of Agriculture, and Mamusha was staying at the farm while working at a mill in town. Therefore, we expected nobody to be at the house in the daytime. And indeed, the house was empty when we arrived dead tired at mid-morning after a sleepless night of walking in the cold. We helped ourselves to food in the pantry and found a couple of half-filled bottles of red wine. We were thirsty and passed a bottle back and forth for a few swigs, then refilled the volume we had removed with a generous portion of water. It was time to get out my .22 rifle and some bullets, and we fired off a few practice rounds.

To my surprise, within minutes Phil Potter came driving up the driveway. He had been laid off from work because the mill was shutting down, and he had heard the shooting. We didn't run. Phil stopped, got out of the car, and asked: "What ya doin', boys?" We had nothing to say.

"Will you wait while I get your mother, Ben?"

"OK," I said simply.

A little woozy from the wine, and with most of our energy dissipated by then, we sat and awaited the inevitable.

We later learned that the state police had been looking for us the first day that we spent on campus under the boat. We also learned that I was expelled from Good Will. The school president, the Reverend Mr. Garrison, told Mamusha that I was not welcome back. My two accomplices, because they had no home to go to, were allowed to return.

I spent my sophomore year at home on the farm, going to the local high school, Wilton Academy, rejoining the classmates with whom I had bonded when Marianne and I first started school in Maine. The best part was joining up with my mentor and adopted father, Phil, who again took me on canoe trips and trout-fishing expeditions all over the state. We even went to the woods near Mount Bald where my buddies and I had planned to spend our lives as hunter-gatherers. I loved the man. He made me feel alive and encouraged me in the things that I most loved to do.

Not Papa, though. Years later he told me that my running away had forever disappointed him. He told me, "You did not do your *duty*" (my italics). My duty was to obey authority—mainly his, in this case. He had wanted me to stay at this school, and I had disobeyed his command. And hadn't I long been taught that there is nothing more despicable and dishonorable than not doing one's duty?

To be honest, I had not even given *Pflichtig* (doing my duty) a thought. I was so far gone already that I had only thought of myself living in the woods with a crow as a companion, looking for bee trees and living off the land as the Indians and the pioneers had done. Actually, I hadn't really run away so much as run *toward* my dream. Hadn't Papa left Borowke to explore the wilderness? But the fact that I had not done as he had directed mattered immensely to him, and he once told Ulla that he could not love me any more because of it. Fortunately, Mamusha did not condemn me so deeply. Instead, she let me know that she had done me a great favor, by permitting me to come home, although she told me she would now have more work—cooking, washing, cleaning up—and less time for reading, writing, and other work that was fun for her.

So I would be an inconvenience to have around. But I didn't give it a thought. I would minimize staying in the house in any case. I was happily at school, in the woods, working, and visiting the Potters. I was ecstatic for a whole year. But my bliss came to an abrupt end: Papa and Mamusha were going to Angola for another two years, and somehow Papa had inveigled Good Will into taking me back.

Things had changed a little for the better when I returned there. I was now a junior in high school, and this automatically put me high in the pecking order at Good Will. But I had one strike against me at the start. I had written to my friend Bob and asked him: "How is the old hen?" The matron had intercepted (and kept) this letter and read it to me as soon as I stepped back into the cottage. I had insulted her behind her back, and she threatened to inform the headmaster if I gave her so much as a dollop of disrespect.

I was terrified. Not only had I been guilty of insubordination, but I had been sneaky as well, because I had not had the courage to tell her to her face what I thought of her. My Saturday afternoons were now no longer free. I was ordered to slosh chemical-smelling gallons of paint remover over the varnished staircases and banisters all over the house, then scrape the dissolving gooey stuff all off and revarnish everything. The fumes from the stuff made me feel ill. Papa had told me that unnatural chemicals would cause cancer even in minute amounts. My dream of being in the great outdoors now grew even more urgent.

Luckily I was no longer assigned cooking duties. I was held captive on Saturdays, but every morning before breakfast I worked outside, at the barns. I helped milk the cows at dawn and again in the evening. I could now pick up a 100-pound bag of Blue Seal feed and lift it up above my head, even though I didn't weigh more than 140 pounds myself. I joined the cross-country running team, where my lightness would be an advantage rather than a deficit. I had a ravenous appetite; in my autograph book a student wrote: "Ben likes to eat and run."

My pet crows had showed me that food is important for building a healthy body. A friend's pet crow died of rickets because he had fed it only bread soaked in milk. My pets had always remained healthy. They had been fed beetles, berries, grasshoppers, and minced mice with all the parts. I could not sustain myself on the identical food, but I needed the equivalent, because like my pets I was also growing. Additionally, I was working hard and running.

We were allowed one orange per week, and this did not satisfy my appetite for fruit. Often I didn't get even that one orange, because our matron kept it for herself. My gums were bleeding a lot.

In a letter to Papa on June 10, 1958, I wrote:

Tomorrow I will have my back teeth pulled. I don't think it is because I don't brush them. I do it regularly. It must be diet. It's not good, because the other kids have the same problem. No use crying over spilled teeth. . . . *Denkst Du, Papa, das ich vielleicht mal auf einer Expedition mit Dir gehen könnte—nach Süd Amerika, Australien, oder Afrika?* [Do you think, Papa, that I could maybe someday go on an expedition with you to South America, Australia, or Africa?]

I dreamed of becoming a prime physical specimen when I grew up, but this was not happening. Something was wrong. I thought it could be due to the lack of fruit, because I craved fruit so much. The matron declared that milk was a "perfect" food to make us "big and strong" and we could have all the milk (and bread) we wanted. That diet was not sufficient even for a baby crow. How could it be for me?

My two buddies—the ones I had run away with the year before—occasionally got a can of pineapple, peaches, or fruit cocktail, and they sometimes shared these delicacies with me out in our hideouts in the woods. They had stolen them from the school's central store, from which food (but not this kind) was distributed to the cottages every Saturday morning. The matrons and other adults got or could buy special treats at the store. Earlier, I had written to Papa, saying that I wished for "a belt, pencils, toothpaste, and postage stamps" for Christmas. He did not send them. I knew as surely as God made little apples that I'd never be granted such a luxury as a can of fruit cocktail, just by asking for it.

My friends had secured a key from a boy who worked at the storeroom, who had got a duplicate and given it to another boy, etc., etc. We did not know who else had a key.

I wanted no part of stealing. Then I realized I'd already done the sneaky, dishonorable thing: I had eaten the fruit while they had taken all the risk. So, to ease my conscience and try to be a little more honorable, I once went into the storeroom with them. But instead of feeling better afterward, I felt even more guilty. I resolved never to do it again, and I absolutely kept that promise to myself. Instead, I roasted squabs from the barns over a fire in the woods and now and then scooped a handful of cow grain out of a bin to supplement my diet.

Things began to look up after I was transferred from Pike Cottage to Hall Cottage and had a new matron, Mrs. Allman. For about a year nothing went wrong, except that the boys with the keys continued raiding the store all the while. Then someone lifted cash out of the cash register. That was noticed, and the thief got caught. He tried to dilute his crime by tattling on another two or three boys who had been in the store, who each fingered a few more, and so on. Eventually that included me. One day the headmaster sent for us all. He lined us all up in a long row in his office and I heard frightening words: he said he would find all the culprits, "even if I have to call in the FBI," and "this will go on your permanent records."

Nobody knew that I had been on the straight and narrow at Good Will for a year, and it did not even occur to me to tell my side of the story. Papa was enough of a Prussian to neither accept excuses nor give them. My explanations would not be of interest to anyone in authority. Their judgment of me was absolute. This was the most traumatic experience in my life so far: our flight from the Russians paled in comparison. (It was not until decades later that I finally got up the nerve to ask Good Will to send me my records. They obliged. There was no mention of the unsettling thing that had shamed me all those years.)

The only person whom I felt I could turn to, if necessary, was Gordon ("Lefty") Gould, the village postmaster. And I went out of my way to try to please him. He, like Miss Dunham, was the rare adult who took an interest in one of my skills, instead of concentrating on my faults.

"Lefty" had only recently been Private Gordon Gould, A Company, 504, Eighty-second Airborne, U.S. Army. When I graduated from barn duty to become the mail boy, I ran the mail route from the Prescott administration building on the boys' campus, through the girls' campus, and on to the village and his one-room post office twice a day, carrying a leather bag with a dozen or so pieces of mail. Lefty emptied and refilled the bag, and then I was free to run back. But I often lingered, as nobody came through the door for ten or fifteen minutes. At those times Lefty leaned toward me at the window with the grate in front of it, and he talked. If he noticed my German accent, it did not offend him.

Lefty had been a boxer before the war. But his boxing days were over; he now limped along with a lot of metal in his legs, after having spent a year in hospitals. He told me matter-of-factly (and I believed him absolutely) that he had been on his way to becoming the welterweight champion of the world before he was badly wounded.

Just one look at his large, grizzled, scarred head, topped with gray curly hair, and you knew that his skull was at least two inches thick. "I was never knocked out," he said. "I'd run five miles every day, and I could do two hundred push-ups in nothing flat." He had been hand-picked by General George Patton to serve as Patton's bodyguard. Lefty liked it that I ran to and from the post office, whereas all the mail boys before me had walked or used a bicycle. He seemed to think I could become an athlete, too. Some of my pride was restored whenever he talked with me. I felt that he cared. Perhaps, as his daughter Linda told me many years later, it was because he thought of me as "his son."

I listened spellbound to Lefty's adventures. Sweat would form on his brow as he took us off into the mountains of Italy, along with his buddies T. J. McCarthy and Eddie Adamzak. One night they were being shelled by "the krauts"; another night they were singing American folk songs and a German soldier approached with his hands up, and said in perfect English: "I'm not coming to surrender. I just love to sing, and I want to sing along with you guys." The soldier joined in a deep clear voice, and afterward they let him sneak back to his own lines.

Lefty's stories seemed improbable to me, and he may have embellished them a bit, but I think that most of them were pretty close to the truth because, unlike fishermen's yarns, his stayed exactly the same with each and every retelling. He had nothing to prove to anyone—he was a mensch, and I did not meet another like him.

Under his approving eye I escaped more and more into running. The sport appealed to me because it is clean, simple, and elemental. The basic formula is to get from point A to point B in the shortest time. It is a pure distillate of concentration, will, skill, and power. It's open to anyone, and the standards are spelled out in discrete distances, discrete numbers. There is no faking it, and little in the way of luck is involved; there is no ambiguity but plenty of justice.

At that time I saw myself almost strictly in terms of what I would become, by molding myself through will into the image I imagined. I had never thought of myself as a passive "product" of genes and environment. I'd had no role models until Miss Dunham had told us about Glenn Cunningham from Kansas. When he was eight years old, Cunningham threw a bucket of gasoline into the stove, thinking it was water, to douse the fire. His brother, who was with him, died, and he nearly died as well; he had such severe leg burns that his doctor said he would never walk again. Instead, he was a star of the Olympics in 1932 and 1936 and became the world's fastest miler and one of the greatest track stars of all time, and he received a doctorate from New York University. With that kind of inspiration set before me, I felt that my puny handicaps could become strengths. I ran a path in the woods once a day, and stopped at an old gnarled maple tree from which we boys had suspended a long rope. I climbed it several times before school, hand over hand. I practiced my swimming strokes in Martin Stream.

As I ran back and forth twice a day to deliver mail and visit with Lefty, I now had a double motive. I wanted to please my new role model, Lefty; and I wanted to become a better runner. After every cross-country meet, Lefty showed me the newspaper clippings about our meets. Our Good Will team competed against the best in the state, and I proudly wrote to Papa with reports.

July 3, 1958: "In two months is cross-country season again. It will be the last one for me. You said you had trained all the time to do handstands; I practice running all the time. My coach thinks I could become state champ."

September 19, 1958: "I won our first cross-country meet against Waterville, last year's New England high school champions. I came in first, a half minute ahead of everyone else. I am doing OK in all of my subjects except physics."

October 5, 1958: "We had another cross-country meet. I came in first again and set a course record. I did it in 11:45. Their coach says it's 2.5 miles, however I don't believe it." He then provided a new twist of Anneliese's words to me in the Hahnheide (that I needed to make up in my legs what I lacked in my head). He said instead that "if you can run like that, you ought to go to college."

However, Papa wrote to me that he was "troubled" by my running. I think he thought I was dissipating my energy. But from what exactly? He didn't actively encourage me to go to college, although he mentioned that he thought I could earn money and independence being a doctor, and continue with ichneumons as a hobby. (I wondered if he was thinking that I could then eventually support him or his collection, or both.) On the other hand, he did nothing to help get me into college—he did not even help pay my application fees. Mamusha was more positive. She pointed out that I could always live on a farm later, and that I'd have other options if I went to college.

Meanwhile, my schooling at Good Will was not yet finished—especially my chemistry lessons. Our chemistry class consisted of going around the class reading aloud out of the book, in front of Mr. Russell, our only science teacher. (He had collected artifacts and therefore called himself an "anthropologist.") I preferred my science with a more experimental bent, and so, with a couple of other students, I dusted off some chemicals in the science room closet, whose door had not been opened in decades. But what to do? We found a whitish powder, said to be "saltpeter." Isn't that what they put into the food? Someone wondered. Maybe, but somebody else thought it was an ingredient of gunpowder. Could I make a firecracker and find out? "Fine," Mr. Russell said. "Just don't tell the principal."

I mixed a teaspoon of the saltpeter with one of sulfur and added maybe half a teaspoon of powdered charcoal, stuffing the mixture into a hollow vial of wood that I picked off the floor, and sticking in a string "wick" to act as a "fuse." I thought it likely that nothing would happen, but I was eager to see.

I lit the vial on top of the concrete abutment that bridged the creek about 100 yards from the school building, during the lunch break. Wrong timing. Unfortunately it was also when Winfred Kelly, the new school principal, happened to come driving around the corner. Instead of a bang there had been a two-foot-long blue flame. (Measuring the ingredients by the teaspoonful rather than weight had been a mistake.) The smoke had not yet cleared, and my running prowess did not help me, as Mr. Kelly identified me sprinting around the corner (third mistake) up Green Road. He stepped on the accelerator. When he caught up with me he ordered me to stop, jumped out of the car, and grabbed me by the scruff of the neck, and I felt the tip of his shoe hit my butt, precisely where he had aimed it.

Later that afternoon, when I got to the post office, Lefty told me that Mr. Kelly had stopped, as he usually did, to shoot the breeze. The gossip that day was that "the little kraut kid was trying to blow up the bridge." It didn't matter to anyone that I had just become a naturalized citizen who had sworn to protect my country. And I certainly never had any thought of blowing anything up, other than a little firecracker.

It all blew over by winter. Cross-country was also over, and my new passion was skiing, At our winter carnival, held with two other local high schools, I won all seven of the skiing and snowshoeing events, and was crowned king of the festival. That required leading the first dance at the ball, but I could not dance a step. Nevertheless I regained the good graces of Mr. Kelly, who told me "Ben, you are college material."

Wow! I could not believe his words. Going to college seemed tantamount to getting a date with Marilyn Monroe: highly desired and nearly impossible. Papa and Mamusha didn't even have money for a telephone or an indoor toilet. But maybe I could get an athletic scholarship. (I never got one; but, ironically, I did eventually go to college, and Good Will gave me a scholarship of $200.) With my successes in running and skiing, and my by-then good grades (mostly B's), I felt confident; and I didn't think about the practical difficulties now, any more than I had considered them when I tried to live in the woods. I went to Mr. Kelly's office and got a bunch of applications: Yale, Bowdoin, Bates, Harvard, and a few others.

The first thing I noticed about the applications was that they all wanted a mug shot. I figured that was the most important thing, since all the schools required it. If I were a college president, would I want ugly students? No way. I'd want the students at my college to be the handsomest of all; then everyone would want to apply there. So I dug up a pretty good photo of myself. It showed a thin smile. Not too much. Just a little, to indicate I was serious *and* a nice guy. I had a hard time finding one that didn't show my pimples, though. They'd want athletes, not prepubescent boys.

The next things wanted for the application were references—presumably people who'd say what an upstanding guy I was. I picked Lefty, of course, and Phil and Floyd. I wanted to show that I was a "regular" guy. Not one of those fancy-pants city-slicker types. You know, a real country gem. I didn't write with a fancy ink pen, either.

I had insufficient credentials for college, but luckily I didn't know it. My report cards from Hamfelde, the school I attended while living in the Hahnheide (which I still have), indicated that on most subjects I was *genügend* ("satisfactory"), although in singing and arithmetic I was *mangelhaft* ("unsatisfactory"). The only "subject" in which I got a straight A's, all through grammar and high school, was behavior. I was a model citizen—quiet, reserved, compliant, minding my own business. My teachers all liked me, and vice versa. But in another regard my records were less complimentary; in looking at them, I learned that on my first IQ (intelligence quotient) test I scored ninety-three. At the time I may have thought this was a good score. The questions were made to look easy, but I was suspicious and thought a long time before answering each one, to avoid getting trapped by the obvious but wrong answers and having my IQ score (to whomever it was revealed) tag me as retarded. Lefty said after the IQ tests that the principal had complained to him that "all" the Good Will kids were "idiots." But I didn't think much about the test once it was over. My anxieties were focused on whether I could be attractive to girls. That was the big question. I did venture to write to ask Fay, the girl who sat behind me when we read out loud out of the book for our chemistry class, if she would "go steady" with me. I got a note back the next day saying, "Yes," she would. But then nothing changed, and she broke up with me the next day.

Around that time, in my exuberance over the approaching graduation and my anticipation of college life as either a cross-country star or a skiing star, I had a sudden inspiration. It came to me that there was one last thing I must do before leaving Good Will. This occurred to me one foggy night a couple of days before the graduation ceremony. I decided that I just had to put a paint mark on the top of our 150-foot-high water tower. I didn't ask myself why. I asked, "Why not?" The only reason I could come up with was fear, and that was not something to be proud of. So I ran out to the barns, found a can of red tractor paint and a brush, tied a long stick onto the brush handle, and then—after dark—climbed all the way up to the catwalk. Over the top of it, in bold letters that were about eleven feet tall, I painted the words: "The Big Bopper Was Here." (The Big Bopper was then a rock star. I knew nothing about rock stars, but I had heard and liked the sound of the name.) As an artistic flourish I added a sketch of Kilroy (a head peeking over a wall—the graffiti mark of American GIs as they passed through foreign towns during World War II). And while suspended up there alone in the night I asked myself, Why not go all the way? And so I climbed onto the tip-top, way above the catwalk, and meticulously painted the entire dome cherry red. There! I was done. Good-bye, Good Will. I'll be off to college now. A tiny and barely visible red smudge, however, would soon cast everything in a new light.

The next dawn I waited with nervous curiosity to see what would happen. The first thing that happened was in the daily morning convocation. After the Lord's Prayer, the Pledge of Allegiance, and Mr. Kelly's unusual announcements, our principal said that the culprit would be found. He ordered the whole senior class, all twelve of us, to stay in the auditorium while the others dispersed to their classes. He lined us up, it turned out, for a fingernail inspection. Indeed, my fingernails had hours earlier been crudely blotched with red tractor paint, but I had found some terpentine, for fingernail paint remover, and had cleaned up thoroughly.

The second thing that happened was that our school president went to our cottages (at least to mine, which was Hall Cottage then) and rummaged through our personal belongings (or at least mine). He would later claim to have found what he was looking for—a little stain of red tractor paint—of exactly the same kind as that on the tower, on one of my bedsheets. He also said he found B'nai Brith literature in my desk drawer. I had applied to this (and several other) organizations for scholarship aid. I had no idea you had to be Jewish to apply. I had just recently become a naturalized American citizen. I'd been taught at school that in America everyone has equal opportunity. But the first, and at the time most relevant, of these facts was not yet known to me when I got to the post office that morning and Lefty told me that he'd just heard from the principal that "the little Jewish kid did it." I thought then they'd accused the wrong guy because I was not Jewish—I had no idea what the comment meant.

It all became much clearer to me a little later when Mr. Garrison called me to his office and told me to take my things and leave by the next day. In that moment I knew precisely who had been meant by "the little Jewish kid."

I LEARNED MUCH AT GOOD WILL. BUT THE MOST important lessons I learned during my six long years there were those not specifically taught in my classes. I did poorly in Latin, which from reading Papa's labels of ichneumon names I assumed was the "language of biology." My failure, it seemed then, would prevent me from ever becoming a biologist. I had, however, for a while sent away for correspondence lessons from the Northwestern School of Taxidermy, hoping to learn how to mount animals so that they looked real, like those in the museum displays. But since I had few Saturday afternoons free to work for cash (one dollar per hour), I could not afford more than a couple of their pamphlets, and the only taxidermy I did get to know of was making cotton-filled skins of the rodents and shrews that I trapped and gave to Papa. I had only a few dollars and little time to spend on these lessons, and soon dropped them. I was, I thought, very poorly prepared for becoming a professional zoologist, and so I had no high expectations. I also realized that my desire to attend college was similarly unrealistic. I can't say I was completely stressed out when I got rejection letters one after another from Bates, Harvard, Colby, Bowdoin, Yale. . . . After all, a large part of me desired nothing more than to return to the farm and live there for the rest of my life.

IT WAS A BEAUTIFUL JUNE DAY AND THE LILACS were in bloom when Phil Potter drove Mamusha to Good Will one last time to bring me home. The familiar, long-missed birds were singing, and some were building their nests. I was finally free, and back in my heaven several days ahead of schedule.

I could have been nervously giving my long-practiced speech on capital punishment up on the stage at the Prescott Building. I could have been stiffly marching down the aisle of Moody Chapel in a black medieval gown, to receive my high school diploma. Instead, I was in the leafy crown of the big elm tree in the yard in front of our house, entranced by a beautiful oriole nest artfully woven together out of gray, weathered milkweed fibers. Just then the Reverend Mr. Garrison drove up and parked his long sedan in our yard. Had he come to see Mamusha and talk to her about her wayward son? He looked up at me in the tree and said "Hi!" I returned the salutation, and then he knocked on the door of the house. He had come to hand-deliver the bill for repainting the water tower.

seventeen

Making the Cut

Of all the sunshine that brightens our lives, there is
no beam more complex and more brilliant than that
distinguished as College Life.

—"PRISM" (UNIVERSITY OF MAINE YEARBOOK), 1894

PAPA'S TALES OF EXPEDITIONS TO THE FAR-OFF
lands had stimulated my imagination with the lure of strange and exotic
nature. But his example sent me a clear signal that one could learn about
nature only by being in it, not out of it at a university. After all, he had
studied only the "classics" (Greek, Latin, and literature) at the gymnasium in Berlin. He had no higher degree, and he held academics in some
disdain. Yet, as a world authority on the classification of a very rare and
little-known group of insects, he considered himself to be at the "forefront" of *Wissenschaft* (science). And I heard nobody contradict him. If I
continued with his ichneumons, as he suggested, I would probably learn
nothing about them in a classroom. He also warned me that one could
not make a living in zoology and biology. Rather, he saw them as a lifelong hobby and something that would help in my old age so that I would
not die of boredom. In order to make a living, Papa suggested that if I had
to go to college, it should be to study medicine, and "then, if you become
a doctor on a ship, you could travel free and collect in Africa or anywhere
else in the world."

I wasn't convinced that his logically formulated plan was right for me, despite his lifetime of experience: I would never be smart enough to memorize the names of hundreds of ichneumons (or to cure diseases). I wasn't even sure I wanted to collect ichneumon wasps for the rest of my life. My dream was at that point a toss-up between becoming a trapper in Alaska and making a tranquil and happy home with a family on a farm, hopefully our New Borowke. I loved every nook and cranny of the house, and I loved the woods and the fields. I envisioned raising bees for honey, collecting eggs in a barnyard full of chickens, watching the woodcock's sky dance over the meadow in the spring, fishing in the brook in summer, hunting deer in the deep woods in the fall, and listening to birdsong every day. The farm was where my compass pointed.

My idyll there after the incident of the water tower didn't last long, though. Shortly after I returned to the farm, I received an envelope from the University of Maine. A crisp typed letter with an official letterhead was inside, like the others. But this one said that I was accepted to enter the freshman class in the fall! "Maine" was the last college I heard from, and also the only one that accepted me. In retrospect I think I could have gotten more acceptances, if I had also penned my other applications in ink rather than in pencil. I had not talked with any "guidance counselor," or with my parents or anyone else, and I had no idea that penmanship could be considered a measure of potential scholarship.

There was no athletic or any other scholarship attached to my admission, but I knew I had to go and at least try to make the cross-country team: one step at a time. And that I'd have to get a job; my slim chance of ultimately graduating from college depended on getting a job to pay for tuition and board.

Close to home in the little town of Weld there was a job opening for someone to empty the trash cans at the Mount Blue State Park summer campsites. Hoping to secure that job, I knocked at the park director's office with some trepidation. He took only a minute or two to size up my small stature before informing me that I did not "look my age." He said the job required "respect from the public." That was the end of the interview. I did not seem competent even to empty trash cans!

The next job I applied for was catching male gypsy moths. This one required a totally solitary existence. The candidate would be isolated and alone all summer in the uninhabited north woods of Aroostook County, near Mount Katahdin, where Papa and Mamusha had collected small mammals. It was some 200 miles from home. It required a driver's license. Thanks to Phil Potter, who had taught me to drive his beat-up Ford pickup all over his hayfields, I had one. I interviewed, and was offered the job on the spot.

In fact, I wasn't expected to actually catch any moths that summer. And I didn't catch a single one. But every day I got the pleasure of driving a brand-new pickup truck issued by the U.S. Department of Agriculture, and I traveled on every back road in Aroostook County. I stopped every mile or so and set out a "pheromone trap" in the bushes alongside the road. Each of these 325 contraptions that I was expected to set out and tend was a tin can painted camouflage green, with a small opening for moths to fly into at each end. The "bait" inside was a wad of cotton dipped in a scent called "gyplure" that mimicked the female moth's sex scent. It would drive males crazy and cause them to fly miles upwind looking for a receptive female. Moths entering the trap expecting to mate would get entangled instead in a sticky glue called "tanglefoot" that was spread onto a strip of paper inside.

Where outbreaks of gypsy moths occur, the forest is left denuded and looking like the winter woods, except that the ground is covered in caterpillar feces. The idea behind the trapping program was to see if a vanguard of the moth population had arrived, so that it could be sprayed from airplanes with DDT. We now know that letting nature take its course is cheaper, safer, more effective, and also more dependable than dropping pesticide from the sky. Left alone, the ichneumon wasps, predators, and diseases build up and then control the pest on their own. Like bombing, which chalks up a huge body count, spraying indiscriminately kills the good guys, too, and it keeps the infestation going much longer.

Day after day I set out and checked my traps. After work I stayed in a dingy little room that I rented in the town of Houlton, scrounging to save as much money as possible from my salary of sixty-seven dollars a week. I had no private vehicle. Therefore I could not travel home except by hitchhiking along the highway outside town. I usually got a ride within an hour or two on any stop along the lightly traveled roads south through the 100-mile woods.

Until this time I had been prepared for disappointment, because I was not saddled with high expectations. Now that changed. I had a pay-ing job and would become a university student at the end of the summer. Running was my passion, and I wanted to become a member of the famous University of Maine cross-country team.

With hands on the steering wheel and foot on the pedal, I sang loudly to myself on the long lonely roads, and every mile or so when I stopped to set and check another trap for the gypsy moths, my feet were itchy to run, and I didn't stop my truck by each trap. Instead I stopped about 100 yards before or past a trap, so I could get in a sprint back and forth between trap and truck. With constant practice, I just might have a chance at making it onto the team.

I don't recall Papa saying anything about my going to college; and when I told Mamusha that I wanted to stay at the farm, she told me that I could do that even with a college degree. That logic made sense to me, since I wanted to run and learn about nature. There was nothing more "nature" to me than the woods, so since we all had to declare a major, I naturally chose forestry. Finally the day in late August came when we freshmen arrived on campus in Orono. I was assigned to Hannibal Hamlin Hall, and we faced a week of orientation before classes would begin. I enrolled as a forestry student and then went immediately to the gymnasium and looked up Edmund Styrna, the coach for track and cross-country. To my great excitement he escorted me to get a locker and showed me to the stockroom, where I got an issue of clean running clothes, a fresh towel, and a pair of thin-soled black canvas running shoes.

The cross-country course started in back of the gym. Eager to run, I changed clothes right then and there, but I wore the boots I had worn all summer, to make the future running seem easier. I met another freshman runner on the course and recognized him as the previous fall's high school state of Maine cross-country champion in the class M (medium) schools. I had failed miserably in the state meet that he had won. Now, as we jogged side by side, I admired Alan's beautiful smooth stride. He speeded up a bit, and to my surprise I kept up with him. Gradually we both accelerated and (as he told me thirty-five years later when we chatted about the incident): "I said, 'Race you to the top of the hill!' You said: 'Let's go!' And when I caught up with you, you were on top of the hill bending over retying your boot laces." I assumed he had chosen not to push, but I had nevertheless gained a tremendous surge of confidence and an intense eagerness to prove myself in real competition. Maybe my summer training had paid off. A little more couldn't hurt.

On my first visit to the weight room, a day or so later, I watched a few brutes work out who obviously had much experience with barbells. Not cowed by anyone's previous achievements, I tried the various things the others did, including lifting a barbell while bending over and keeping my back horizontal. I had never tried anything like that before, but then I had no previous weight training, either. Indeed, I had never lifted a barbell in my life. Suddenly I experienced a sharp pain in the small of my back.

It didn't get better the next day, or the day after that. I wrote to Papa on September 17, 1959, telling him that I was barely able to hobble to classes, much less run well in cross-country:

> Everything is challenging here and that is what I like; a feeling that you're out of the rut and going someplace. I have a job in the cafeteria, working three hours per day at seventy-five cents per hour. I am taking seven courses and like chemistry, but it is difficult. After graduation I could probably make more than two dollars an hour. You say you could not give me money because you have to save for your old age. Could you invest it in me and then I could pay you later when I start earning myself?

He didn't.

A week later I wrote to him again, telling him that the pain had increased and saying that I was "completely 'down' and shaken."

I had gone to the infirmary, telling the doctor that my back and also one leg ached. He didn't examine me, saying my problem was "just muscular." That seemed an odd diagnosis because I had pains all the way down my left leg. Still, I went out for cross-country and endured the pain, desperately trying to make the team. I wasn't performing well and told my coach, hoping he would have consideration. He sent me back to the infirmary and told me to say that I was a runner hoping to make the team. The doctor examined me and then told me that I was through with running, for good. He referred me to a neurologist, and I hitchhiked to Bangor to see him, to be told the chilling news that I had a ruptured disk in the spinal column pressing on the nerve going down my leg. He said the chances of a successful operation were "fifty-fifty" and recommended that I reconsider forestry as a career. I was devastated, thinking that all my plans had been shattered by a boring afternoon when I tried to become more physically fit!

My studies were not going so great, either. I got my first English essay back—a C. On my first quiz in engineering drawing I got a fifty-six. I hoped I would not fall into the unlucky twenty percent who flunked out of the University of Maine in freshman year. My academic adviser, Professor Frank K. Beyer, was moved to write Papa (on November 13, 1959) warning him that my midterm grades "are only an indication of progress and nothing more." He expressed regret that my spine injury "has prevented him [me] from taking part in athletics."

By early December the pain was "no better and no worse," as I wrote to Papa; but I got an eighty on my zoology exam, whereas the class average was fifty-four, and I was now "doing OK" in chemistry. Papa answered that he wanted me to come home for a visit, but I said that I was unable to come home on the weekend, as I had "too much to do to catch up." He was, of course, with his ichneumons, so he was always unable to come up and see me. Also, he had a car. I didn't.

By the end of the first semester in January my back had improved slightly and I was able to come home to the farm on semester break. I could move around sufficiently well to run a trapline for catching weasels. Along with the weasels, I also caught smaller mammals for Papa to sell as museum skins. Red and flying squirrels, mice, voles, and shrews also went into the traps. Not wanting to waste them, I skinned and prepared them as museum skins. I happened to tell a wildlife biologist, Dr. Malcolm Coulter, about some patterns I'd observed in my catches. He encouraged me to write them up. I did, and to my surprise this piece of research and writing was accepted in the *Maine Field Naturalist*. I was a freshman in college, and it was my first scientific publication. I suspect I felt how Papa may have felt when he told Stresemann at Berlin University about his bird observations from the Dobruja. I also talked with my mammalogy professor, Dr. Al Barden, about the problems of identifying shrews and gave him one of my weasels for the university museum's collection. "They were quite interested in the skins although they do not have the funds for a collection," I wrote to Papa in January of my freshman year. "We came to the conclusion that the shrew definitely is not (Sorex) cenarius. It is Sorex fumeus. Should I send you the rest?"

Miraculously, by spring I was occasionally getting A's and B's in some of my courses. I even once scored 100 on a chemistry quiz. I had pulled my overall grade-point average slightly above a B, and made the dean's list. Good Will then sent me the $200 scholarship it had originally intended to give me at graduation—maybe the staff realized that I never meant anyone any harm, or maybe they were happy to see one of their wayward youths succeed. Despite the doctor's negative prognosis, I did my best to take care of and strengthen my back. I was running again, however tentatively; and in July I was offered, and took, a job as a "timber cruiser" with the International Paper Company in northern Maine. Our crew of five forestry students from the university spent all day out in the woods. We were on our feet most of the day marking trees for selective cutting. We were living in big lumber camps built of unpeeled logs and populated by a couple of hundred woodcutters and a fair number of horses that they used to twitch the logs out of the woods. We got up at dawn and came back to a lumbermen's supper in the evening with all the loggers, most of whom were French-Canadians from Quebec Province. Contrary to my doctor's prediction, I was thriving (even though I could not bend over). I again saw a ray of hope for my running, and so every day after supper I went out for a seven-mile run, somewhere down a dusty logging road. I combined my runs with setting traplines for small mammals, which I sold to the Turtox biological supply company (which sells biological material for teaching labs) for supplemental income.

I corresponded with Papa throughout the summer.

July 19, 1960, from the town of New Sweden, in northern Maine: "The bog where I caught the black-banded ichneumon males is quite a way from camp, but I will go there to see about catching the females. . . . How many mammal skins do they want?" (I don't remember the particulars of this, but Papa must have had an order, and I must have offered to stuff mice and shrews for him.) "Papa, I was going to tell you that there are a lot of hummingbirds deep in the [spruce-fir] woods here. The sapsuckers make their licks on big lone yellow-birch trees and then there are always a few hummingbirds that congregate there, also squirrels. The cedar swamps are getting tiring. My back has not got any worse from all this, but it is not pain-free by any means." I attached a sketch of a Tennessee warbler, which I had never seen back home in western Maine, along with a descriptions of its nest, eggs, and voice, and noted: "Found very many in spruce-fir woods; especially where the trees were so spaced as to allow a little sunlight to reach the ground. Trucks are hauling out pulp now. Have read everything in camp. Could you send me something?"

July 31, 1960, from New Sweden: "Today I pulled in my traps." I had eleven *Blarina*, six *S. fumeus*, one hairy-tail mole, and one "small unknown (to me) shrew. The habitat of the jumping mice seems to be in grass along water by evergreen woods; *fumeus* in mixed (woods) with lots of leaf mold and moss."

I was surprised to learn that Anneliese was coming to visit Papa and Mamusha (and maybe me?), traveling by bus from Chicago, where she was living with Ulla. I wished there was some way to come home for a weekend visit, but it could not be done. It was a two-hour drive on a dirt road through totally uninhabited woods to get from our lumbering camp to Ashland, and another four hours on the tar from the village of Ashland downstate to Wilton. I felt guilty for not being able to see them all.

Once we got into the northern Maine Allagash region, I wrote:

> Trying my damnedest but there just aren't any wood jumping mice here. Finally, I did however catch a smoky shrew. Very proud of it. One more, *Microsorex* [pygmy shrew] and two more water shrews. Can catch all kinds of *cinerius*. *Blarina* are very abundant only along the riverbank (no water shrews there, however). When we get back from work there is never enough daylight for ichneumon wasp collecting—only a little for trapping. Will try next Sunday.

I enjoyed trapping the small mammals. In a way it reminded me of the Hahnheide, where Papa had taken me out on his trapping rounds. Now I had names for the tiny "gnomes" that I found under cover of damp green moss, in the labyrinthine recesses of rotting logs and under undercut banks along trickles of water. They were called *Microsorex hoyi* (pygmy shrew), *Sorex palustris* (water shrew), *Blarina brevicauda* (short-tailed shrew), *Sorex cinerius* (masked shrew), and *Sorex fumeus* (smoky shrew). I wrote to Papa that the smoky shrew is well named because in the New Sweden camp where they were so plentiful, one could still see the charred stumps of a forest fire, which consumed the whole area about ninety years ago.

. . .

THE CAMPUS OF THE UNIVERSITY OF MAINE IN
Orono at the end of the 1950s was centered on a large quadrangular mall
lined by tall elm trees. It had a huge library at one end and an athletic field
house several times larger at the other, with basketball courts, a swimming pool, and an indoor track. The mall was surrounded by brick buildings filled with classrooms, and radiating out in all directions for over
100 acres were a half dozen dormitories, eight sorority houses, seventeen
fraternity houses, a bookstore, a barbershop, a farm with apple orchards
and livestock, an outdoor athletic stadium and running track, endless
lawns, tennis courts, and fields and forests extending down to the Penobscot River. I could hardly begin to imagine something so huge, grand,
and magnificent devoted solely to learning and unbiased by religious, national, or other prejudices. Despite its seriously rigorous loftiness, the
campus was comfortable. The student union building next to the library
had a reading room where I often browsed in the freely available magazines. It also had a theater, a bowling alley, and Ping-Pong tables; and
downstairs was the "Bear's Den" where probably most of the 5,000-
strong university population stopped at least once daily to drink coffee
(out of *real* cups) and eat doughnuts made fresh in the kitchen.

We walked on two roads under the elms along the side of the quad
on our way to classes, and we were appropriately labeled in categories.
The engineering students had slide rules hanging from their belts. Then,
unlike nowadays, there was never a backpack in sight on campus. The
women invariably carried their books tucked in the crook of their left arm
while the men carried theirs on the right. That's just the way it was done,
and I doubt if we were conscious of it then. The men wore tan or black
pants (chinos), but women were not allowed to wear pants. Summer
and winter, they had their legs bare, except for knee-length socks. They
wore pleated skirts and sweaters. By their junior year many of the girls
were "pinned." They wore the icon of a particular fraternity attached
to the front of their sweater to signify their engagement to a guy in that
fraternity.

There was almost always love "in the air." I heard it in almost every
song we played in the jukebox in the Bear's Den. The Beatles sang "I Saw

Her Standing There," and "I Want to Hold Your Hand." We heard Roy Orbison's "Pretty Woman," Chubby Checker's "The Twist," Hank Williams's "I'm So Lonely I Could Cry," James Brown's "I Feel Good," and Bobby Vinton's "Roses Are Red." I especially liked "When," a song written by Paul Evans and sung by the Kalin Twins. All these songs induced euphoria in me, but the last one can still make me spring into the air. If it didn't make me want to leap ten feet into the air, then I didn't like it. "Good" music was to me that which exceeded my aerobic threshold while dancing.

Most of the men wore bright-colored uniforms. For the most part these were short jackets that identified the fraternity they belonged to: the pea-greens of Alpha Gamma Rho, the dark greens of Phi Eta Kappa, the maroon of Kappa Sig, the purple of Sigma Alpha Epsilon. The school jocks proudly wore their navy blue "M jackets" that they had been awarded at the fall athletic banquet, for winning a varsity letter. These were emblazoned with a tall, light blue "M" attached to the left side on the front. Only upperclassmen could wear an M jacket, since freshmen were not allowed to play on a varsity team. Indeed, school pride, the engine of our famous "Maine spirit," was due to the fact that we earned our stripes. I hoped to wear mine proudly.

Biology on campus was much different from what I had experienced during my time out in the field. In introductory biology labs we dissected pickled worms and frogs. It was not fun. I studied morning, noon, and night, writing down what I did not have solidly in my memory, making notes from my notes, and reviewing them while slowly advancing in the sometimes endless chow lines at the cafeteria so as not to waste any time.

Several other introductory courses stick out in my mind. One was "Surveying," which was considered an easy or "gut" course, but I had nightmares about it. I dreamed I took the entire course, and when I was going to take the final exam I realized I'd missed all my Thursday classes because they had been held at a time and place that I had not put on my schedule. It was a secret to me what this course was all about even while I was awake. Teams of three of us were issued glass tubes on tripods to look through. We set our tripod up on the campus mall somewhere over some very specific spot. And it was very important which spot. Indeed, the tripod dangled a brass weight with a point, and the point had to coincide with that spot. Then one of the other guys looked through the glass to the second member of the party who was holding a rod with gradations on it. He read off the numbers he saw to me, and I entered them in columns in a little notebook that would later be handed in to the instructor for grading. It didn't matter to me whether the distance from the library to the gym was 1,658 or 1,600 feet, but this was a required course in the forestry curriculum. When we got our final grades, the instructor had been merciful. I was saved. I had received a D-minus, a passing grade.

Words were my nemesis. The information that I received through reading and hearing was in English, but in my mind I processed the world in German. Then I had to pass it on in speech and write in English. I was frustrated because my words were inadequate. They intruded and dulled my sensations. "English 1" was supposed to correct that, and all underclassmen took this course. We were required to write weekly or biweekly "themes." These writing exercises were usually on an assigned topic, often about a poem. I can say two things with confidence. The first is that I seldom got a B on one of these themes. The second is that I deserved all my C's.

Most of my time out of class was spent at the field house, on the outdoor track, or on the cross-country course. But occasionally on weekends I would shyly call a girl and ask for a date. We usually went to a movie in the student union. The girls' dorms closed at ten o'clock at night, but if there was still time we went to a coffeehouse, where we all sat down on the floor together and sang Woody Guthrie's songs. I dreamed of someday seeing the "redwood forests and the Gulf Stream waters" of this beautiful land. Our nostalgia was stoked by the Maine "Stein Song." This lusty, boisterous, bacchanalian song that had originated at the University of Maine had been a national and international hit. Singing it made me feel alive. We sang it like an anthem at athletic events and other get-togethers. The words, to a catchy tune, were: "A toast—to the trees, to the sky, to the eyes, to the lips, to the girls who will love us someday—drink to the spring in its glorious happiness—stand and drink a toast once again, let every loyal Maine man sing."

I had sung and drunk a few toasts and become a loyal Maine man in my first eventful year at the university, and I returned to the campus eager, confident, and filled with anticipation for my second year. My financial situation had improved because I had gained residency in "the Cabins"—

self-sufficient units where six of us lived together and did our own house-keeping. I once figured out what I spent for groceries—it averaged $3.32 per week. In part the food bill was so low because I had killed a deer that fall, and a friend gave us the meat of a bear (it tasted awful). It now looked as if I could "make it." Also, I had made the varsity cross-country team, and in early October 1960 I wrote to Mamusha and Papa:

> This has been an exciting day for me in cross-country and I will tell you about it. Today we had a meet with University of New Hampshire, which is recognized as a strong club. We lost to them by only two points, one of our runners was sick. One of the men on their team, Doug McGregor, is supposed to be near the top in New England, as is our Mike Kimball. During the whole 4.3-mile race, Mike dogged right on Doug's heels and I stayed with Mike. Two hundred yards from the finish, both Mike and I sprinted past him to finish together in 21:51. That was a full minute better than I had run the course a couple of weeks ago. Next Saturday we will run in Lewiston.

Lewiston was about twenty-five miles from home. Only Phil Potter came to watch me run on any meet in my college career.

At a later meet, at the University of Vermont, Mike and I were first for the third time, and we ran the course 25 seconds faster than it had been run that season. I felt I could have run faster, but it seemed wrong to me to try to beat a teammate. Keeping up with the great Mike Kimball was more than enough glory and far more than I ever expected.

My studies were going well, and I believe it was not in spite of my running, but perhaps because of it: our cross-country team got the high-est grade-point average of any of the dozens of clubs on campus. But Papa felt I was taking unfair advantage in mammalogy. As one of our course assignments the professor, whom I liked, had us trap small mammals, and told us that we'd get a bonus if we could collect rare species, includ-ing the water shrew and the red-nosed vole I had captured while working in the north woods with the International Paper Company, so I wrote Papa asking if he would please return them to me so I could in turn give them to the university collections. Papa did not want to give them back, and accused me of brown-nosing. I was disgruntled and told him that I did not see why someone who had never seen a smoky shrew, or what have you, should get a better mark in the course just because he could memorize the book.

In early December, as I was completing my third semester, I wrote to Papa: "Today we had our fall athletic banquet. Mmm! Very delicious steak dinner. The varsity football and cross-country teams received their M letters and sweaters and also chose next year's captains. I feel very proud and honored to have been chosen the captain of our team for next fall!"

I was hoping Papa would relent and be proud of me in my studies and in athletics. But if he was, he never let me know. Perhaps he felt that if he admitted pleasure, then he would make me complacent so that I would let up. Maybe he thought that all the pleasure should be mine so that I would be doing it for myself and not anyone else. Alternatively, he probably thought that he should not encourage my running because he felt it would be a dissipation of my energies. As for me, I expected little from him, and I felt no lack. Instead, I was "on top of the world."

I was getting all A's and B's in most of my exams. I liked Mack McClain, my boss at the International Paper Company, who contacted me and offered me a raise to come back that coming summer and work again up in the north woods. I had a (requited!) crush on a wonderful girl who had seen me run at the track and with whom I had gone to watch a movie at the student union. My back injury was miraculously improving. I felt resurrected. I was on top of the world. I'd made it. I felt like the student prince—proud and exuberant. I was no longer, as at Good Will, chafing to escape to far-off lands. *Here* was for me the best of times in the best of places.

eighteen   Year on Safari

> The great journeys of youth mark a man, showing
> him possibilities others never see, uncovering po-
> tentials that stagger the young mind and monop-
> olize an entire life in attaining.
>
> —FROM *THE COVENANT*, JAMES MICHENER

AFTER RETURNING FROM THEIR SECOND EXPEDI-
tion in Angola, Papa spent months away in Ottawa working on the Canadian ichneumon collections, while Mamusha stayed at the farm in Maine.
Papa (who always required a woman's adoration in his life) must have
been lonely in Canada. At the age of sixty-five he had entered an apparently "serious" flirtation with a "maid" he met at a party where he had
played the role (and wore the costume) of an "entomologist." He made
no secret of the fact that he was married, and he made no secret of this
"new love" to Mamusha, either. Indeed, he appeared to flaunt it, induc-
ing Mamusha not only to stage a fit of temper, but also to refuse to go
with him on a newly planned expedition to Africa that Yale's famous
S. Dillon Ripley had agreed to sponsor. Ripley wanted, in particular, a
bird called *Alethe montana*, which Papa and Mamusha had discovered in
Angola on their previous expedition and delivered to the Field Museum in
Chicago. I had no knowledge of the drama unfolding around the expedition until one day during the midwinter semester break in my sophomore
year. I was home, warming myself in the kitchen by the big black wood-

burning stove, when Papa came in and said he wanted to talk to me. I knew by his tone this was going to be serious. We walked into his study and I stood expectantly. He did not sit down either. He looked me in the eye and told me that Mamusha had decided that she didn't want to go on the expedition with him. Then he asked me: "Do you want to join me for a year on my *last* expedition—to Africa?"

All those expeditions over the years had kept us apart, but they had also provided the stories he told that brought us together, and I had long dreamed of going with him. Ordinarily such an offer to join him, especially to my land of dreams—Africa—would have been more than a wish finally come true, but the timing was all wrong. I had become thoroughly engaged in my own new adventure. I was a university student and athlete. I had found my own glory! I had not only survived my incredibly grueling freshman year but had also miraculously recovered from my debilitating back injury, and I'd been chosen captain of the cross-country team for the next year. I liked my professors; I liked my friends and loved going out on Saturday nights with them to drink beer at Pat's Place in Orono; and I had a crush on at least one beautiful coed. I had been offered a dream job for the summer. It was impossible for me to go, but just as impossible not to.

Perhaps noticing my hesitation, Papa opened his desk drawer and pulled out a picture of an attractive young woman of about my age who he said would be joining the expedition. I was immediately more interested. Papa was probably right to say that this would be his last expedition; I would never again have another chance like this. The university and my studies would be there when I came back. I agreed to go.

When Mamusha found out about Papa's new plan, she hit the roof. She also changed her mind, and went down to New Haven to tell Ripley that Papa had decided to replace her, the veteran of three expeditions, with "two inexperienced teenagers!" Not only that; the teenagers Papa wanted to take were both his own children.

Both his children? He had (unknown to me, but not unknown to Mamusha) a daughter, Christel, living in Germany, the product of another wartime liaison—with a woman named Brigitte Wanner. Gitti, as Brigitte was called, was the relative of a neighbor near Borowke who was a very close friend of Anneliese and Ulla, and she had been their guest at Borowke while Papa was off to war. Papa met her when he came home on leave from Russia, and while Mamusha was hiding in Berlin with the newborn Marianne. Papa became smitten with Gitti, wanted to marry her, and wrote to Mamusha that he would be "a fool" not to. This pronouncement had predictably been devastating to Mamusha, who had two kids by him and had received no such proposal. And it came on top of her other traumas, especially that of her being considered Polish and hence officially an "enemy of the state." Gitti, in turn, had been shocked when Papa demanded that she give up their child (presumably to "Lebensborn," the Nazi establishment for unwed mothers where both Christel and I were born, and where children who were abandoned by their mothers would be trained and indoctrinated as good National Socialists). Gitti flatly refused Papa and struck off on her own, hiding in southern Germany and working on the land with a farm family. So, in the end Papa ended up with Mamusha and me and Marianne, rather than with Gitti and Christel. In retrospect, I shudder to think what might have happened to Ulla, Marianne, and me if Gitti had accepted Papa's offer of marriage.

On December 1, 1960, Papa wrote Gitti a long letter from Ottawa (where he was working on his ichneumons), trying to determine if she thought Christel would be up to the rigors of a year on an African expedition. "This is no 'safari,' " Papa wrote; it requires "work from dawn into the night every day for a year" and with no pay. He declared, "Hilde has been skinny and nervous of late"—maybe because of their tension over his affair in Ottawa—and he wasn't sure if she would or could come, "but still I must give her preference because she has become *Weltmeister* [the world champion] in this area."

Six weeks later he wrote to Gitti again, praising Mamusha for her "fantastic finger capacity" and bragging about her being one of the best assistants he has ever had, although he acknowledged that he still needed at least two bird stuffers. He continued:

I have considered the situation back and forth, with the result that it would be best if Bernd came also as an additional preparator. This way, one or the other could eventually improve in efficiency enough to get a few hours off to go into the woods and catch insects— the boy has made something of himself recently, and is a superb hunter and collector, but as a taxidermist he is a null (except for small mammals).

Papa then brought up the potential danger that Christel and I might fall in love, but assured Gitti that there would never be any energy left for distraction—the hard work would exhaust us, and anyway, "The boy would be informed of the truth."

Of course, at the time in Papa's office, and even after I left it, I knew nothing of "the truth," nor any bit of the background leading up to or preceding it. I agreed to go. But then Mamusha told the whole saga to Ripley, who said in no uncertain terms that Papa could *not* replace Mamusha. So Papa now had to take her along, or cancel the expedition. The young woman—who later told me that she had dreamed of going to Africa with her father "since I could think," and who had then prepared for a year at a zoological institute of the University of Köln (Cologne) to learn *baelgen* (the taxidermy involved in making museum study skins) and been certified by her professor, and had gotten a visa, immunizations, and a contract with Yale to pay for her passage—was summarily bumped. Papa wrote to her (as I found out years later):

> Even though I asked Hilde twice if she would come, she chose not to. Only then did I ask you. In the meantime, and behind my back, she told several people that I want to go on the expedition with two young inexperienced people, even though she is available. She is trying to make me look foolish and dishonest. I know you have your contract [with Yale] and could maintain your participation, since you fulfilled all of your obligations. But if you love me even a little, I ask that you please willingly step back from your contract. Otherwise I will have no chance for a peaceful life at home.

Christel relinquished her right, and I signed a contract with Papa, saying that I would follow his orders throughout the year of the expedition and would receive no salary.

PAPA, MAMUSHA, AND I LEFT FROM NEW YORK Harbor aboard the freighter *African Moon*, on July 29, 1961. The day before we sailed, we stayed at the Hotel Lincoln Square and had time to kill, so I paid a visit to the Museum of Natural History. As I entered, I saw the elephants that Carl Akeley, the great African explorer, had collected. I had read Akeley's books at the Good Will School, and they had made quite an impression on me. Only forty years earlier (in his book *In Brightest Africa*) he had written about spending "a day and a night in the Budongo Forest in the middle of a herd of seven hundred elephants . . . which had got my wind and were determined to get me."

By the early 1960s, much was known about elephants, but little was known of the many African birds. We needed to learn something about them in order to collect them; and Papa, being one of the world's best bird collectors, always prepared meticulously for every expedition months in advance. On our six-week trip across the Atlantic and around the Cape of Africa (with stops at numerous ports), Papa and I observed daily study hours in a long engagement with the 836-page *African Handbook of Birds*, by C. W. Mackworth-Praed and C. H. B. Grant.

We had heard that thousands of people had recently been killed in Angola (this was why the expedition had been rerouted to Tanganyika instead of going to Angola, as originally planned). Perhaps Papa's friends from the previous expeditions were among the victims. But East Africa was also in turmoil. One of the ship's crewmen gave me a book to read, *Flamingo Feather*, a lurid account of disembowelment and cannibalism during the recent Mau-Mau uprising of the Kikuyu in Kenya (which borders on Tanganyika). The leader of that uprising, Jomo Kenyatta, became the new president of Kenya. I knew we'd be isolated for months in "the bush," where we would be at the mercy of local people. I hoped it would not be true, as Elspeth Huxley wrote, that: "Africa is a cruel country; it takes your heart and grinds it into powdered stone and no one minds."

I spent endless hours of our Atlantic crossing at the ship's rail, fascinated by great schools of flying fish that erupted out of the water, sailed for many tens of yards on their outstretched pectoral fins, and then dived back down into the ocean. At night I sometimes went to the bow to face the breeze and feel the ship cut through the waves; the wake left an eerie green iridescence in the water. Spots of illumination twinkled on the water and then were quickly extinguished. The sounds of roaring water mingled with the muffled small talk of cardplayers coming through a nearby porthole.

One day we crossed a sharp divide between the bluish-green open ocean and the brown Congo river water laden with floating weeds and the silt washed off the African continent. I grew more restless in my bunk at night in the cramped cabin I shared with Papa. (Mamusha had her own cabin.) When I came on deck the next morning we were far up the river, with swamps and green walls of jungle on both sides. Occasionally I saw a dugout canoe along the distant shore being poled by one or two mostly naked black figures. Vultures spiraled lazily overhead. Eventually the steaming swamplands gave way to bare, rolling yellow hills. Flakes of ash from fires in the interior drifted like snowflakes onto the deck, and I wondered what it would be like to hunt alone in the thickets among elephants, snakes, and wild and savage peoples.

On August 8, 1961, only a week after leaving New York, Papa had written to Stresemann from the freighter:

> Why Tanganyika? you will ask. It was to be Africa over all else, because my specially outfitted truck from the last trip is still in Luanda [Angola] and waiting for me. The truck will now be shipped from Luanda to Dar es Salaam. This time there is a cloverleaf: my son, who has been for two years a successful forestry student of the University of Maine, accompanies us. He is an excellent bird observer and bird hunter, as well as a useful taxidermist. Cross your fingers that we will not be massacred on this, our last expedition, which can nowadays happen everywhere on the black continent. One year is a long time and much will change during that time in this region.

The *African Moon* eventually stopped far up the Congo River at the docks of a little town called Matadi to unload its cargo of 101 United Nations three-ton trucks. Each one had an opening for a gun turret on the top of the cab and seats for twelve soldiers inside. I did not know what conflict they were intended for, and didn't ask.

It was here in Matadi that I took my first steps onto African soil, Papa at my side and insect nets in hand. With two Angolan expeditions under his belt, Papa already had collected many African ichneumonids, and he now wanted to enlarge his collection. We would collect around the perimeter of the continent wherever the freighter stopped. Papa would undoubtedly fulfill his obligation to collect birds for the Peabody Museum as he was commissioned to do, but his personal mission lay with the ichneumons.

After thirty-five days we reached Cape Town. I went into the city and at a post office noted one window labeled "Whites and Europeans" and another "Non-Whites." I wrote in my diary that I went by a movie theater featuring *The Magnificent Seven* with Yul Brynner. There was a crowd in front of it carrying placards that read "Down with Segregation." This didn't seem like the Africa I had imagined.

After Cape Town, Papa and I next disembarked at the port of East London. Here in a wooded valley by a small brook we hit on a bonanza where Papa caught "more ichneumons than ever in one place." Papa was attempting to master the Ichneumoninae of Africa, and after this third African expedition his collection would be comprehensive enough to enable him to write a full-fledged volume on the African wasps. Like an ichneumon parasite inside a big fat caterpillar, this private work of his was growing within the bowels of the officially sponsored ornithological collecting expedition funded by public institutions with public support.

Finally, on September 7, we reached our official destination port, Dar es Salaam, "Haven of Peace," but neither the specially outfitted truck from Angola nor all our collecting and taxidermy gear had arrived. Unable to travel into the wilds without them, we stayed at the Oceania Hotel at the edge of the Indian Ocean. Papa was kept busy trying to get all the necessary permits, and I was free to collect ichneumons as well as other insects in the area. On one of my daily jogs I went four miles out of town and found a patch of tropical tangle. Everything up to that patch had been grazed bare by the humpbacked cattle that were on the loose everywhere. But here in the little patch of shaded trees I found ichneumons. I saved an especially handsome one to give Papa later, for Christmas. I do not recall ever having a disagreement with Papa during the entire trip. Given the common and mutually taxing mission, we all pulled together, and even his relationship with Mamusha seemed tranquil.

After we had been in Dar es Salaam for two weeks, our baggage finally arrived from Angola in a dozen big green wooden crates stamped "Yale University." We now had guns and ammunition to hunt birds, and the cotton, arsenic, and potato starch needed to prepare them. Papa had also just received the good news from his agent in Luanda that the truck was finally at the dock, and would soon be loaded onto the *African Planet* bound for Dar es Salaam. In the meantime we could begin to collect birds nearby.

Our first outing for birds was into the Pugu Hills, only sixteen miles from our hotel. We left in the gray dawn in a rented Volkswagen, passing by streams of people walking into town to work in an American cigarette factory. After getting to the Pugu Hills, we drove onto a logging road and parked at the edge of the forest. Papa counted out fifteen shotgun shells for each of us, and then we went into the forest in separate directions.

We never carried things like binoculars, compasses, or a canteen of water. Papa did not believe in lugging inessentials. We carried only our shotguns and shells, and each of us also had a small green army shoulder bag for carrying birds. That's it. Simplicity allowed focus. We agreed to meet back at the car for lunch at noon. Papa promised a cookie for the one who got the best bird. I didn't need to be bribed, but I nevertheless liked getting a rare token of appreciation from him.

Entering the dense growth, I felt entirely swallowed up. The trees were covered with vines. The forest floor was bare. By lying on my belly I could see a dik-dik (a type of tiny antelope) grazing here, a rat scurrying there, a thrush scratching in the fallen leaves. I crawled and encountered a troupe of birds in the lower branches and vines. I saw a glimpse of yellow, perhaps the belly of a forest weaverbird, a flash of white—perhaps a bush shrike. Identification was difficult, and it was hard to get the correct range and an unobstructed shot. I was hot, and the humid air smelled musty. Sweat dripped off my back in rivulets, and my khaki shirt was quickly soaked through. I heard a crashing above me—monkeys flinging themselves through the green sea above. The bird I had been trying to follow vanished.

I felt as though I was being reborn into a new and uncharted world, one where I knew nothing and expected anything. It was a world of new odors and strange sounds. I was an outsider. I did not know which stimuli were relevant or which signaled danger, and I had to be open to them all. Whistles, groans, ticks, scrapings, and sibilant melodies come to me from all directions. I would eventually learn to identify the bulbuls that answered each other with a loud whistle, pigeons that boomed like owls, sunbirds with twittering little ditties, the loud barking of turacos, the monotonous clucking of the *Camaroptera* hidden in tangles, and countless others. I had at first imagined vipers, charging elephants, and bloodthirsty tsetse flies infected with sleeping sickness. But the fear that I had anticipated gradually declined and was replaced by a sharp alertness that left little room for worry. I became a hunter, a role for which my senses had been honed throughout most of our evolution as a species. I felt as though I was awakening from a long sleep. Slowly, I became less conscious of myself. To be engaged in the hunt is really to become one with the animal and the landscape and the cycle of life in which every creature is food for something else.

The ground that I walked and crawled over was covered with decaying leaves, as little vegetation can grow in the gloom of the forest floor. Occasionally a bird would flit by up ahead, but I kept my eyes on my feet—a curled-up viper could blend in anywhere. Like the snakes, I hunted by movement and by sound, and in this dense multilayered forest, the birds in a flock needed to be noisy to keep in touch. Their voices told me where to go. When I heard a chirp ahead of me and saw a speck of orange and black, I took aim with trembling hands and pulled the trigger. I saw my bird falling through the branches and strained to keep it in sight until it landed. Then I crashed through the underbrush to try to retrieve it. I had brought down a forest weaver boldly marked in orange and black: *Symplictes bicolor*. I felt a surge of excitement at handling a bird I had never held before. Quickly I pried open its bill and inserted a small wad of cotton (so its feathers would not be soiled by regurgitate or saliva). I wiped off a spot of blood, then carefully rolled the bird up into a precut piece of paper and put it in my shoulder bag. Success!

The birds of the troupe had fallen silent for several seconds, then resumed their babble farther down the slope as though nothing had happened. I pursued them, alert for what other wonders the flock might hold. There in the twilit forest, a drab little bird clambered among the lianas. I patiently waited for an opening, and after another shot retrieved a forest bulbul, one of the Pycnonotidae.

Hours passed in a flash and by noon I had expended my fifteen shells and gained about a dozen birds. Among them were several more bulbuls. They did not seem alike to me. Were they perhaps several species?

By the time I eventually found my way back to the car I was grimy, tired, hot, and hungry. Papa was not there. I waited an interminably long time before finally hearing a shot. I hollered and heard no reply. Finally I went to search for him and used our distinctive seven-note Heinrich whistle, which Papa had invented and which had served our family well over the years. Like a bird's, it is distinctively specific. Nobody else would ever dream up such a strange sequence. It therefore stands out above the babble of all competing calls, whether they be from bird, insect, or human. I heard Papa's identical answering whistle and then quickly located him—he seemed to be confused and admitted having been lost. He had followed a path down into a valley. The path got smaller and smaller until it became a dik-dik trail. Then even that ended. When I found him he said he'd been trying to find his way out for about an hour by crawling on his hands and knees.

"This way is out," I said, pointing from where I came.

He answered, "No—I just came from there." So I agreed to go first and call for him when and if I reached the road, and did so. He answered, and I waited. No Papa. I hollered again, and heard him answer—from twice as far away. So now I hollered steadily so that he could follow my voice, and when he finally reached me he was thoroughly beat. "I'll never again go without my *panga*," he said, as though taking the machete would have been the solution to his problem.

Back at the hotel that evening he had intense muscle cramps, and complained that he could not sleep at night—symptoms of dehydration and salt depletion (as I would later learn). "This will be my last expedition," he had said. Now, in my bones I could feel the truth of this statement. He seemed fragile and vulnerable, no longer the indestructible he-man I had always thought him to be.

I did not know then how fortunate I had been to come on this expedition, because it would be not just the last of *his* expeditions. It would be the last of the classic zoological expeditions, the end of a tradition that stretched back over 100 years through the Victorian era, and it encompassed my heroes—Darwin, Wallace, Humboldt, Audubon. Such older fieldwork was giving way to the beginning of modern biology. In a few short years there would be virtually no new birds to discover, except by new methods of DNA analysis of already collected museum specimens in closely related species. Then in only a few more years, the unimaginable would happen: people would stop talking about finding new species. Instead they would be talking of ecological destruction and the extinction of even well-known species on a global scale. But at this moment in time in the Pugu Hills, the natural world still seemed endless, even if the personal world seemed so finite.

Our main assignment for this expedition was to visit the many high mountain cloud forests scattered throughout the vast African savanna in order to obtain a series of about six specimens each of all the bird species confined to each individual forest island. I did not know why we needed to collect what seemed like the same species from different mountains. But later I learned that the seemingly similar birds found were in fact potentially different species. The underlying points of difference between the species concerned ecology and species formation, thus requiring exhaustive collecting at each locality.

Tropical forests are rich in species, but the closer one moves to the poles, the fewer species one finds. One theory holds that tropical forests contain such a wealth of species because long periods of thermal stability allowed time for adaptation and differentiation into new species, each specializing in its own habitat niche. Another theory asserts the opposite: that habitat *instability* is a key factor in diversity. In the past, habitat has been closely related to climate, and climate can be reconstructed from a variety of (sometimes conflicting) evidence. Because there was no glaciation in central Africa, it had been assumed that the African climate had remained stable. However, East African geology shows that water levels in the lakes of the Great Rift Valley were much higher in the distant past (roughly 20,000 years ago) than now. While we were collecting birds in Tanganyika, Daniel Livingstone from Duke University was examining the fossil pollen deposits in sediment cores of East African lakes. The kinds of pollen in these cores (which could be dated by their radiocarbon content) revealed what kinds of plants grew when, and hence indicated the specific habitat at the site at different times in history. The end result of this and subsequent pollen studies showed that the currently forested regions in East Africa were once grass-rich and drier. The surprise was that East Africa was, in contrast to North America, drier in the glacial period, and warmer in the postglacial period, resulting in a wetter climate and forestation. One can deduce that during the fluctuating glaciations of the Pleistocene epoch (which lasted from 1.8 million years ago to 10,000 years ago), Africa experienced tremendous fragmentation of its habitat. During past glacial periods (the most recent of which started about 20,000 years ago), the mountains became virtual islands, separated from one another by oceans of grass, cammiphora bush, and acacia savanna. Populations of forest animals were thus stranded on each mountain and isolated. These populations were then free to evolve into new species, provided their population size was small enough to prevent random but evolutionary neutral changes from being diluted into the gene pool. (Imagine a flock of chickens in which one chicken has a neutrally selective random mutation for a long bill, and in which offspring of that bird have intermediate-size bills. If there are ten chickens in the flock, then soon all the birds of that flock could have bills that are longer by one-tenth, no matter how many birds the flock eventually grows to. But if you start with a flock of 1,000 and one has the mutation, then the bills of the offspring will be lengthened only by one-thousandth.) When warming episodes between glacial periods allowed the forest islands to coalesce and the animals to leave their refugia and renew contacts, they could at that point interbreed, or diverge further (by "character displacement") to accentuate their differences, or displace one another through competition. The birds we collected in the Pugu Hills, and the various isolated mountains named the Ulugurus, Usambaras, Ukugurus, and Pare, and the Meru crater, would help to tell the evolutionary story of East Africa.

The cycles of forest growth and regression occurred repeatedly in the last 2 million years, and the present is a relatively brief, warm interlude in what has been a predominantly cold Pleistocene period. These natural climatic cycles of instability (that will be devastatingly accentuated or even overwhelmed by massive releases of greenhouse gases) are now thought to provide the engine for bird diversity. Our collecting would provide the material to refute, modify, or support these ideas, while also creating a reference for future studies yet undreamed of. Indeed, there is now a renewed interest in some of the chats and thrushes that we collected using new DNA technology for investigations of species' relationships. Perhaps even more important, as habitats now change or disappear entirely, owing to disturbance by humans, museum specimens become an ever more valuable record as bio-indicators for many phenomena, including toxins spread in the environment.

My job description was "hunter-taxidermist." I had skinned and stuffed small mammals since I was a small boy. Mamusha had already taught me bird and small mammal taxidermy, and in Africa bird preparation took on an assembly-line routine. After Papa and I got back from hunting, usually around noon, we'd have a quick lunch, and then I'd sit down to start skinning. The whole process took me about ten minutes for a small bird, perhaps an hour for a fat duck. I placed the accumulating skins under a wet towel to keep them from drying out. Mamusha would take them in turn to work on them further. The next step was to apply the arsenic solution (which we mixed from a dry powder into a square, yellow, unmarked bottle) with a paintbrush. The solution moistened the skins to make them pliable and stretchable, and it would ultimately preserve them forever from insect pests in museum collections. Mamusha then did the stuffing—delicate work that requires sensitive fingertips. She never used rubber gloves. The stuffed birds were handed off to Papa, who held each of them up by a protruding stick that served as a temporary scaffold. He primped its feathers with a pair of tweezers, tied a data label to the crossed legs, and placed each "skin" in an appropriately sized cotton-lined cardboard bed where its feathers and its form were held in place until it dried and the stick could be pulled out. The successively stuffed birds were placed on a tin tray, and one of our hired African helpers maintained a slow-burning charcoal fire and tended to the drying process. (Insects did not need to be stuffed. Most of their mass is exoskeleton, and in most adult insects, this external skeleton does not collapse as the internal organs dry.)

By September 30 we had already established a rigid operational schedule. Papa and I got up at five o'clock in the morning, and after our oatmeal we drove out to hunt. We got to the Pugu Hills at about six-fifteen. By ten o'clock we returned to the car to leave the birds, eat a banana, and replenish the exact amount of ammunition that we had used, so Papa could better tabulate "who shot what with how many." Then we went out again and came back at noon, by which time we hoped to have twenty-five to thirty specimens. The next day, Mamusha and I prepared skins from dawn to dusk. After supper we cleaned the guns and got the ammunition ready so we could start the cycle all over again the following day. By the end of October, as the African spring approached, we had collected 170 species of local birds, just by making short forays into the hills outside Dar es Salaam.

On October 30 our truck finally arrived from Angola, and by November 16 all our boxes were finally cleared through customs. We packed up and left for the interior. My diary entry on November 18 reads: "Made it to Morogoro. We will camp in the mountains, far above town. Here, near the lower slopes, there is a steep winding path. Swarms of kids, staring, shouting, following. Some wear rags, others nothing. Multicolored goats graze on the narrow shoulders of the 'road.' "

We began our climb up the mountain with the help of porters. Big tin boxes that Papa had specially constructed months before the voyage were full of supplies and were balanced on the men's heads on little circular woven-grass cushions. Barefoot, the sweaty porters ascended the slippery, winding paths cut into the hillside to the *shambas* (fields) and grass huts. We followed them, and the swarms of chattering children joined the procession. Brown wren-like cisticulas chattered from grass stems. Coucals with short broad wings and long tails, flushed out of the grass up ahead, flew laboriously in front of us, and then disappeared again into the sea of grass before us.

By afternoon we finally reached a flat area with bare trodden earth where thatched huts stood next to small gardens surrounded by banana plants. From there we had a panoramic view down the mountain and up to the forest. Barefoot women in long dresses with brilliant red, blue, and yellow patterns walked past us, coming down the mountain as we went up. They were loaded with huge bundles of wood that defied gravity by remaining balanced on their heads. They murmured *Jambo* (hello), as they furtively cast their eyes toward us, but they did not and probably could not move their heads under their burdens.

Finally we reached a suitable camping spot. On level ground in a notch of the mountain by a brook, we set up our two tents, though there were also a few mud huts nearby in which to quarter our crew. Primeval forest lay above us. Below us the maize *shambas* clung to the slopes. White clouds drifted down from the dark green summits, descending like diaphanous veils enveloping first the trees and then us and our piles of supplies. Swarms of wood starlings swirled through the low-hanging clouds, and giant black and white hornbills flapped by in ones and twos. Green wood pigeons settled in groups in the nearby trees. Strange whistles, deep croaks, and maniacal chatter echoed down the mountainsides in random volleys, and I wondered if I was hearing leopards or monkeys.

The clouds that enveloped these mountains rose daily into the cooler higher air, driven by the moisture-laden warm air coming up from below. The water in this air condensed, and we would hear a roaring sound from the forest above, followed by the pattering of raindrops, which itself was followed very quickly by pounding rain and then rushing rivulets that sweep over the ground.

From our campsite (at 5,000 feet elevation), Papa wrote a letter to Stresemann, who published it in 1962, in the *Journal für Ornithologie*. Papa confided, "Maybe this is the most difficult research area that I have ever experienced—it was hard to reach, storms, rains, wind and cold, and fog without end, and steep slopes for difficult hunting." He described thickets, tree ferns, lianas, moss forest, and giant trees.

I saw huge numbers of termites in our clearing: they came tumbling out of their subterranean nests as soon as the rains let up and sunshine warmed the ground. They rose into the air like plumes of smoke. Chattering swifts then swooped in, scooping them out of the air. Swallows flew back and forth catching more of the bounty coming up out of the earth. Some of the more lucky termites managed to return to the ground and to pair off there. Once paired, they almost instantly dropped off their wings, like a kid casting off a Batman costume. The wings snapped off along predetermined lines of weakness. The pair would now dig into the ground and try to start a colony. Flight, which had been a dangerous but necessary part of their life during the nuptials, would now be superfluous. Wings would be a burden in the pair's lifelong union in a clay castle rising up out of the earth.

The now wingless termite couple would, however, have to bury themselves in haste. Ants, their terrestrial predator, are plentiful. In some places ants seemed to form a carpet over the ground. I saw big black columns of them, and most individuals within it carried a dead termite. Columns of these army ants were so long that they took a half day or more to pass a given point.

Every dawn as I left the camp to hunt—rain or shine—I saw something new. And I continued to keep an eye out for leopards and snakes. In the evening, after Mamusha and I had prepared the birds from the morning hunt, and our supper of rice and fried birds has been eaten, I would relax at the fire with our helpers, Mohammed, Waziri, Baccali, and Odilo. We talked about America and Africa in a mixture of English and Swahili. When the coals died down and Mamusha and Papa had long ago retired for the night, we drew closer in our confidences.

"Have you noticed how these people are afraid of us?" Odilo asked.

I had noticed. They gave us all a wide berth.

And he continued: "They notice you speak German, but they do not believe that you have come all this distance for such tiny birds."

"Then what do they think?" I wanted to know.

"They think," Odilo said softly, ominously, and with only a hint of a smile, "that you have come to get blood for making medicine. But since birds are too small for getting blood from, you will get your blood from elsewhere."

"From where?" I wondered out loud.

"They have told their children to be careful and stay away."

We later learned that British and German doctors and their staffs had indeed once come to take blood from the people, but only to sample for malaria parasites. (Indeed, malaria was still prevalent, and all three of us eventually contracted it.)

But I was not at ease, especially after I found out that the local people had recently murdered a prospector picking at rocks in the hills, because they thought he was disturbing ghosts. Another time a man found one of the white gauze Malaise insect traps that I had set up in the forest to catch ichneumon wasps. This trap is a big white tentlike contraption. It is almost transparent, and the man reputedly fled out of the forest in terror, claiming to have seen "a house of ghosts." Luckily he had not seen me in it.

Papa helped to diffuse the situation by inviting the "king" or headman of the area to come for a visit and have tea and cakes. The headman came barefoot, wearing a tattered old suit jacket and shorts, and carrying a black cane. He was friendly and spoke a few words of German (Tanganyika had been a German colony before World War I).

A week later (on November 28) we were struck with a fierce rainstorm that nearly dislodged our whole camp from its mountain perch. Our work tent, where Papa and Mamusha also had their folding cots, was torn. My small pup tent was spared, although my bedding soaked up a lot of water and the rope strands holding the tent down were yanked out of the ground. The headman immediately came back to our aid. He gave orders and sent men into the jungle, who soon reappeared carrying bundles of vines, poles, and huge leaves that looked like banana leaves. In an hour or two they had built us a solid wind- and waterproof structure that was far superior to our western tent.

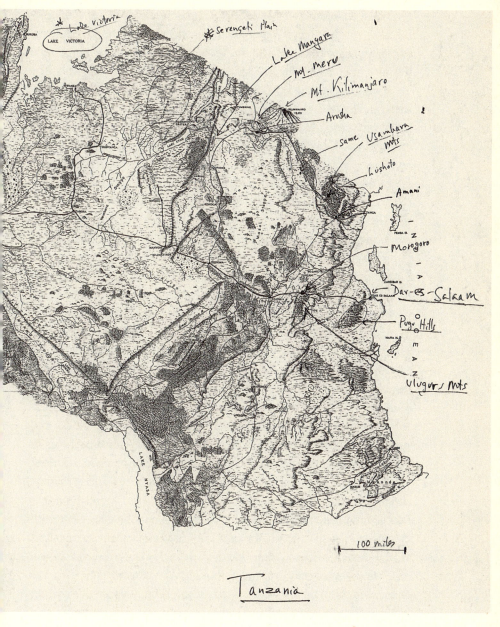

Lake Victoria

Serengeti Plain

Lake Manyara

Mt. Meru

Mt. Kilimanjaro

Arusha

Same

Usambara Mts

Lushoto

Amani

Morogoro

Dar-es-Salaam

Pugu Hills

Uluguru Mts

100 miles

Tanzania

A map showing the topography and isolated
mountains and place-names where we
stayed during our year in Tanganyika (now
Tanzania).

We did not try to explain to the local people that we were not collecting birds for their blood, but we had to justify what we did to ourselves. There was, of course, our faith in science—what we were doing was routine for explorers over the past century, many of whom were our scientific idols. I did not question them. However, I found it hard if I had wounded a bird and had to kill it by squeezing it in my hand, the way a lion or leopard throttles prey. After all, as a hunter I became aware of the lives of these creatures and I liked them very much for their physical beauty, their song, and their intense passions. Ultimately I reasoned that none would have an easier death in the wild. I also knew that knowledge is valuable and that it does not come cheap. My ethical imperative was to try to minimize the costs. Papa made it very clear to me that I was never to shoot unless I was sure what I was aiming at and sure of getting it. If I lost sight of something after shooting, I was not to give up until I found it. His kind of hunting reminded me why he had disdain for "sport" hunting and fishing that is just for bragging rights. We stuffed every bird, no matter how tattered it was or how much work was required to make it right. A final lesson was never to waste any meat. The flesh of even the tiniest bird was fried to be eaten.

By December 21 we were fairly certain that we had collected representatives of "all" the species (about sixty) existing on the top of the Uluguro Mountains, and we were ready to leave for the next mountain island. However, first we celebrated Christmas. We were invited for Mass and dinner with Padre Salzmann at the Catholic mission down below us in Morogoro. Mamusha also baked a cake at our campfire, and I surprised Papa with a present of four new, pretty ichneumons, including the one saved from Dar es Salaam. He gave me a note: "Mr. Bernd Heinrich, hunter-taxidermist of Yale-Tanganyika expedition 1961–1962, is herewith entitled to cash up to twenty shillings as Christmas present," signed Gerd H. Heinrich.

There was no time to spend it, though. We had work to do. This was an expedition! To my diary, my sole confidant, I complained that I was not being duly appreciated. After all, I was investing more than a year in this expedition, and losing two full summers when I would normally have been earning the money for my college expenses.

Not all the Africans who worked for Papa remained with us for the entire expedition. Adam was fired on the spot for the unpardonable sin of toasting some of the bird skins in the drying box he was tending over the hot coals. I was especially sorry for Mohammed, a small thin man who looked nutritionally stunted and whom I never saw eat anything except corn mush. He was merely slow, and unlike Adam he was given "one more chance" to improve his work habits or leave. By that point he had accumulated 320 shillings on his account, enough to buy 150 bottles of "Tusker" beer, so he decided to leave. Waziri claimed on May 15 that it was a Muslim holiday, and asked for and received a gift of five shillings from Papa in the morning. He promptly wanted to go to town to celebrate and was given permission, but told to be back by two o'clock that afternoon. He arrived back at three. Papa promptly told him to leave and take off the rest of the afternoon, "and tomorrow, and the next day as well." Without pay, of course. But Waziri apologized and offered to work anyway, without pay. Later, when we were driving to the next camp and had stopped along the roadside to make some oatmeal for Papa and coffee for Mamusha, Papa drew him aside and told him he was fired. Papa paid him 500 shillings for his six months of work. To our surprise Waziri said "Auf wiedersehen!" and I answered "Qua heri" (good-bye) and "Uhuru na kazi" (the independence slogan: "Freedom through work"). Odilo and Baccali laughed uproariously as we drove on.

Our heavily laden truck now lumbered up steep dirt roads to Lushoto into the Usambara Mountains. It was dreamlike here; cool, with clear, sunny skies, and malaria-free. We had an avocado tree next to our quarters in a little house, from which we could pick all the fruit we wanted. There was a small pond near us, and bright yellow weaverbirds nested in the reeds close to the water. Here we got the *Alethe montana*, an inconspicuous little bird that inhabited dense underbrush only in these mountains that our boss, Ripley, at the Peabody Museum wanted. But what I remember most were the women. Two in particular.

Both were bare-breasted, with colorful cloth wrapped around their hips and flowing down around their legs. I would see them in the morning facing each other near our hut, where they crushed maize by pounding it in a wooden pot with the end of a heavy wooden pole. They worked as a team. One smashed her pole down just as the other was lifting hers out. Thud, thud, thud, thud it went, in a steady monotonous rhythm. But then I heard thud, clap, thud, clap, or variations. While the pole was lifted out to its highest elevation one or both let go for a second to clap hands

A baobab tree on the bush steppe by Same.

before grasping it again to smash it down. They were making music! I could dance to that. But I could not talk to them, and I was too shy to try. One of the African workers from near the coast, Baccali, had his own ideas. "Lushoto no good. Women no Swahili." He could not talk to them, either. I longed for companionship and missed my friends at the university.

At our next camp, near Same (pronounced "sah-may"), a village nestled in a dry open acacia steppe, we finally got a break from the closed-in forest. Here I could see all the way to the snow-covered peak of Mount Kilimanjaro, about ninety miles away. It was a great relief to be able to walk on solid dry ground without being drenched and without stooping to squeeze through brush. The bird sounds were all new. Each species was a novelty, and I noted in my diary that we collected forty *new* species here in only six days.

In the bush steppe, the ground was hard-baked red clay exposed in patches interspersed with sand and gravel. Etched into the sand were the tracks of dik-dik, guinea fowl, rodents, dung beetles, various small birds, and mongoose. The grasses were kept cropped by antelope, and on the tall cactus-like agaves perched steppe eagles and chanting goshawks. They searched for lizards, rodents, and hares hiding under the cover of grass and brush. The flat-topped acacias were home to colonies of noisy weaver finches. Here and there a giant baobab tree dominated the landscape and provided an apartment complex for bees, barbets, and bats. Under blue skies and gentle breezes I heard the chatter and twitter of finches in the trees and cisticulas in the grass, and above this background music rose the raucous crackle of the guinea fowl. Sometimes the frog-like "kwok, kwok, kwok" croak of the gray desert turaco, and the shrill "kee-kee-kee-kee" of a hornbill joined in. I was especially stirred by the sounds of the dawn. In the hour before the sun peeked over the horizon I awoke to a gradually emerging concert of yellow-legged spur fowl, hornbills, weaver finches, cisticulas, bulbuls, bush shrikes, guinea fowl, turacos, frogs, katydids, crickets, and buzzing bees. The whole show gradually burst forth its music, and from far and wide the sounds rang out. It started with the spur fowl; then more and more birds chimed in, until, in predictable sequence, they again dropped out as the sun rose. The individual sounds and the pattern were ageless, and all of the participants of this drama were performing for their lives. They did not know it, but they were announcing their presence and their vigor to potential or actual mates; they were proclaiming their space to live and reproduce to their rivals.

So far we had not seen any elephants; nor had we seen the large charismatic megafauna that Africa is known for, because Papa deliberately skirted every game park. But that changed at our next camp, near Momella on the slopes of the Mount Meru crater, even though this area was not yet a park. Camped in the forest close to a series of alkaline ("soda") lakes filled with ducks, geese, flamingos, and other waterfowl, we saw bushbuck, Cape buffalo . . . and elephants and rhinos! Nearby, the forested slopes teemed with herds of the big black buffalo, the m'bogos. Black and white colobus monkeys roared in the moss-hung and cloud-veiled trees on the mountain slopes, and their bellows reverberated over the hillsides in the morning mists. Occasionally we heard the high-pitched scream of an elephant. It was an intoxicating paradise, but Mr. Nagy, a safari leader whom we visited in nearby Arusha, warned us: "It is dangerous in the mountains—the rhinos charge on sight."

Out in the forest alone every day from dawn to noon, I had to be constantly alert. I followed deeply worn game trails made by buffalo, magnificently tusked elephants up to fourteen feet high at the shoulder, and lumbering tanklike rhinos. I saw their fresh tracks and dung daily. I saw the signs of their browsing and felt the vibrations of their footsteps. I had to get into the mind of every animal, and I became so alert that my attention was occupied every moment, and time seemed to stand still. I was an intruder in a kingdom that did not belong to humans. Papa wrote (on June 26, 1962) to Stresemann from our forest camp on Mount Meru that collecting here was life-threatening, since the forests were full of elephants, Cape buffalo, and rhinos, of which especially the last two "turn around and make the hunter the hunted."

I found out the hard way that this was literally true. Mr. Nagy, having already warned us about the rhinos, nevertheless offered to show me the way up to the Mount Meru crater. We had brought along his African helpers as well as his taxidermist, Bodo Muche. We walked along a deeply worn game trail and had not gone far when I heard a gurgling bellow, the bushes appeared to explode, and a big Cape buffalo bull impaled Mr. Nagy on his massive horns, thrashed him against the bushes, and trampled him within an inch of his life. He had ambushed us after lying flat on the ground, hidden behind some bushes. He jumped up at us with such blinding speed that the armed hunter didn't even have time to raise his rifle. We carried Mr. Nagy down the mountain on a stretcher made from our clothes. As we were doing so a herd of about 100 more buffalo thundered past us down a gully. We did not encounter a rhino on this day, but after this episode the mountain forests became even more forbidding, especially for Papa, who was no longer as nimble on his feet or as adept at jumping behind rocks or climbing trees as I was. But the harder the nut, the sweeter the meat; it seemed unlikely that many bird hunters (and especially ichneumon hunters) had ever ventured all the way up and into the crater, and Papa felt certain it would contain riches.

I had brushed the buffalo attack off as a freak accident, and if anything, an expedition to the top of the mountain seemed more exciting. I would try again.

About a week later, I teamed up with Bodo once more. He was a boy my age who had come from Germany for adventure and worked for Mr. Nagy. With Mr. Nagy now in the hospital, Bodo was free. He had connections with local tribesmen and organized a group of carriers to take us up with enough supplies for a week. We hired a mzee (respected old man) who liked chewing tobacco, and his two rugged, spear-toting sons, Karino and Mirisho, as guides. These three led the advance; Bodo and I and the carriers followed. A full day's march brought us through the forest of the mountain slopes, and into grazed areas that looked like the cow pastures back home. These were interspersed with groves of low trees (which we skirted). We again followed game trails that had been worn deep into the ground and, judging from the fresh dung on them, were in current use by elephants, rhino, and buffalo. One trail that we followed took us into a dense thicket of giant heath. I felt nervous, because visibility was limited, but we slipped through and then went down into the crater. Now we were surrounded by a dark and forbidding forest of thick gnarled trees with broad but low spreading crowns. Dark green cushions of moss and small vermilion orchids covered the limbs, and long pale green lichens hung from them. I had never seen anything like it. The air smelled of sulfur. An eerie silence prevailed in a thick, opaque fog. There was no chatter of birds except for the occasional tweet of a sunbird or the raucous cry of a turaco. The only other sounds to break the silence were the occasional scream of an elephant, the rumbling of a buffalo, or the crack of a broken limb as one of these animals ambled through the forest.

The floor of the crater itself was flat and heavily grazed, like a meadow. We saw blue irises, pink impatiens, and buttercups like those found at home. We set up our two tents on a small knoll—an island of trees next to a shallow pond whose shores were trodden into a quagmire by the animals coming to drink. We heard them from our tents at night.

I don't recall collecting a single bird the whole week. But the crater, where the exotic was next to the familiar, turned out to be a Shangri-la for ichneumons. I found them everywhere on the foliage under the spreading gnarled trees. There were big bluish wasps with dark wings and smaller ones richly colored in russet. Everywhere I saw these prized creatures, and all of them were easy to catch because they were too chilled to fly. I made a large collection, anticipating Papa's happiness when I came back to show him my catch. Until then, I had Bodo, the mzee, and his two sons as company. On the cool foggy evenings we hunkered around the fire, and drew close together and talked. The mzee and his two boys kept the fire going next to the tents all night, and we spoke in hushed tones when we heard splashing at the pond and the snapping of twigs in the nearby woods.

In the five-volume publication (1,258 pages) on the African Ichneumonidae—with the long-winded title *Synopsis and Reclassification of the Ichneumoninae Stenopneusticae of Africa South of the Sahara* (Hymenoptera)—Papa acknowledges his gratitude to Mamusha and me for our participation in the expedition: "To my wife, Mrs. Hildegarde M. Heinrich, who took part in the three African expeditions and assisted in collecting, and preparing the typescript of this work; and to my son, Bernd Heinrich, who gathered in Tanganyika a valuable collection of Ichneumoninae—in some instances under difficult or unpleasant circumstances."

In this work Papa also provided a description and a hypothesis relevant to the ichneumons I had collected on my excursion to Mount Meru:

> During July 1962, Mr. Bernd Heinrich camped for a week near the summit of Mount Meru, Tanganyika, collecting birds at altitudes of about 2,700 meters. There, at the timberline, he observed numerous Ichneumoninae, mainly large specimens, sitting motionless upon and under leaves of lower bushes and plants. He was able to pick up dozens of these usually very swift and nervous insects by hand. This was just about the middle of the dry season and the coolest time of the year. The ichneumonids were too cold to be able to fly. As they belonged to species which usually occur in median altitudes, my explanation of the phenomenon is that the ichneumonids flew up (for reason unknown) toward the summit region of the crater during a warmer spell in the lower altitudes and were then paralyzed and caught by the cold air gliding through the northeastern rift of the crater-hole. In a similar way great numbers of ichneumonids have sometimes been found paralyzed on the surface of snowfields in high altitudes of the European Alps.

I was a little surprised to learn that Papa had given me such handsome credit, since I do not recall his ever praising me in person, or telling me that he was glad to have me along. I think he felt he didn't need to, since he assumed it was a privilege for me to be there with him. And it was indeed.

For me, the African safari ended on the evening of August 13, 1962. I sipped a cup of hot coffee with sugar and cream while reclining in a jetliner five miles up, on my way from Nairobi to Maine via London. I looked out of the window at the campfires below, orange specks in the black African night. At each one I could visualize dark faces gathered close around the fire. My mind was busy wandering through images of surging herds of Cape buffalo, delicate emerald sunbirds, and gorgeous reddish ichneumon wasps with purple wings, when I was interrupted by an airline stewardess with a powdered face and red-painted nails and lips. She came as a representative of my culture to refill my coffee cup. I was eagerly returning to my studies at the university, but I was not quite sure whether I was awakening from a psychedelic dream, or about to enter one.

nineteen   War: Prospects and Retrospect

Lacedemonians, when we go out of here,
let no person see us weep or conduct ourselves
in a manner that is unworthy of Sparta.

—"COME, O KING OF THE LACEDEMONIANS,"
CONSTANTINE P. CAVAFY

IT WAS "RUSH WEEK" IN THE PHI ETA HOUSE,
near the beginning of the semester. Everyone was walking around with a
beer in hand.

Everyone was talking, too, but I didn't know what to say. One of the
brothers came over to check me out, maybe so that he would have something to report at the house meeting, which would be held after we would-
be pledges left. I knew that the impression I made would be important.
But I didn't know how to be cool, or even what sort of impression to
make. What were the magic words that would open the door to fraternity
society? I assumed fraternity life to be synonymous with college life, and I
didn't want to miss out. I'd been away with no real peers for a whole year,
and being part of a brotherhood seemed like a good way to make friends.
Phi Eta is noted for its track and cross-country athletes, and the brother,
named Harold, himself a track star, recognized me as the cross-country
runner who had been elected captain, but then didn't bother to show up.
He looked at me accusingly and said, "Maine lost the Yankee Conference
by only four points. Where were you last year?"

"In Africa." What else could I say? I felt a little guilty.

"Yeah!?"

"Yeah—I was hunting birds." I needed to explain.

"Aren't there enough right here?"

"Sure. But these weren't just any birds. These were rare birds in jungles on isolated mountains. I skinned them and stuffed them. But I ran, too."

I told him how I tried to keep in shape by running barefoot like a native. Coach Styrna had sent me a newspaper clipping about Marchukas, the Kenyan runner at Cornell, who had won the championship of the InterCollegiate American Amateur Athletic Association (IC4A) running barefoot. So I thought I'd try it, too. When we were in the Meru mountain forests I usually ran about four or five miles a day, going from our camp down a dirt track to a little pond, and then straight back. When I finally tried it barefoot, my feet started to hurt a little after two miles. I didn't want to quit—I can be obsessive about these things—so I kept going. By the time I got to the pond I noticed blood between my toes. On the way back the blisters started to break, and they hurt so much that I wanted to crawl. But then I had to run really fast because I had misjudged the time. It was getting dark and I knew that the buffalo, elephants, rhino, and lions would be starting to stalk around, and they would be hard to see. I escaped back to camp by the skin of my toes, but I lost the soles of my feet. I had to crawl on my hands and knees for about a week or two after this experiment.

A couple of other brothers came over to listen. I continued:

"Another time I set a personal record in the fifty-yard dash—when a spitting cobra chased me."

Even more brothers came by, and all were taking frequent nips from their cans of Black Label beer.

"I had number two birdshot in the left barrel of my .16-gauge, but I didn't use it because I didn't want to damage the cobra's skin. So I used my right barrel with the .32-caliber auxiliary barrel inside that was loaded with a cartridge of bird dust shot instead. It only made the snake mad. It reared up and looked at me with beady black eyes and then it lunged at me in a flash, probably wanting to spit me in the eye to blind me. I sprinted like an SOB but it kept right on my tail."

I rattled on about my year in Africa, and by the time I got to the part where I dodged behind a tree to avoid the rhinoceros charge, and the bit about the buffalo attack on Mount Meru, I figured I'd wowed them sufficiently with my adventure stories that they would want me. They didn't. (I got about the same response when I tried to woo my date by telling her about eating fried mice in the Hahnheide.) None of them ever called me back. I dropped the idea of becoming a frat boy and contented myself with being a track man.

I found kindred spirits among the runners. We were a tight-knit group, and I would grow to love every one of those guys. In my junior and senior years, we ran down the opposition to win numerous state and New England track and cross-country championships. We were the only team in the east who startled, and hopefully flustered, the opposition by yelling out in unison, *Lassunslaufengehn!* immediately after the starting gun went off. That was our secret code meaning, "Let's go—run!" During the race we worked hard to catch up to opposing runners just to be able to stun them from behind by yelling, *Forbeigehen!* ("Let's pass!") Maybe it worked, because our team won the IC4As in a meet held in Van Cortlandt Park, in New York City. Only one team from Maine had done this before, and none has done it since.

Papa didn't show interest in my college running, any more than he did in my high school running. Nor did he explicitly disapprove. He just made it clear that he considered it unimportant by never once attending any meet during my four years of almost year-round competition. I ran cross-country in the fall, indoor track in winter, and outdoor track in the spring and earned three varsity letters per year for three years. Papa never did understand my passion. But it didn't matter. I wasn't overly concerned about his opinions on my running at the University of Maine. On the other hand, I certainly would have welcomed some financial support for my studies. While I was at the Good Will School sending out my applications, a bill for four dollars that I had failed to pay with a college application was sent to him by mistake. Papa sent me the bill. I got the message. College was my thing. Not his.

Now, returning to the University of Maine after a year away, I again took a part-time job in the campus kitchens. I would fund my studies myself; and when I made it through the university, the success would be mine alone. I even felt that I had career options. The year in Africa had reinforced my fascination with nature. I had been startled and transfixed by giant beetles I had caught in mist nets set for birds, had seen brilliant butterflies and birds, had watched army ants on the march, and had seen magenta jewelweed in a shaded mountain forest. My heart lay in the study of biology, not in forestry, which had turned out to be some sort of job training. I was uninterested in vocational training for any job. But I felt handicapped in biology because I had taken only half a year of Latin at Good Will and I had assumed Latin to be the language of biology. I now for the first time sought professional advice; and in a meeting with my faculty adviser, the geneticist Benjamin Speicher, I said that I loved animals and plants and wanted to learn about how they worked, how they stayed alive and survived. I admitted that my Latin was weak, and he replied: "Ben, first let me put your fears to rest. You don't need to take any Latin." Bliss! Sweetness! Relief! Then came the "but": "But you will have to take lots of math and chemistry." This was followed by some hope: "You have a rare background and keen interest, and you have the potential to make a fine biologist." (Hurray!)

When I switched my major from forestry to zoology, I elected a minor in biochemistry. Some of the students in my courses were now at the graduate level and my biochem course required a thick green textbook full of turgid chemical formulas. I was prepared to flunk, but instead became fascinated that an infinite variety of substances in nature could be described by such formulas. Knowing them, one could predict all sorts of possibilities, as Watson and Crick had done with DNA. Every day little lightbulbs went off over my head, and in the end, I even received an honor grade, B.

With the exception of the ornithology course (which I didn't think I needed to take), I took every biology course offered—general biology, embryology, parasitology, cell biology, general and comparative physiology, genetics, and mammalogy, as well as entomology, botany, bacteriology, anthropology. I was getting straight A's in these courses and making new friends.

But through all this a scary prospect hung over my head: the threat of war. When I was younger Papa had told me, "You, too, will experience war." I was reminded of his words almost constantly, and particularly at the athletic events. Before each ball game we all had to stand up and sing in unison about flags waving in the breeze and bombs dropping. I'd heard about the Nazis, who did a lot of flag-waving, and I wasn't too keen on it. To me the dropping of bombs had a negative connotation, so rather than pride welling up in my chest, I felt a chill inside. In October 1963 President John F. Kennedy and his entourage arrived in two military helicopters at our football stadium, where he gave a speech to the students and Mainers who'd driven from far and wide to hear him. It was on a Saturday, and unfortunately I chose grouse hunting over the president. Later I read that he had said: "While maintaining our readiness for war, let us exhaust every avenue of peace. Let us always make clear our willingness to talk, if talk will help, and our readiness to fight, if fight we must—until the world we pass on to our children is truly safe for diversity and the rule of law covers all."

Kennedy had talked about asking not what your country can do for you, but what you can do for your country. Tyrants make the same appeals, but Kennedy had stood up for the people of other countries threatened by force. Several years earlier, in 1953, Russian tanks had rumbled down the streets in Berlin, and the people there had attacked them with their bare hands and by throwing stones. Kennedy had recognized their bravery when he stood on the Berlin wall and shouted, *"Ich bin ein Berliner."*

In 1948, when the desperate Berliners were blockaded, the United States had airlifted 2,380,794 tons of cargo in almost 300,000 flights until the Russians opened the land route that supplied the Berliners and kept them free. In 1950 the communists invaded South Korea, and the United States again sacrificed for the victims. And in 1961, American and Soviet tanks faced off in Berlin across Checkpoint Charlie. The United States did not back down, even though we had nothing to gain in Berlin. To my mind, America was different from any other country on earth, because it stood for and defended the freedoms of others. I was proud to be a U.S. citizen. Only a month after Kennedy's visit I was sitting in an advanced biology seminar. Ken Allen, our professor, had just asked me a question about the famous experiments Harold Urey had done, relating to the biochemical origin of life on earth. Before I had a chance to answer, we all heard shouts through an open window: "Kennedy was shot!"

Kennedy's death sent a shock wave through the entire nation, and our safe little university was not exempt. Instead of enhancing peace, we continued to stockpile ICBMs until there were enough to blow up the world several times over. Bob Dylan then sang: "The times they are a-changin'." And he was certainly right.

Lyndon B. Johnson assumed the presidency, inheriting a little skirmish that was brewing in Vietnam. It didn't affect us, yet. Despite the theme of our famous Maine "Stein Song"—"drink to the careless years"—the chilling undercurrent of the cold war did not go away. It was especially evident to us male students, who were all issued army uniforms. In order to remain enrolled at the university, a state institution (and few of us could afford to enroll elsewhere), we were all required to take a two-year course of the Army Reserve Officers Training Corps (ROTC). We wore our khaki uniforms and made sure our black shoes and brass buttons were shined, really shined. We each wore a flat olive-colored hat with a black visor that required a mirror polish. We carried our M-1 rifles braced across our chests. During dress drill, once a week, thousands of us organized into squads, platoons, and companies and filled the dusty field house. We'd look straight ahead out from under our shiny visors and march in step as squad and company leaders shouted "Hut, hut, hut, hut, right h'ace, forward h'arch, left face, hut, hut, hut . . . , company halt, present h'arms!"

My back ached during these drills because my ruptured lumbar disk didn't adjust well to a continuous ramrod-straight posture. But no matter how much I hated it I knew I had to do it, because it was my duty. That much Papa had instilled in me. The idea of trying to get out of military service didn't even occur to me. When I was nearly finished with my undergraduate studies I knew that I had to enlist as soon as possible to get my military obligation over with. I'd heard that enlisted men could choose where they wanted to go and what they wanted to do. I trusted authority absolutely, and it never occurred to me that any official could be so foolish as to undermine their authority with a promise not kept.

The question was what assignment to seek. I decided I would like to be in Europe, preferably in Germany, and I wanted to be a paratrooper, like my old friend Lefty. Jumping out of airplanes and landing on the ground with a heavy pack might be hard on the back, but I had overcome obstacles before.

Around this time I had long discussions with Papa when I came home to visit. I asked him if I should enlist, and to my surprise he no longer believed that one needed to defend one's *Heimat* (homeland), regardless of the regime that happened to be in power at the time. This time, he told me that the "United States is the only state on earth without an ethnic foundation. It is an incorporation of all nations of the world with the common goal to make as much money as possible—and at the same time to enjoy as much 'liberty' (meaning license) as possible. One does not like to sacrifice one's life or limbs to protect the profit of others." On hearing this I got angry. I took it personally, because I had indeed worked hard at making money. I had to, in order to finance my education, because he couldn't or wouldn't do it! Instead of being with him, collecting ichneumons, I had to take summer jobs and work in kitchens. I wrote him an incensed letter in which I half-jokingly suggested to him that he wanted to keep me around only in order to catch ichneumons for him. Here is his response to me:

> Dear Bernd—I would like to answer the point of your previous, bellicose letter, a point which concerned me particularly. You wrote with reference to the fact that I had advised you not to enlist for Vietnam, that you believe I did this because I wanted "to have you around to catch a few ichneumon flies" etc., etc. This I could call jumping to conclusions. I certainly like "to have you around" and also am glad if somebody catches a few ichneumons for me (in the right place). But this egocentric aspect was by no means the real reason for my advice. I will try to explain this to you:
>
> I was a coward by birth, as most animals are, by the strongest of all instincts for self-preservation. I feared death so much as a boy that I went out of my way to avoid to see a dead human being. I never had seen one until I was seventeen years old. War was in my imagination sheer horror and hell. I learned about it in the history lessons, but I thought this was something of the past and nothing that could ever concern me personally. But I grew up in a close-knit society, based on common love for our country, on common endeavor to make it strong, beautiful, and successful. The idea of "duty" was almost born with us. To defend our country in danger was considered as one of the basic duties. High or low, rich or poor, king or worker—everybody believed in that duty. The number of individuals who questioned it was next to zero.

He then wrote several pages recounting his own experiences of enlisting at the beginning of World War I at the age of seventeen, and how much competition there was to get into the military because everyone went to enlist. He continued:

I also was not accepted at first in the uhlans. But I telephoned my father, who knew some general of influence and by "nepotism" and a cable of the general I had the honor still to be accepted as an enlisted man. And from this day on I started to become a man and a soldier; I came to the front with the first replacement and went out to the regiment about four weeks after I had enlisted. And I can tell you: it was not easy for me. I, the born coward, learned to defy death completely. I ceased slowly to hope or think for a return. Later, when in 1916 I volunteered for the Air Force, I knew that my chances for survival were next to nil. And yet, when I think back today, I knew this was the greatest time of my life. Why—is beside the point here. But the point is that I believe that I could not have been able to do what I did: to sacrifice my life, to offer it every day anew to destruction and torture by burning to death (the most common way of dying in an air crash then)—if this war had not been the *common* cause of an entire nation, of front and homeland together. In a situation as it is here, it must be sheer hell to experience war, if nobody cares, if the people at home think and say that what the soldier stands for at the front is all nonsense, etc., etc. And this I wanted to spare you. I wanted to spare you the agony of dying or being maimed for nothing, without knowing that everybody at home is proud of you and grateful for what you are doing. This makes the sacrifice easier. Without it, it is beyond human capacity. My advice was based on experience, which you could not have. Under circumstances similar to those which I have described above, I could not have given you that advice, as much as I had liked to know you in safety, as much as I had needed you as my successor in Borowke. All—or nothing—that is the only alternative of war. The people of this country are perhaps more "sophisticated" than we were. The ones more "educated" perhaps know better and talk everything to pieces, but they lack the only thing which justifies the sacrifice of war: real love for their country, real pride in it, solidarity, and the sense of duty.

Books could be written about all this. But I suggest that you do not try to discuss all this with me. So I say: take it or leave it—but do not try to "talk it to pieces." *I* have experienced it, and it was something very great for me, something almost sacred. I anticipate you would say: just mass suggestion. Maybe. But what means a word like this? The truth, the essence is felt, not talked. And for the two of us the point of my story is to prove to you that I thought about you, not about me, when I advised you *not* to enlist.

Kind regards,
Gerd

I could appreciate that Papa had gone to great lengths to try to make me understand German society as it was before World War I, his sincerity in volunteering to serve in the war, and his wish for my well-being. I noted, however, that as always, he understandably made no mention of World War II. He did not serve the Hitler regime because he wanted to—he had no choice. I did have the luxury of choice. Was he still stuck in the pre–World War I past where he went willingly, out of "honor"? I felt he was trying to make the point that enlisting now would *not* be honorable, and I was angry at his remarks about American society. The society he referred to was my society, just as the pre–World War I society was (still) his. Papa was a pessimist who favored enforced order. He pointed out "moneygrubbing," "pornography," crime, and civil disobedience, all of which were on his list of corruptions. He told me about the regime of Frederick the Great (r. 1740–1786), the Prussian soldier-king. Frederick was down-to-earth. He made laws to protect the peasants, and laws with a reliable legal system fair to all. There were no special privileges. To him, the function of man was to work for the common good, the fatherland. He was an extraordinarily gifted man who considered himself a servant of the state, who ruled by himself and decided everything himself. He decreed, and he expected action. Those who produced were promoted; those who made mistakes were fired on the spot. When Frederick sent soldiers to war, he personally led them into battle. He stayed with them throughout the campaign and experienced their discomforts. Frederick the Great had no place for pomp, ceremony, or luxuries. He said, "It is not necessary that I live, but I must do my duty."

So the honor of a leader was equated with the goodness of a regime? Hadn't Frederick, precisely by his virtue, made people dependent rather than independent? What if some charlatan came along who said, "Trust me! I have the answers," and bamboozled the people into placing themselves under his authority? Could, then, not a scoundrel achieve power? "What about Hitler?" I wanted to know.

Papa took my cavalier comparison of Frederick the Great to Hitler to mean that I was being a wise guy, as if he were saying, "How dare you compare my hero to the ultimate lowlife who destroyed my fatherland?" My problem was that I was by now starting to become like a professor— one who, according to a definition I once heard, "is someone who thinks otherwise."

Papa saw our vaunted individual freedoms in America as "license" and a shirking of responsibility. He was not even convinced of democracy as an ideal. He thought of it as an "experiment," as if it were a luxury that could exist only because of our unlimited resources. Maybe this was because he had grown up in a world of limited resources where free choice had led to chaos.

I decided to go against the old man's advice. I did not want to be guided by fear. I was proud to be an American, and I trusted my president absolutely. Hitler lied, and was ignorant besides; but I had faith in the absolute goodness and honesty of every president of our country. And none of my friends were moneygrubbing.

I knew what I had to do. My draft card (all of us were required to carry one) said "1-A," meaning top choice for military service. So I hitchhiked to Bangor to see the army recruiter.

What happened next was totally inconceivable and certainly unanticipated. My draft card, which had been issued to me at age eighteen, when I became a citizen, had always identified me as prime military material—the first in line to be called. But now, soon after I went to enlist, I was sent a new card, which said "4-F," meaning "unfit to serve." The change came after the army asked for and received the doctors' reports on my back injury. No doctor had anticipated that I could recover, and the army did not think me tough enough to jump out of airplanes and land safely.

I'm not sure if I should thank my four-leaf clover or be grateful to Papa. His "anti" stance made me "pro." In part because of him, I had gone out to enlist at a time when there were only a few "advisers" in Vietnam (who were thought to be sufficient to help the South Vietnamese against the North Vietnamese aggressors) and there was no desperate need for recruits. The military was rejecting anyone with a possible disability. Had I *not* tried to enlist then, there is no doubt that I would have been drafted a year or two later, when the military was no longer looking for excuses to reject young men.

I was spared another jungle tour, but some of my best friends and cross-country teammates were not—Fred and Kirk flew helicopters; Bruce commanded a PT boat; others carried a rifle. Their wounds, and in some cases their ultimate sacrifice, colored my perceptions. I resented the protesters. If I had chanted any slogan at all, it would have been "LBJ—all the way." My infinite faith in America, my country, had prevented me from seeing the truth, and now it scares me to think that had I been born twenty years sooner in central Europe, I might have been similarly blinded.

In retrospect, the Vietnam War was unjust, and those who saw that and resisted it showed uncommon courage. But to be honest, I tried to stay clear of politics. I felt that I was not qualified to know all the facts necessary to make a decision. I felt that this was the job of those in government. They had professional agencies, which existed to gather the appropriate information. It was their job to get it right and not for me to question them. Some claimed that our government had lied. To me that seemed impossible! Little did I know that even Robert McNamara—the secretary of defense and the war's main architect—would later admit publicly that the war was "a big mistake." So, Papa was right. The country was not behind that war, and my sacrifice would have been not only much harder, but also in vain. Miraculously, I had escaped unscathed and unscarred; I was free to be what I could, and maybe even what I *wanted* to be. This was a precious gift of fate that could not be squandered.

twenty   A Biology Education

I stopped by the bar at 3 a.m.
To seek solace in a bottle or possibly a friend.

"CLOSER TO FINE," THE INDIGO GIRLS

I REALIZE NOW, MANY YEARS LATER, JUST HOW prophetic Papa's words were. At the end of the war, thousands of young men had sacrificed their lives, their bodies, and their sanity in Vietnam. When they returned, the America they loved did not love them. People were deeply uncomfortable with the "defeat," and so the veterans were not hailed as heroes, but rather seen as suckers or, even worse, murderers. Papa had seen it coming, perhaps, as he told me, because of his vast experience in life and in warfare. I had reacted to his advice like a child, seeking to prove him wrong by doing exactly the opposite of what he advised. (I can thank my lucky stars that we both got what we wanted!) I was busy thinking myself a rebel because I was not pursuing his very narrow path of expertise with the ichneumon wasps. I had made it clear that I was *not* going to be like him. But in fact, there were many more divergent paths I could have chosen. I could have become a businessman. That would have really irked him. I would have been part of the American "moneygrubbing" machinery. Even though it was clear that my interests, abilities, and passions were not dissimilar from his, I would never, ever admit such a thing.

When it became clear that I would not be joining the army, I dived back into my studies. Perhaps believing that I now needed to serve in some capacity, I applied to medical schools. I was rejected by every one, despite my application essay, a thunderous tirade about the stupid practice, then current, of X-raying pregnant women. Sir Peter Medawar, a British Nobel laureate biologist once said, "One does not have to be terribly brainy to be a good scientist," and, "One would do better for owning some of those old fashioned virtues—application, diligence, a sense of purpose, the power to concentrate, to persevere and not to be cast down by adversity." I found these words reassuring. After all, they are the same qualities that make a good runner.

I had a part-time job washing glassware and generally trying to make myself useful in Professor James R. Cook's lab in the basement of Coburn Hall, the zoology building. I had not the slightest notion, nor did I care, whether or not he was a famous person in his field (he wasn't). Dick, as we called him, was raising one-cell protozoans, *Euglena gracilis*, and had a method involving light cycles and temperature changes that made hundreds of millions of these cells divide more or less at the same time. They would grow and divide in gallons of a watery medium that looked like a green broth. He could then harvest the cells from the water by spinning them down in a centrifuge at any point in the cell-division cycle that he chose. This allowed him to extract specific cell contents, which in turn enabled him to deduce processes involved in cell division. Knowledge of the mechanisms of cell division would have fundamental applications. Cancer research is one example, because cancer is basically an unrestricted growth that consumes all the resources it can grab.

When I finished washing the glassware, Dick let me help with ever more varied and technical aspects of his work, including measuring concentrations of enzyme molecules, determining metabolic rate, doing radiocarbon labeling, and making cell counts with his Coulter counter machine. This work made me feel that I was doing "real" science, and it was a heady feeling. I liked it and worked harder. Medawar was right, because one morning while I was counting the population density of the euglena growing in one of the many 250-milliliter Erlenmeyer flasks in the lab, Dick pulled the pipe out of his mouth and remarked in a nonchalant way: "Ben, you ought to make a master's thesis with this work."

Just like that, in an instant, I not only had a full-time summer job but also had an exciting project and goal for the next two years. Me, become a graduate student? I had never dared to aspire to something that seemed so lofty, or so daring. To be a doctor you had to apply what was already known. But to earn a degree in biology meant discovering something that nobody had ever known before.

An unforeseen revolution had recently started in biology. In 1944 the physicist Erwin Schrödinger had elegantly posed a bold but pertinent question: "What is life?" In his little book of that title, which I had excitedly read, he had refined the question by asking: "What maintains the specific characteristics of organisms for millions of generations?" He reasoned that there *must* be instructions, and these instructions *had* to be passed from cell to dividing cell. He deduced further that there *had* to be a replicating molecule, which encoded the instructions. The "secret of life" as he defined it thus boiled down to discovering and understanding this molecule, which was as yet unknown.

Only six years before I started as a freshman at the university, James D. Watson and Francis Crick had published a bombshell paper in *Nature* that addressed Schrödiger's question. At the end of their paper they penned the immortal (if you're a biologist) words: "It has not escaped our notice that these specific pairings [of nucleotides] we have postulated immediately suggest a possible copying mechanism for the genetic material." They had been less guarded off the record, reputedly running into a pub to shout that they had "discovered the secret of life." Cracking the DNA code was, next to Wallace's and Darwin's insights, the most momentous discovery of the ages, and the impact on biology was revolutionary.

I wondered if there were any discoveries left to make, but if Dick Cook was willing to take me on as a graduate student, then there had to be reason for hope.

Dick had noticed that a cell's respiratory rate (the amount of oxygen consumed) varied greatly depending on the specific point within the cell-division cycle where it was measured. It also varied according to what source of carbon—say, glucose, fructose, or acetic acid (the main component of vinegar)—the euglena cells had as a "diet" for their energy. But the cell cultures' overall rate of growth (i.e., the cell-division rate) remained constant, despite changes in respiratory rate. More specifically, I found out in the experiments Dick had me make that the euglena cells consumed about four times as much oxygen when we grew them on acetate (a 2-carbon molecule) than when we fed them glucose (a 6-carbon molecule), but that they maintained the same rate of cell divisions. It didn't make sense: respiratory rate is supposed to be mirrored by metabolic rate, or in this case, reproductive rate. Here was an enigma. I was hooked. Dick had showed me a problem to wonder about.

Dick Cook was a quiet, private person, competent and of real integrity. He became my model of what I thought a scientist is, or should be. He stood for Professor Gottlieb, the fictional scientist depicted by Sinclair Lewis in his novel *Arrowsmith*. I loved Cook but knew little about him outside of academe. About forty years after I left his lab, I asked one of his two daughters, Karen—who had become an architect and designed important buildings in London, Berlin, Frankfurt, and other cities—about her father. She told me that earlier in his career he had encountered a problem that is almost epidemic in academe: a major professor taking credit for his student's work. I believe it was sensitivity derived from this hurt that allowed Dick to hand me a great thesis problem on a silver platter. How good it was I would learn only later, when I happened to look into a three-volume opus, *The Biology of Euglena*, and find that my master's thesis work was cited twenty-eight times. I had performed a large part of the labor, but since he had played the major role in the concept, ideas, and writing of our three papers, I did not think that I was entitled to any intellectual credit.

Yet after two years of work our names had appeared together on three scientific papers that we published in the journals *Cell Biology* and *Protozoology*. At the end of this work I gave a seminar to the assembled faculty, and got to show off what we had discovered. It became the title of my master's thesis, "The Physiology of *Euglena gracilis* during Substrate Induction and Repression of the Glyoxylate Bypass." I may not have discovered the secret of life—but then, there are many secrets worth discovering. I had helped discover one of the many tiny steps involved in how the protozoans metabolize vinegar.

The euglena metabolism that we elucidated gave me my first glimpse of cellular mechanisms. It was very different from my previous experiences in natural history, and it hinted at layers within layers of existence. In 1928 the famous British biologist J. B. S. Haldane said, "Now, my own suspicion is that the universe is not only queerer than we suppose, but queerer than we *can* suppose." I was in complete agreement. It was exhilarating for me to discover that one could, with the right methods, occasionally see important processes at work in a cell so tiny that it is barely visible to the naked eye. Nature is a magic show of the highest order. I can hardly think of a greater grandeur and satisfaction than the process of finding out, and ultimately understanding, how life really works. Haldane also said that science is "vastly more stimulating to the imagination than are the classics." How could it not be? It's about real life!

My professors were mesmerized by the new findings that had been revealed by clever experiments; and experiments became my holy grail—they were the means for understanding. The insights derived from molecular biology dovetailed with our growing understanding of the mechanisms of evolution. Nobody could have any idea, for example, how closely or distantly organisms are related until one could decipher their genetic blueprint: DNA. Life, which was previously thought to be a phenomenon that could be understood only from the standpoint of history, now also became linked with biochemistry. Names of obscure researchers who had provided the building blocks for these momentous revelations and who had languished in oblivion suddenly became my household words. In my mind these researchers were the new explorers of nature, heirs of people like Papa who had traveled to distant lands to discover and document life's diversity. Here was a new frontier, and I was drawn there to try to make my own discoveries.

My master's thesis work was more than adequate to satisfy any PhD requirements, but there was then no PhD program in our department. Dick therefore suggested that I go on at another university and get PhD certification while broadening my education. One of the other members of my thesis committee, Professor Charlie Major, suggested that I go to the State University of New York in Buffalo to study respiratory physiology. I applied and received a letter from Professor Hermann Rahn to "come look us over," all expenses paid! When I got there, his graduate students took me out for lunch. The burning question for them was whether the shells of birds' eggs exchanged gases with their environment at rates of x versus y. I didn't quite understand their excitement. The embryo had to inspire oxygen and expire carbon dioxide. What was the enigma?

Finally, I was admitted into Professor Rahn's office. He looked at me. I looked at him. Being just a student (and still true to Anneliese's teaching in the Hahnheide), I was not going to speak unless spoken to. I waited politely. He still didn't say a thing; he just kept looking me over. What to do? If I started telling him about myself, that would be presumptuous. If I asked him to talk, that would be bold. What did he want? He wouldn't say. After maybe five minutes that seemed more like hours, he rose from his chair and declared, "The interview is over."

I slunk out of the office to face my next tormenter. This one was a little chattier. He asked me about my math background. The interview was very short.

Next I was shuttled off to another professor, who asked me why I wanted to study biology. I eagerly told him about my enchanting visions of bees buzzing on flowers, birds in the marsh, caterpillars, and moths, and perhaps even watching the grass grow—all the things that turned me on. He stared at me and, after a long pause, finally snorted, "You're a *naturalist!*" It wasn't a compliment. He quickly escorted me to the door.

There was, in the aftermath of Watson and Crick's discovery, little room for naturalists in academe, at least not at the State University of New York in Buffalo. However, it did not seem to me that being a naturalist would be an impediment to becoming a biologist. Studying life requires many different levels of analysis, from the biochemical and molecular processes within a cell to the physiology, anatomy, and behavior of whole organisms to the dynamics of populations, interactions between species in ecosystems, and even global processes that support life and enable life to exist. The discovery of the DNA molecule was momentous, but it did not trump any of the other levels of biology!

After I returned to Maine I told Dick about my fiasco, and he suggested that I look into the University of California at Los Angeles. He explained that UCLA (where he had recently taught biophysics in the medical school) had a great faculty in biology, one that would not be troubled by my "naturalist" leanings. He offered to write me a letter of introduction.

In the summer of 1966 I sent UCLA my application and he sent a letter of reference. Within a couple of weeks I received a reply. Eagerly I opened the crisp clean envelope and unfolded the letter . . . Accepted! Not only that; I was offered a research fellowship. Within days I was packing my sleeping bag into the backseat of my white Plymouth Comet and heading west, singing all the way. For the first time, I'd get to see the prairies, the fields of amber grain, the deserts, the majestic mountains, and the redwood forests to which Woody Guthrie and Pete Seeger had introduced me. To Papa, I was just "going away," maybe because I told him little if anything, since I expected him to be less interested in what I was going to do than he was critical of my leaving. Unlike me, Marianne was attending the teachers college in Farmington, only six miles from home.

On my first day in California I visited Sequoia National Park, with its giant trees, and I knew I was in an exotic and exciting new world. I proceeded directly from there to Malibu Beach to try my hand at surfing. I rented a board, and to my great disappointment and surprise found out that I could not even stand up on it when it was anywhere near water. Surfing was nothing like the canoeing I'd been used to on the Penobscot River. I decided to move on.

I drove several more miles down the coast highway to Santa Monica, and after following the signs to Los Angeles' Westwood neighborhood, where UCLA is located, I parked at the campus. It was exotic—red- and white-flowering oleanders along the roads, eucalyptus and orange trees, palms, and tall brick-facaded buildings. The air was hot and dry. I'd never felt or smelled any air like it. Everyone I saw looked fit and healthy and tanned. The women were especially beautiful.

I had not yet thought about where I might spend the night, but within minutes of finding the biology department I met a graduate student who just happened to be vacating an apartment at 1666 Greenfield Avenue, an easy walk from campus. I'd found my home for the next four years. I met my PhD supervisor and the four graduate students in his laboratory. They were African-American, Japanese, Jewish, and Irish Catholic, all men, all natives of Los Angeles. I, the only Maine-German-Polish naturalist-woodsman, was surely the most minor of the minorities, but I fit in just the same. Within a week, one of them, the lab clown, decided to take me out on the town. We went to a bar and he bought me a few drinks. He studied me, grinned a lot, and finally asked if I'd noticed anything peculiar about the girls in Los Angeles. I shrugged; not really—where upon I was informed that the "girls" were really men in drag.

About a week later I went to a social at the International Club on campus. I was nervously standing around, not knowing what to do, when a beautiful, slight dark-eyed brunette approached and shyly asked if I would fetch her a drink at the bar because she was not yet twenty-one.

The ice was broken and I felt a warmth radiating from her that I suspected originated from her having faced difficulties. I sensed that she had depth and I immediately felt an affinity for her. Kitty had chosen me as the lucky guy to get the drink for her. What an amazing coincidence that we would meet, I thought as her friendly eyes met mine. I was in love. We soon wanted to be with each other all the time. I showed her my little apartment, and she loved it even though it was modest in the extreme. She was as short of money as I, and we decided to set up housekeeping together. We soon had more roommates—a young raven and also a long-eared owl that we got on one of our outings into Owens Valley, near Lone Pine.

Everyone in our lab was working with flagellate one-celled protozoans (either tetrahymena or euglena). What might they tell us? I faced the universal problem of all PhD students, to get started on a research project that would yield a discovery. Discoveries in science go to the clever, the persistent, and the lucky. It's like an Easter egg hunt . . . you have to know where to look. There actually does have to be something there to uncover.

It had recently been discovered that DNA exists not only in the cell's nucleus (on the chromosome with its sequence of genes) but also in some of the cell organelles (such as in the mitochondria and chloroplasts). Mitochondria release energy from food, using oxygen and liberating carbon dioxide, while chloroplasts capture energy from the sun's rays (and power the earth economy of nature) and absorb carbon dioxide while liberating oxygen. In many protozoa (such as euglena and tetrahymena) the chloroplasts could be either present or absent. When the cells are grown in the dark and fed sugars or acetate as an energy source, they have no chloroplasts. But when they are grown in light, these protozoa turn bright green because they fill up with chloroplasts that capture energy from the light to make sugars, proteins, etc., to grow on. Some protozoa ingest and digest bacteria and single-celled algae. About two billion years ago some of these ingested organisms had evolved to resist digestion. It had been proposed that both mitochondria and chloroplasts are highly evolved organisms. Mitochondria, which look and divide like bacteria, are thought to have been derived from them, and chloroplasts were derived from one-celled blue-green algae (now known as cyanobacteria). Following the general evolutionary pattern of parasites who become symbionts, both had become highly specialized to live in and then become dependent on the environment in cells. In turn, their host cells had become dependent on them. Thus stable partnerships are formed.

I tried for a year, and failed, to get any DNA that I could confidently identify as being other than that from the cell nucleus. I was probably not skilled enough to make a proper sucrose gradient, or to load it, or to read it. Alternatively, the DNA of the chloroplasts may have had a base ratio too close to that of the nucleus to be differentiated from it. If so, then no matter how skilled my techniques were, I would never get it. Not by this method.

I didn't trust my lab techniques and ended up looking at the chloroplasts themselves (through a special "phase-contrast" microscope). I was surprised to see that the chloroplasts under the lens looked like tiny specks, as well as large rounded structures, and I began to wonder if each chloroplast grew from a single primordium, or whether the population of chloroplasts in any one cell arose by division. I became excited, photographed them through the microscope lens, and deduced from my observations that they increase both in size and in number. I wrote a paper about my results, "The Activation of Protoplastids in *Euglena gracilis*," and submitted it to the *Journal of Cell Science*. It was rejected because of criticisms by established scientific peers, and since I trusted their judgment, I immediately threw away my manuscript and dropped the project.

I dreamed of the beauty of simple, elegant experiments like those that Karl von Frisch had done to reveal the totally unheard of and unimagined wonders of bees, using only sugar syrup and dabs of paint. They earned him a Nobel Prize. I had worked with expensive machinery, and after applying it with unrelenting force and determination I had achieved not one scrap of useful data in a full year of trying.

As I had been on track for a biology degree, I had to study for my very stressful comprehensive examinations. I had to read widely, because we would be tested about anything and everything in biology. The Mojave Desert, where Kitty and I went camping to smell the air, feel the sun, and enjoy the crisp coolness of the night, inspired me. I thoroughly loved the deserts here, which were so different from the landscapes I knew; I loved the animals that survived and thrived here. Kangaroo rats burrowed in the sand next to clumps of yellow-blossomed creosote bushes. Pale blue flowers of mint were visited by hummingbirds, and red-tailed hawks circled in the sky. I began to read about desert animals and I was captivated by studies done by Knut and Bodil Schmidt-Nielsen on how camels survived in desert heat where other animals died, and how kangaroo rats were able to live in the desert without taking a drink of water their entire lives and eating only dry seeds. I had heard about studies by Donald Griffin, who proved that bats could "see" inanimate objects in a totally dark room with their ears. My idol was Kenneth Roeder at Tufts University. He had discovered that some moths can hear. He had illuminated their audio abilities in field experiments, and by doing electrophysiological recordings of individual neurons, he discovered how they could detect the bats' echolocating cries, which are beyond our own senses.

Like ancient mariners looking at the maps of the world, these pioneers discovered a blank space in the world of biology, and filled in a thing of beauty. That, too, was my ultimate dream. But I did not know where to look or how to proceed. Only those studies that would sprout into peer-reviewed articles in the scientific literature could count.

At no time in my graduate career was there any formal discussion with peers about the "big picture" of what we were doing, how to proceed, what to expect, or why or how it might fit into life. We were supposed to know. Mostly I didn't. Not knowing isn't always bad, because it makes one behave like a horse that has blinders put along the sides of its head so that it can proceed only straight ahead and is never diverted. I think someone forgot to put the blinders on me, because I often turned my head, stumbled, and got bruised. In retrospect, I don't think I learned the ropes until it was time for me to retire, but I would not have had it any other way. Nevertheless, while I was at UCLA I sometimes wondered—especially when I had a glass of beer in my hand at the Night School—if there was another way.

This tiny pub on Wilshire Boulevard was close to our apartment. It was a cozy place with four small tables surrounded by chairs, a dartboard, and an aquarium with one large fish. I stopped there sometimes on my way home from the lab in the evening, and Kitty would join me after her day of teaching kids in the Watts ghetto. I had been taking remedial courses at UCLA in calculus and physical chemistry. Both were an agony for me, and I could not see how they were helping me solve the DNA problem. Graduate school was no longer fun. Here at the pub, with a beer in front of me, I could relax and talk. I can't remember our exact conversations, but I think one went something like this:

"You look kind of peaked," Kitty probably said.

"I am. I don't know what I'm doing or why. I get distracted by the red-tailed hawk soaring over the canyon."

"Look, if you're not happy, why not do something else?"

"Easy for you to say—you haven't spent almost three years on the road to becoming a molecular biologist."

"Well, why don't you talk to some of the other grad students and see what they are up to? Maybe there is something else going on that is more your style. It's a big university."

Indeed it was, and my answer actually lay under my nose, downstairs in the biology department. There, in the discipline of "physiological ecology," people were working on birds, snakes, lizards, and desert rodents. They went out into the Mojave, studied the behavior of animals there, and then brought them back into the laboratory to focus on their physiology. I had never heard of anything like physiological ecology while I was in Maine. It was like seeing the world through a new window, and I liked the view.

So I went to my major adviser and told him that I needed to leave the lab. I then went downstairs, and approached Professor George A. Bartholomew. "Bart," as he was called, had an impressive record of studying desert rodents, reptiles, and birds. He had a flock of students and didn't need any more. He was about to leave for a year on sabbatical to study bats in New Guinea and Australia, and as I had washed out with another adviser, I must have seemed to him more a potential liability than an asset. We chatted only briefly, and he suggested that I try a colleague of his who had fewer students but similar interests. I suspect this professor also didn't want to be the one to accept a lost soul, but he generously agreed to give me a try. He asked me to acquire data comparing rates of water loss between two species of quail, one of whom was native to "xeric" environments, and the other more partial to "mesic" habitat (what he meant was that one lived in drier conditions than the other). He had a new weighing scale called the electrobalance that was so sensitive it could detect water loss from a bird one breath at a time.

A couple of weeks later, after a good deal of effort, I managed to get some data and proudly showed them to him. He pulled open the drawer of his desk and showed me his own data. Yes, my results were correct—they were in agreement with his! Water loss by panting, and the amount of body cooling resulting from it, had already been measured to three decimal places. He had meant to compliment me, but I was irked to think that I had spent two weeks trying to find out something that was already known. The discoveries that I might make didn't have to be "important," like Ken Roeder's revelations about moths, but they had to be interesting to me (i.e., unexpected); and I didn't just want to be accumulating data for another professor. I went back to Bart and told him that I wasn't satisfied measuring water loss in quail. He then made a very unusual demand—and one that was almost impossible for me to meet. "Make a list of six potential research projects that you might want to solve," he said, "and then we'll talk."

As with evolution, so with academic research: the more options there are to choose from, the more possibilities for progress exist. After spending several weeks, morning, noon, and night, in the library, I had that list of six potential questions. In retrospect, posing the questions themselves was perhaps the most productive time I ever spent as a grad student. All but one of my potential research projects involved insects, but it didn't occur to me that nobody in Bart's lab had ever worked with insects.

I sat down in Bart's little office and showed him my list. We talked it over, and as I recall he was not dismissive or discouraging of any of my proposals. He said he would leave it up to me to decide which would be best. I didn't want to blow the chance to make a discovery, so over the next three years I worked on five of the six projects on that list. And the sixth, on hibernation, I later wrote about in a book called The Winter World.

The first potential thesis question that I explored involved tiger beetles. They awoke in me pleasant memories of chasing them in the Hahnheide. The question itself was prompted by a student in Bart's lab, who was examining whether color in birds affected their body temperature and energy balance under the direct heat from sunshine. There were both relatively dark as well as light-colored tiger beetles in California, and I wondered if the "white" ones were able to venture into hotter habitats from which the darker beetles were excluded because they would overheat. I went to China Lake, California, on February 5, 1968, to visit Norman L. Rump, an "amateur" entomologist and collector, like Papa. Rump showed me his collection of 16,500 tiger beetles, and when I told him what I wanted to do, he said, "I like your approach. The information you propose to obtain is very important to better understand insect behavior and response to environmental pressures." He helped me by pointing out where I might find various species, such as the half dozen nearly white ones at Furnace Creek in Death Valley, and the darker ones on the salt flats near Balboa by Huntington Beach, California.

One of my major study sites soon became the edge of salt flats by a coastal marsh at Huntington Beach where stilts and avocets nested. It was a beautiful spot, and I loved to linger there monitoring the beetles' activity while using the lab's equipment to continuously record temperatures and solar radiation to find if and when the beetles took refuge in their cool underground burrows. I brought beetles back to the laboratory and observed their behavior in temperature gradients that I built for them. I assembled equipment and got data on what wavelengths of light were absorbed versus reflected from their dorsal elytra (wing covers). I was making progress, but then I found a recently published paper showing that my idea was not original.

I next turned my attention to caterpillars. During my forays into the Mojave Desert to study tiger beetles, I had found green patches of jimson-weed—the plant of the tobacco family that the Indians had used as a hallucinogen. To my surprise and pleasure I found that the big green caterpillars of the tobacco-tomato hornworm, *Manduca sexta*, fed on these plants. I had already seen these caterpillars on the tomato plants in Phil and Myrtle Potter's garden and at the Good Will School. But here in the hot desert they seemed anomalous. They must be losing water very quickly in this hot sun, I thought. Do they *eat* more to make up for the water loss? This seemed like a simple enough question to answer. It wasn't. I worked on this project for two years, and gathered reams of data of all sorts. But the picture got ever more complex as I went along and it became evident that many variables were interacting in complex ways. I felt that I might get an answer, but it would end up being a description rather than a discovery of a new phenomenon. Since there was no discovery to report, I called it quits and restricted the scope to an investigation of a tiny side project: to try to find out how a caterpillar could consume a large flat leaf whole without wasting any of it, without moving from the spot, and while using the leaf as a sunshield at the same time. The caterpillars did this routinely, but it was not obvious how they did something that seemed "intelligent" and that had to be the result of a series of programmed responses. The caterpillars, while remaining attached to the base of a large leaf (I experimented with different shapes and sizes), fed by bringing the leaf blade to their mouths by pulling it in with their legs at their front end. Pulling down and eating forward resulted in bending the leaf and feeding to its tip, and then to the base, where they had been shielded by the bottom portion of the leaf the whole time. I published the results and conclusions of this simple little project in the British journal *Animal Behaviour*. But I did not assume or intend that this study would suffice as a subject for a PhD dissertation.

I turned to bees next. I was walking down Santa Monica Boulevard on my way home through Westwood, when I noticed a swarm of honeybees hanging from an ornamental Australian bottlebrush tree along the sidewalk. People obliviously walked by this swarm, one foot wide by a foot and a half long, of about 10,000 bees hanging barely ten feet above them in the branches. When I saw it, I had to have it. I went home, hurriedly hammered together a wooden box with a volume of about three or four cubic feet, and then went back to the scene of the swarm. There I climbed up and snipped off the branch that the swarm was attached to, carried it down in my free hand, and secured the bees in the wooden box. I knew these bees would be docile, since each of them would have tanked up on honey before leaving the hive, and in any case, they had no hive (yet) to defend. Thanks to my experience lining bees and keeping bees in Maine, I had no hesitation about handling these bees, even without gloves and a bee veil. I brought the bees into the lab, hoping to measure their metabolism with the oxygen analyzer that the students in Bart's lab were using to measure almost every other critter around.

My boxed bees did indeed give me totally unexpected results. The water produced from their metabolism soaked into the wood. The wood then warped enough to spring a crack big enough for the bees to escape. A portion of the 10,000 or so bees streamed out into my temperature-controlled room where I had been working with the caterpillars, and when I opened the door to the room they also spilled out into the halls. I took it with a smile; unfortunately, almost everyone else did not.

It had taken me a little over a year to strike out with projects on chloroplasts and chloroplast DNA. It took another two years for strikes three and four, with the tiger beetles and the caterpillars. Other students didn't seem to have my problem—maybe their professors thought them worthy, because they got assigned great problems that were being worked on by the professors themselves. The results of the honeybee episode, strike five, were obvious to me in about one day.

twenty-one   Making a Discovery

To him who in the love of Nature holds
Communion with her visible forms
She speaks a various language.

WILLIAM CULLEN BRYANT

What we know is a point to what we do not know.

RALPH WALDO EMERSON

THE 1970S WERE HEADY TIMES OF BIOLOGICAL
discovery. I tried to educate myself by almost daily reading of the current
issues in several biological journals and going to weekly talks by famous
biologists who had been invited to the campus, much as Papa had done
when he went to visit Kleinschmidt, Jordan, Rothschild, and Stresemann.
New concepts were being revealed that would revolutionize how we saw
nature work, especially the role of evolution as it applies to social adaptation and ultimately to humans. Papa could not have had the slightest notion of them, and indeed when he was leafing through his *Journal für
Ornithologie* (which he was receiving free as an honorary member of the
German Ornithological Society), he said that he did not like the "new biology." There was too much emphasis on numbers for him to understand
and feel comfortable with. I found the new ideas, such as those on fit-
ness, sexual conflict, optimal foraging, and altruism in social evolution
interesting and informative, but a bit trendy and not fertile ground for

359

making a clean discovery. I was undoubtedly as old-fashioned then as Papa had been, not being anxious to jump on any bandwagon. I felt more comfortable trying to find a little problem and working it out to see where it might lead, rather than having a destination in mind and trying to figure out how to get there.

Papa always had a destination in mind: describe the ichneumons of Burma, Africa, North America. After he'd completed a survey of one place, he could proceed to the next new project. I, on the other hand, could start from a quite random observation (but usually at least a casually informed one) that seemed at odds with some current theory, and chances were good that I would learn something new, which in turn would lead to the next observation and perhaps to the next set of ad hoc experiments. To me this was like tracking a weasel through the woods. I had no idea where it might go, but I knew some of its habits and I had tentative expectations or "predictions" that could help me checkmate the beast. It didn't really matter if the weasel was caught. What mattered was that the journey be fun and that I encounter new territory.

As is required for science, the only observations suitable for dissemination are those legitimized by publication in the peer-reviewed research journals. Thus there was a certain formalized structure that had to be adhered to in taking and presenting observations. Unfortunately the presentation leaves almost no trace of the journey. I give in this chapter a highly abbreviated and selective sketch (which may be either too detailed or too broad for some readers) of my adventurous and often serendipitous path.

KITTY AND I CONTINUED TO LIVE IN OUR FUNKY little apartment complex at 1666 Greenfield Avenue after we married in 1968. It was a little Shangri-la, complete with a patch of lawn in front and some yucca and oleander along the sides. We continued to make outings into the nearby deserts to keep an eye out for tiger beetles, caterpillars, lizards, hummingbirds, and other animals. I usually took along an electronic thermometer from the lab to do a few "grabs and stabs" of anything that crawled or flew within reach of my butterfly net.

On one of our trips to the Anza Borrego Desert we packed our sleeping bags and a few cooking utensils and went up a dry wash that mean-

Hyles lineata, the lined sphinx moths.

dered into the rocky cactus-studded hills. We camped on the sand behind the first bend. Pale blue flowering mint, yellow composites, and *Belloper-one* with long bright-red tubular flowers were all around us. We gathered some dry paloverde twigs and started a small campfire while the sky darkened and Costa's hummingbirds, *Calypte costae*, finished their evening feeding frenzy. The purple iridescence of their throat feathers as they hovered before the red flowers changed to black as the shadows started to fall. Then I saw something even more surprising under the darkening sky, in the chilly air: white-lined sphinx moths.

Superficially, sphinx moths resemble hummingbirds. They are close analogues, similar in size, swift and agile in flight, and colorful. The lined sphinx (*Hyles lineata*) has a soft "fur" of green and tan scales and its pink underwings are framed in black while the abdomen has patches of white.

I wanted to catch one of these moths, and when one stopped nearby to hover and sip nectar from a *Belloperone* flower (where the hummingbirds had fed minutes earlier) I had my chance. With a swoop of my net I had one fluttering inside against the gauze. I grabbed the moth between my thumb and forefinger, stuck the point of a tiny thermometer probe called a thermocouple into the flight muscles of its thorax, and quickly read the instrument's dial. The needle jumped to indicate a whoppingly hot body temperature of 42.5 degrees Celsius (108 degrees Fahrenheit)—and it stayed there for several seconds until the moth, now stationary, gradually started to cool.

The temperature on the dial—ten degrees Fahrenheit higher than human body temperature—startled me. Maintenance of a high body temperature was considered an advanced trait of the "higher" vertebrate animals, mammals, and birds. Others, like fish, amphibians and reptiles, were thought to be "cold-blooded" because they are unable to generate sufficient internal heat to sustain a high body temperature.

The idea that an insect could maintain a high body temperature not behaviorally, by engaging in an on-off activity like some "cold-blooded" animals, but by internal physiological mechanisms, like a bird or a mammal, during continuous and sustained activity was (when I started at UCLA) almost revolutionary. I had previously caught sphinx moths that had been hovering while feeding on flowers on warm and even hot, sticky nights. But the moth I had just now measured had flown on a cool evening when it could not have been heated by any external factors. Yet its muscle temperature was a phenomenal fifty-eight degrees Fahrenheit higher than the air temperature! The moth's body heat could have resulted only from muscle activity, and the only muscles that are available to a moth for heat production are those used to power its flight. I knew the moths flew even at high *air* temperatures, such as at least thirty degrees Celsius (eighty-six degrees Fahrenheit), and in sunshine. If under those conditions they produced the same amount of heat that the one I had just captured apparently produced, then according to my calculations (eighty-six plus fifty-eight) they should have heated up to at least 144 degrees Fahrenheit! In short, they would *fry*.

I was baffled. It seemed that since the moths are *continuous* flyers that do not regulate temperature behaviorally by stopping either to heat up (bask) or to cool down (rest), they must somehow regulate their thorax temperature using a physiological mechanism. If so, that would essentially make them *more* warm-blooded than a mammal or a bird!

I had no a clue as to what mechanism might be involved, but I felt that such a mechanism could be something new. Maybe I had finally found a thesis problem; if I could solve it, I would make a discovery. My quest would therefore be an adventure, and I resolved to do at least something every single day to solve the puzzle. I would use the common garden variety "tobacco hornworm" sphinx moth (*Manduca sexta*) whose caterpillars I had found on jimsonweed in the desert and which I was then rearing in the laboratory.

As soon as I got back to campus, I tried to duplicate in the lab what I had seen in the field, because in the lab I could control numerous variables. For example, I could eliminate the possibility that moths had, at high temperatures, simply stopped flight to prevent themselves from overheating.

I allowed the moths to shiver and warm up in Bart's temperature-controlled room, which was big enough for them to fly in. After they lifted off, I let them continue to be airborne for at least two minutes (sufficient time for their body temperatures to have reached an equilibrium). I then caught them in flight and took their body temperatures within five seconds (before they had a chance to cool or heat). I did the same with many moths and at many air temperatures, and the results were clear. The moths' thoracic muscle temperature remained at around (104 to 109 degrees Fahrenheit) over the range of air temperatures that I could produce in the room and that the moths could be induced to fly in (about 60 to 93 Fahrenheit). It appeared they were physiologically regulating their thoracic muscle temperature. The temperature of the abdomen, on the other hand, was not regulated. I next measured the moths' metabolic rate in flight (in a two-and-a-half-gallon pickle jar) over the same air temperatures and found that it stayed constant regardless of air temperature, even though the difference between muscle and air temperature varied threefold. That is, the moths were *not* exercising their muscles more at low temperatures in order to produce additional heat when flying than they were at high air temperatures.

I wondered, Could my body temperatures be wrong? I felt that my "grab and stab" technique of measuring body temperatures might be criticized for being too simpleminded and unsophisticated. After all, there had to be heat flow between my fingers and the moth when I grabbed a moth to be measured. So I designed a "flight mill": a rotating arm of thin, stiff metal syringe tubing at the end of which a moth could be tethered so that it could fly in a circular orbit (of 12 feet). Each moth flown on the mill had a thermocouple implanted in its thorax; the leads from the thermocouple were strung through the hollow syringe tubing to a sliding contact at the base of the mill, and the electrical temperature signals were then recorded at ten-second intervals on a recording device and displayed on a chart recorder. This "toy" (which the biology department's shop built to my specifications) worked beautifully! Except for one thing: it was a total surprise to me that the body temperature of the moths attached to the mill stayed, on average, a meager 20 degrees Fahrenheit above air temperature over a wide range of air temperatures. I now worried about my grab-and-stab measurement. Did this prove that the moths *didn't* regulate their body temperature after all? Or had I discovered a critical variable that would help me understand?

It turned out to be the last. Various additional tests showed that moths which were suspended by a tether, and thus "assisted" in flight, as much as halved their energy expenditure. They were working only hard enough to support themselves in the air, and the heat they generated was all a by-product of that energy expenditure to support their flight effort. That is, once in flight the moths were expending energy not to keep warm but to stay airborne. Thermoregulation in free flight therefore had to involve a physiological mechanism for regulating heat loss.

I had no clue how heat loss might be accomplished or regulated. Insects don't sweat as we do when we run; nor can they adjust their insulation by varying the thickness of their fur or feather layer, as mammals and birds do. Sir Vincent Wigglesworth, the father of insect physiology, had even published a paper in which he concluded that heat balance in insects is unaffected by evaporation and blood circulation.

For me the problem was somewhat narrowed because I had convincing proof that as long as flight was absolutely uninterrupted, there was no greater heat production at low relative to high air temperature. Therefore, the data left no other conclusion than that there had to be a heat-loss mechanism. The problem was that I had no idea what the heat-loss mechanism might be. I had clues, though. I had been taking abdominal temperatures of my flying moths and found that at high air temperatures, when the difference between thorax and abdomen temperature was the least, the difference between abdomen and air was the greatest. This observation suggested that the moths were dumping excess heat from the working (wing-moving) muscles in the thorax (the "flight motor") into the abdomen. To me it looked as though the abdomen might serve in a manner similar to the radiator of a car. This hypothesis looked even more likely in an experiment where I tethered moths on a board and focused heat (using the beam of a lamp) onto the thorax and simultaneously measured thorax and abdominal temperature. I found that moths allowed their thoracic temperature to rise, but only to about forty-two degrees Celsius (108 degrees Fahrenheit). Abdominal temperature then suddenly increased sharply, even as thoracic temperature stabilized while the heat was still being applied to the thorax only. I rubbed the fur (or scales, as they're called) off the back of the abdomen of a moth, and through the transparent cuticle (the hard external "exoskeleton") I saw that the moth's heart, in the abdomen, was pumping blood forward into the thorax just as the temperature of the abdomen started to heat up. It looked as though the heart (in the abdomen) reacted to the temperature in the thorax, but if so then the moth must have a neural sensor or sensors in the thorax, which sends nerve impulses to the heart telling it to respond by pumping blood forward. Cutting the neural connections between the thorax and the heart should abolish the communication, and the heart's forward-pumping action, and it did as I predicted.

With a sensitive motion detector and thermocouple I also simultaneously measured individual heartbeats and temperature pulses. They were correlated—one heartbeat resulted in one high-temperature pulse arriving from the thorax to the abdomen and one low-temperature pulse in the thorax. Abolishing the action of the heart while leaving the neurons intact should have likewise abolished temperature regulation (and of course also the animal's life, were it a mammal or a bird). I therefore resorted to invasive heart surgery to confirm my conclusion that the moths were using the abdomen as a radiator to prevent their thoracic muscles from overheating.

Operating on moth hearts would, I thought, lead to a big breakthrough. It would be the sweet experiment that nailed the solution to this problem on the head. I had already read all the published papers on blood circulation in insects, all the way back to William Harvey's treatise of 1628, "*Moto Cordis et Sanguinis in Animalibus.*" Harvey, the father of circulatory physiology, had proved that the blood makes a circular path through the body and that the heart is a pump that pushes the blood. He had written on the insect heart as "a throbbing vessel which is the mainstay of life."

But how can one remove the "mainstay of life" to see how it affects some *select* aspect of a very active life? It might be possible. William Harvey didn't know that insect blood, unlike the blood of vertebrate animals, is not part of the respiratory system and therefore does not function as an oxygen carrier.

I became one of the first insect heart surgeons, if not the only such surgeon. Using a curved sewing needle and a thread of my own hair, I tied off the moth's heart, making it unable to pump blood. Moths usually flew well after this operation, but at high air temperatures, where unoperated moths flew for long durations, these crashed to the floor after a minute or less of flight. Significantly, their thoracic temperatures, at a whopping 110 degrees Fahrenheit or more, were above normal in this species. The moths had therefore apparently stopped flight because of overheating, due to inoperative blood circulation. To find out if this overheating was indeed the reason for their crashing to the floor, I did a second control experiment at the same (high) air temperature: I again flew the operated moths, but this time I *also* removed their insulating "fur coat" from the thorax, to remove the necessity for active heat dissipation. The moths now flew much longer, and didn't overheat, *despite* the heart operation. Hence the heart operation, *as such*, had not been the cause of the earlier flight cessation. All the numerous independent lines of evidence were thus consistent with the same conclusion: healthy moths actively regulate their thoracic muscle temperature during continuous flight by using their blood circulation to dissipate excess heat.

I was close to finishing my research and writing a dissertation. However, there was still no guarantee in my mind that I would get a PhD. I knew that my results were original enough to satisfy the requirements for a doctorate, but there was one potential problem. Another researcher, an established biophysicist whom I'll call Dr. H., had been working on temperature regulation in the lined sphinx moth, and he had recently published a paper in the prestigious international journal *Nature*. He had come to exactly the opposite conclusion. His punch line was that the moths regulate thoracic temperature in flight by changing their rate of heat production, whereas I said that they produced the same amount of heat and got rid of the excess. I naively expected Dr. H. to be pleased by my data and what they revealed—the object of science being to further knowledge and understanding, etc.—and I excitedly wrote to him to tell him all that I had found out with Manduca. The response was not what I had expected.

He wrote back (on December 16, 1969): "Thank you for informing us of your interesting studies. I have not yet worked them over in detail but one or two quick suggestions might be helpful." He did have a couple of suggestions, but most of the two pages of single-spaced typing was devoted to criticisms. Six comments most particularly caught my eye:

1) He and two graduate students had "also been working with this animal [*Manduca sexta*] for the last five years and "we have never observed the thoracic temperature to exceed 40°C."
2) "Our animals were never 'forced' to fly."
3) "They only rarely hover spontaneously for more than a few seconds."
4) "You do not mention the time of day that you performed the experiments."
5) "The moths are sensitive to their condition; do you have sham-operated [cut, but heart *not* tied shut like the experimental animals] controls?"
6) "What the animals can do is adjust the period of active flight to gliding or the airborne to grounded period. This the animal does do, although I am not certain these activities would emerge under forced or restricted flight."

I had also written a similar letter to the coauthor of Dr. H.'s paper in *Nature*. The coauthor responded: "These are very sensitive animals and must be handled with sympathetic understanding, or they fail to perform."

The two senior researchers had given me, a mere PhD candidate, a response to frighten off a bear. They were forceful and detailed enough to perhaps make me reconsider before continuing my research. But with this project—unlike my previous project on chloroplasts—I felt strongly that I had hit a jackpot and had to continue, even though I had not expected anyone else in the world to be working on thermoregulation in the sphinx moth, *Manduca sexta*.

Perhaps to put him on notice, I wrote to Dr. H. immediately (December 23, 1969): "I am finishing up work on a comparison of free flight and flight on a round-about flight mill. I am also working on the physiology of preflight warm-up, and other aspects of the circulation."

I did not hear back from him for months, and I started to get worried. I realized that I had told Dr. H. exactly what I was doing and what I had found out. Finally it dawned on me that I might be scooped.

I wrote to Papa about my concerns, and in a letter dated January 25, 1970, he replied:

> I was rather concerned with what you wrote about your thesis and the negative response you received from one of the researchers to whom you had sent a copy. Your personal judgment about the work of that person sounds about as negative as his has apparently been about yours. You may be right and I hope and trust you are, but if so, then I imagine that you could be up for trouble. In this country, where there has, so it seems to me, never been any education on "nobility" of behavior and thought, almost everything is possible. For instance, I could well imagine that one pseudoscientist steals the findings of a graduate student, to strengthen his own glory, if the opportunity arises. And if you give your results to somebody inclined to such behavior, *before* they are printed, the opportunity is there and a good one at that. Be on your guard! If you are sure of yourself, I could think it best to have a short, very short, summary or extract of the essentials, published as soon as possible, no matter how short, ahead of the thesis. Then your priority is protected.

For once, I took Papa's advice. As he suggested, I quickly wrote an "abstract of essentials" aimed at the journal Science, summarizing my key findings. My paper came back with harsh criticisms from one reviewer, and glowing praise from another. But it was rejected. Science, like Nature, publishes only highly acclaimed papers. The editors are objective, and manuscripts with a fifty percent approval rating are not good enough.

Uncharacteristically for me, I was outraged. I wasn't going to just let this work die. I knew I had made a worthwhile discovery. I took the unprecedented step of writing a letter to the editor explaining why I thought my work was significant. To my surprise, he reversed his decision and agreed to publish the paper.

When Dr. H.'s students' papers (all of which he co-wrote with them) came out shortly thereafter, my worries about theft were over, but my disappointments were just beginning. These papers were clones of H.'s previous work. Now "they" suddenly had a total of four papers in the *Journal of Experimental Biology* claiming to prove that the moths regulate their thoracic temperature "in flight" by varying *heat production*. I had proved that the regulation of the thoracic temperature in flight was entirely due to a mechanism of *heat loss*; my specifically aimed tests showed not a single shred of evidence for any variation of heat production for temperature regulation during flight. The contrast between my conclusions and theirs could not have been more stark.

I later had discussions with one of the coauthors of the moth papers, asking him: "Do you think it is possible that your moths regulated the *amount* (not rate) of heat production precisely because they were not really in flight? That is, maybe the moths were shivering to warm up *in order to fly*." He admitted what I had suspected, that they had indeed been "less than clear to distinguish continuous activity from flight." The moths had been confined in tiny flasks where their metabolic rates could be easily and accurately measured. But it was assumed that movements of wing stubs in grounded moths (their wings had been clipped) were equivalent to flight. Moths "in flight" with clipped wings? Sometimes the details make all the difference. This detail explained *everything!*

At this point in my work I was invited by Bart to help him in the field in New Guinea catching bats—an offer I couldn't refuse. When I got there I told Bart about my exciting results back at UCLA with the moths. He was silent for what I thought was longer than usual; and then right there, while standing in the shade of a palm tree in Madang, he offered me a postdoctoral position back in his lab at UCLA the next fall. He said my work with the caterpillars was already enough for a PhD thesis. He said all I had to do was write up my water-balance data, adding my already published paper on feeding behavior. I felt he was right, and I also wanted to come back to UCLA and work with him on my unfinished study of moth shivering. So here was another offer I could not refuse, except that because I didn't want to shift the focus off the diamond. I decided to write my PhD dissertation on the moths instead, because it seemed to me that this story was clear, clean—and thus easier to write. It turned out to be short and to the point, titled, "Flight Energetics and Temperature Regulation by Blood Circulation in the Sphinx Moth, *Manduca sexta*." Unlike most PhD theses (and unlike my much lengthier master's thesis) this one was quite intelligible, and at a total of only eighty-nine pages, including references, it probably set a record for thinness for PhD dissertations. My results were so clear and unambiguous that I could write most of them in "plain" English, and I was immensely proud. But when I submitted it to the secretary for filing at the registrar's office, she thumbed through it, noting its flimsiness and spotting a number of white-outs. She seemed disgusted and declared pointedly: "The quality of theses is going down all the time." I did not have time to retype the pages; nor did I think form mattered. It was the ideas that were relevant, and of them I was confident.

As it happened, my dissertation defense to the thesis committee of five professors was less a test than the joyful formality of the hunter showing off his prize to the clan. The clan liked mine, and officially declared it "with distinction."

As the Indigo Girls sing, "Got my papers. And I was free."

Without further ado (such as walking down an aisle wearing a medieval cap and a long black gown), Kitty and I jumped into our new green Plymouth Duster; put our beautiful five-month-old baby girl, Erica, into a baby carrier on the backseat, along with our beloved year-old shepherd puppy, Foonman; and headed down the highway. On our first day out we camped on a deserted dirt road in the Mojave Desert. The next morning as we were boiling our coffee over a small campfire, I heard the begging calls of baby ravens. I found the nest nearby in a Joshua tree and could not resist climbing up to look in, and then picking up two of the six youngsters, who were just starting to get feathers. With the ravens in a cardboard box on the floor in front of us, and the rest of us packed in around them, we were soon again on our way across the desert. Later on we stopped to camp in the Sierras and Grand Teton and Yellowstone national parks, then went across the prairies and home to Maine for the summer.

The ravens slept most of the way across the continent, at least while we were driving. But stopping the car for gas was their cue to start yelling for food. Two hungry ravens make themselves heard. In those days there were service station attendants to pump gas. Every time we stopped, the attendants heard the hoarse loud cries, but when they looked in the window all they saw was little Erica in the car seat. Once we were out of sight again we opened the cardboard box and stuffed a few spoonfuls of dog food down the ravens' opened gullets. Each then squirted a huge dollop of liquid feces that I tried to intercept before it landed on the upholstery. A human baby is of course not so regular and predictable.

We got to Maine just in time, because the ravens were starting to hop about and explore in the car. They needed their freedom, and at the farm they got it. Here they romped with us through the fields along with Foonman and Hector, Mamusha's fifty-fifty shepherd-collie mix and 100 percent family member. I was surprised by how well the dogs and ravens adjusted to each other. Much better, it turned out, than Mamusha adjusted to my two black-feathered friends. And to how both Mamusha and Papa adjusted to Kitty. It appeared that Kitty could do nothing right. There was, I believe, greater justification for them to be critical of the ravens. When Mamusha or Papa went down the rows of the garden meticulously pulling out the tiny sprouting green plants, the ravens followed them and did likewise. They also landed on the clothesline, and assisted in pulling the clothes off, but not without first walking all over the clean laundry with their dirty feet. I learned later that they mysteriously disappeared the very day in August that we left to return to California to start my postdoctoral appointment in Bart's lab. I felt it wise not to ask too many questions. Mamusha volunteered that although my ravens disappeared and were never seen again, she saw "Nr. 2" (one out of my marked bumblebees, so I could see them as individuals) at Beans Corner a mile up the road, where it was "loaded with pollen *and* was collecting nectar."

My original plan that summer had been to relax, celebrate, go fishing, and maybe play with bumblebees. I had brought along tags to glue onto the bees' backs so I could identify individuals in the field. I also had the electronic thermometer for taking their body temperature. There were many bumblebees on the wildflowers in the meadow next to the house, and with each measurement I took, my excitement mounted. Otto Plath, who had written *the* book on bumblebees (*Bumblebees and Their Ways*) in 1934, had noted, "Like all cold-blooded animals, honeybees and bumblebees have no means of regulating their body temperature, and their exposure to cold invariably results in lethargy, and often death." He was wrong, very wrong, on several of those "reasonable" assertions. Even the best of logic can stand impotent before a smidgen of solid data.

I got up at dawn while there was still dew on the meadow and temperatures were barely ten degrees Celsius, and I saw a few bees even then. It was with great anticipation that I captured them and measured their body temperatures, taking many precautions to make sure that I was getting their "instant" temperatures, instead of what they might heat up to or cool down to seconds after capture. Thanks to my experiments with my moths on the flight mill, I knew that my grab-and-stab method of taking measurements was not only the easiest, but also the most reliable. No doubt about it, the bees regulated their temperature—they had close to the same thoracic temperature of thirty to thirty-five degrees Celsius (eighty-six to ninety-five degrees Fahrenheit) early in the shady dawn at ten degrees Celsius as on a hot day in the midday sun.

Like the sphinx moths, the bumblebees had abdominal temperatures that varied with external air temperature or sunshine or both. That is, like the moths, they did not regulate the temperature of their abdomen. It seemed logical that, like the sphinx moths, the bees had been heated exclusively as a by-product of their flight metabolism. However, bumblebees do not hover to feed. They land on flowers, and as they pause momentarily, they stick their tongues (proboscises) into the flowers to extract the dilute nectar. I noticed that their abdomens, especially on cold mornings, pumped rapidly while their wings held still. Since rapid abdominal pumping movements indicate "breathing" to ventilate highly active thoracic muscles, this suggested that heat production (i.e., exercise) of the flight muscles was *continuing* after the bees had *stopped* flying. But how to prove it in the field if I could see no external trace of shivering? To find out if a live bee replaced the heat it lost in order to keep itself warmed up, I needed to measure body temperature of a bee that *didn't* shiver. A dead one! Heated (with Kitty's hair dryer) to the same temperature as a live one, the bee cooled at fifteen degrees Celsius per minute in an air temperature of ten degrees Celsius. From cooling rate and body mass I could then calculate the rate of heat production of a live bee experiencing the same rate of heat loss (and the same rate of heat production to oppose it to maintain a steady body temperature).

I don't think Papa was interested in watching me blowing bees with a hair dryer. He asked no questions. He spent the days at his desk by the window facing the meadows and the old maple trees. I suspect he would have liked to be outside with insect net in hand, and me alongside, catching ichneumon wasps with him as I had always done. But I now had no time for that. I had to stand with a stopwatch and time bumblebees' flight durations and stops at the flowers. I was trying to find out if the bees stopped longer because they needed more time to shiver to produce more heat, to warm up for the next one or two seconds of flight. And then I had to grab them and stab them to take their temperatures to find out if they were shoring up their body temperature during their brief stops at flowers. But the stops were normally too brief—I needed the bee foragers to stop much longer than they normally did, to see if they would maintain their body temperature or mimic my dead bee's temperature response and cool down passively. I could do that with sugar rewards.

Normally when a bee visits a blossom (of fireweed) it can swipe the nectar reward with a few quick licks of the tongue and then is off in a second or so of flight to the next flower and another second or so of stopping. So I put sugar syrup droplets into fireweed flowers. Then when a bee stopped at one of these enriched flowers it took sometimes a minute or more to swipe the flower clean. Now I could determine how much heat loss each bee *should* experience as it paused at a flower for a given duration at a given air temperature. From that data and the bee's weight I could then calculate its rate of heat production when alive.

Success. The live bees did not cool at all—if anything, they became hotter after pausing than they were immediately after they had landed. Here I had proof that the key to thermoregulation for foraging bumblebees was heat production (to maintain close to a minimum muscle temperature for flight) after landing (while in the hovering sphinx moths the key had been heat loss during flight to maintain the muscle temperature and prevent it from exceeding the maximum tolerable).

It was exciting to learn new things about the physiology of bees free in the field, and I quickly wrote a third paper for *Science*, where it might be read by hundreds of people. This one was immediately accepted. Meanwhile, Papa continued to labor on his manuscript about the Oriental ichneumons, which he'd been working on for decades. He was experiencing tremendous difficulties in trying to get it published, and even if it was published it would be read only by another entomologist working with ichneumons from that region of the world. I suspect Papa was dismissive of what I was doing because he did not need to know any physiology for his taxonomy, but I needed to know insect and plant taxonomy for my work. Each of us concentrated on his own passion.

At first I measured only the temperature of workers of the same species of bumblebees (*Bombus vagans*), on the same species of flowers (fireweed, *Epilobium angustifolium*), but at different air temperature and in the shade (to avoid complications of heating by sunshine). Later, after Phil Potter took me up-country on a fishing trip, I found that my routine scientific protocol was more justified than I had imagined.

Phil and I went way up north into the practically endless woods near Kokadjo. We stopped in a small clearing in the woods about forty miles out on a dirt road to the west branch of the Penobscot River. After we got out of the car and stretched, I noticed some goldenrod in bloom nearby, and it was covered with bumblebees. They were *Bombus terricola*, and they were all slow in their movements on the large inflorescences (many small flowers massed into one larger unit), which have hundreds of tiny florets each. Almost all of these actively forging bees were unable to take off in flight—their muscle temperatures were too low, even though the air temperature was moderate. These data were exciting because they indicated that something unexpected was going on. I had stumbled into yet another puzzle!

Uncharacteristically, Phil, usually so full of energy and cheerful impatience with my naturalist sidelines, seemed tired on that trip. Weirder still, he was content to hang out while I spent a couple of hours in Kokadjo totally absorbed with the bumblebees. The dramatic difference in body temperature compared with my previous results turned out to be related to different types of flowers. The species of flower, as such, didn't matter—I found the same body temperatures and bee behavior later on spirea flowers, which have a low nectar content and architecture similar to the goldenrod flowers. Back home at the farm the bees had much choice, and they did not bother visiting such low-reward flowers. Here in the north woods there was little else available, and they had to visit them or starve. But if they had been expending as much energy per unit time as the bees I examined previously, they would soon have been dead, owing to energy depletion. Yet they were *not* depleted—I pulled the abdomens off some bees and found that many had collected a stomach full of nectar. Here in the north woods they had reduced their energy expenditure by reducing their energy output for heat production. In this way, they managed to exceed the profit margin despite the low resources. I had found evidence for bees adjusting their profit-loss balance sheet in order to continue making a profit!

As fall approached I was already thinking of starting my postdoc with Bart. I was so excited by my new story with the bees and I felt it was so important that I had all my data and paperwork photocopied and sent ahead, in case we got into a car accident along the way. Fortunately I was not, like Papa, saving my data from invading armies, and I would not have to resort to soldering my papers into metal boxes and burying them in a swamp. Now, as he worked at his desk, engaged in the mind-numbing hours of pinning and spreading his wasps, I occasionally sat across from him, to try to chat. He talked of his ichneumons, of "tradition," and of his collections that were lost in Poland. He seemed to have almost nothing to show for his years on expeditions. And there was no future in sight. I began to understand how he must have felt to have his important life-work threatened. Meanwhile, I seemed to have lucked into scientific riches and my career was practically meteoric, despite my not really caring about anything but what to him seemed like having a trivial good time. He was not far wrong.

I had barely begun my postdoc with Bart at UCLA, when the insect physiologist Franz Engelmann, who had been the cochairman of my PhD thesis committee, declared, "There is a job for an insect physiologist in the entomology department at UC Berkeley. You should get it."

At first I resisted. "No, I want to go back to the University of Maine," I told him. Late that summer, when it was clear that Phil Potter had cancer, and shortly before his death as we propped him up at his daughter and son-in-law's camp, facing Sand Pond, I told him I would be leaving soon to go back to California. He looked at me and said, "Ben, you belong in Maine." On the other hand, Franz said that I'd be "crazy" if I did not apply at Berkeley, because it had the best entomology department in the country.

I applied and was invited to come for an interview and to give a presentation. I went to Berkeley on September 21, 1970, and enjoyed myself, telling the professors and graduate students there the story of the moths and garnishing it with things to come from the bumblebees. The position, which was offered to me almost immediately, was indeed a dream job: it came with a research stipend, promised equipment, and only one course—insect physiology—to teach per year. Furthermore, I would be allowed a one-year reprieve to get the research lab up and running before starting to teach. This was critically important because I had more data from my field research on bumblebees in Maine than I knew what to do with. I would have time to step back, mull it over, and read broadly to synthesize the larger picture.

I read up on pollination, because I had discovered that the bumblebees fine-tuned their foraging behavior, and their associated thermoregulatory physiology, with respect to the flower architecture (they could cool down and walk on large inflorescenses but not on single isolated florets) and the food rewards they found in flowers. It seemed to me that from an evolutionary perspective it was the plants (through their flowers) that were manipulating the bees' behavior, even while the bees were physically manipulating the flowers for their own benefit. This was new and exciting stuff for me. It was as if lids had been removed from my eyes, and suddenly I saw both the bees and "their" plants in a new light. I could do empirical studies on the bees' behavior and physiology, and at the same time frame it all within the larger picture of plant ecology and evolution. I felt that what I was seeing was so bright and luminous that everyone else would recognize it as well. I had no time to lose to stake my claim.

In a grant proposal that I made to the National Science Foundation a little later, I didn't need more than about six pages to explain what I wanted to do and to ask for $20,000. It was funded. It was also chosen by Senator William Proxmire of Wisconsin for his infamous (to us entomologists) "Golden Fleece Award," to publicize how science was "squandering the taxpayers' money." In an article in the *National Enquirer* Proxmire was quoted as saying that I was given money "to find out how bees make honey." At Stanford University, meanwhile, the winning entry of a trivia contest was "the rectal temperature of a bumblebee." I believe I was the only person at the time measuring temperatures in that general area.

Proxmire's vigilance to protect the public from scientists like me luckily didn't stop the National Science Foundation, National Institutes of Health, or National Aeronautics and Space Administration from spending on us. Neil Armstrong had landed on the moon, finding that its surface is barren dust and rocks under a dark sky; and that project had cost a lot more than my bumblebees. Soon NASA spent billions to land Viking 1 on Mars, finding that the surface there was "barren sand and rocks under a reddish salmon sky." Spending a pittance to reveal some of the mysteries of little-known life on earth is also worthwhile.

Although I at first had no interest in the abdominal temperature of any insect, the newly discovered mechanism in the moths had suddenly given this body region interest, if not significance. It almost immediately provided insight into the function of a curious series of loops in the petiole (the narrow "waist") of honeybees, about which all sorts of weird speculations abounded. A few measurements soon convinced me that those loops function as a countercurrent heat exchanger that helps retain heat in the thorax so that the bees can fly at low air temperature. (To prevent overheating, honeybees regurgitate nectar and water, rub the liquid over their heads, and cool by evaporative cooling.) Examining the anatomy of bumblebees, I was struck by the fact that they *lacked* this anatomical arrangement and soon tried to understand why. Bumblebee nests are started by individuals—"queens"—who *incubate* their brood (eggs, larvae, pupae) by applying the abdomen to it and heating it much as a bird heats its eggs, by direct contact. In order to do that, the bees must be able to shunt heat from where it is produced—the thoracic flight muscles—into the abdomen, from where it flows into the brood. But bumblebees also need to greatly *reduce* heat leakage into the abdomen so that they can forage out in the field on cool days, which requires that they be able to keep up a high muscle temperature in order to fly. How could they do both, since the processes seemed to be mutually exclusive?

Here was another worthy scientific question. To me it was, I suspect, like Papa's search for the rare forest rail. I knew what had to exist, but I was not sure what it would look like. It had to be in one thicket or another, and so I fooled around, getting all sorts of anatomical and physiological data to get a feel for where it might lie, to be able to narrow my search for what I knew would be a thing of beauty.

Timewise, science is 99.9 percent dog work. But the 0.1 percent inspiration, like the final shot that bags the rail, is at least as essential as the boring, repetitive, mechanical steps required for getting there. Those 0.1 percent moments make all the tedious hours worthwhile. I remember once drinking coffee at the Three C's Café near my lab at Berkeley, just half-listening to the background music and chatter, going over and over in my mind the graphs and diagrams from my months of collecting data on another puzzling phenomenon—bumblebees increasing the temperature of their *abdomen*, when they could have produced the heat only in their thorax, yet the anatomy between thorax and abdomen was such that there should be a countercurrent heat exchanger that would *prevent* or at least greatly reduce heat flow to the abdomen. When two "truths" collide as opposites, as occurs almost always in reality, then the answer almost always depends on seemingly insignificant trifles, which ultimately make all the difference. Doodling with a pencil on scratch paper—a glimmer, something taking shape, a glow, and a flame . . . eureka! The pieces fell into place: a model I called the "alternating current" heat exchange. Eventually it would be prepared like the bagged rail, stuffed into a presentable scientific product, in this case a paper for the *Journal of Experimental Biology*. Instead of going to a vault deep in a museum where only researchers would ever get to see it, this product would similarly be entombed in scientific libraries where it would be "forever" available for all to examine, to challenge, and to modify or expand, yet just as unlikely to ever be read by anyone other than a very rare specialist.

IT HAD IMPRESSED ME GREATLY THAT THE ANswers to the puzzles in moths, honeybees, and bumblebees lay in minute and precise details of anatomy in conjunction with precise physiological mechanisms (i.e., organ behavior). All the details were functional. There was nothing "extra," and all of what one saw had meaning. If anything seemed extra, then it generally turned out to be harboring something exciting that was still unknown. I think it was with these thoughts, and a few cups of strong coffee, that I was also induced to daydream about the wild nature that Kitty and I had admired on our trips back and forth across the continent, between California and Maine.

The most spectacular sights to me were always the vistas that included the diversity of colorful flowers—the amazing flora of the deserts, the high alpine meadows of the Rockies, the remnants of prairies, and the meadows and bog near home in Maine where I studied the bumblebees. Given the almost incalculable precision of the physiological mechanisms of the bees' apparatus for temperature regulation, I could not help being impressed with what might otherwise seem as trivial as the rectal temperature of a bug: the diversity of flowers in any one habitat. For the "physiology" of pollination, wasn't there *one* form, color, or scent of flowers that was "best" for bumblebees? And if so, would it not then replace all the other flowers so that it would be the same on all plants? Why hasn't one ideal type evolved, with the brightest color and the simplest, most easily accessible nectar?

In habitats where bumblebees are the major pollinators there is instead a huge *diversity of* flowers. Something weird seemed to be going on. Some of the flowers used exclusively by bumblebees are designed so they also exclude these bees. The orange jewelweed blossoms have their nectar tucked into a tiny hidden spur. The white turtlehead flowers can be entered only by a slow process of forcibly prying apart a "mouth." The pink cranberry blossoms have to be vibrated by the bees before they can collect the pollen. The blue monkshood flowers provide a huge amount of nectar, but it is far away from the flower center and hidden in the tip of a specially adapted petal.

As my daydream of the bees and the flowers unfolded I saw not only individual flowers, but meadows of them in their full glory of stunning variety splashed across gorgeous natural wildernesses. I also saw my observations of how the bees were foraging in response to the different food rewards the flowers offered. I started to formulate a simple hypothesis regarding how the flower diversity must have evolved. According to this hypothesis the diversity is not an extra—it is instead a necessity that has arisen from the competition of flowers for bees, and the competition among bees for flower rewards.

I started writing to organize my thoughts. The more I thought, the more holes were revealed in my information. So more reading was required, and more research as well. Helen Jackson, our department secretary (secretaries did typing in those days) retyped my manuscript so many times that I grew embarrassed to hand it to her again and again. After nine drafts I was still by no means satisfied. There was just too much to consider.

Finally, however, I honed it down to what I thought was the essence of the idea. As with my PhD dissertation, I had done everything possible to trim the fat and make the core argument lean and clear. The main idea was that energy economics is a key factor—the flowers must provide enough food to "pay" the pollinator, but not too much, or else the bee will visit only a few flowers and not transfer pollen to many. Since the pollen the plant receives and passes on needs to be of the same species and not that of another concurrently blooming species, the flowers of different concurrently blooming plant species need to be different from each other, or else the foragers would visit them indiscriminately. However, foragers should try to maximize profit and switch to more high-reward flowers, when they can. But if everyone has high rewards, then all lose. Blooming when others are not in flower would avoid the competition for pollinators, and blooming synchronously would reduce the amount of rewards required (because flight times and foraging costs would be reduced). On the other hand, given the great number of plant species, overlap in blooming times is almost unavoidable, but variety of flowers at any one time would then also promote flower-constancy and hence pollination efficiency. And it is less important that particular kinds of flowers are visited by specific species of pollinators than that the *individual* bees are flower-constant. Here then was an ecological problem that could be solved only by studying the behavior of individuals.

Papa cared nothing about individual insects. His mind set was on "the species." His eureka moments involved discovering new ones; and except for the collecting, his work was drudgery whereas mine was purest joy. Coming so close on the heels of the discovery about thermoregulation, this insight—the energy key that had predictive power in animal-plant coevolution—felt to me like what bagging the Snoring Rail, *Aramidopsis*, may have been to Papa. I doubt, though, that either of us saw it this way at the time.

Around this time I was starting to have some conflicts with Papa that stemmed from our different perceptions of and approaches to science. Papa was of the "old" school. Despite his enterprise, his dedication, and the difficulties that he fought to overcome, he was still ignored if not derided and dismissed by some who saw little merit in his kind of specimen-collection science. Seeing me as a "modern" biologist, he may have thought that I felt as they did. I did know what they thought, and in that he was correct. He still felt the pain of having been relegated to the status of a "salaried drudge" when Professor Hall at the University of Kansas forbade him to publish his findings on gopher speciation. Now, in conversations with me, he talked of the "stuffed shirts" in academe, who were "trying to make hay with big ideas," while he was describing new species in a very difficult and little-known group that he alone had mastered.

Papa's low opinion of my colleagues made me mad. After all, I valued good "big ideas." My academic colleagues were heroes that I looked up to. In Dick Cook I saw the Professor Gottlieb that Sinclair Lewis raised almost to sainthood in his novel *Arrowsmith*. I thought Papa was overreaching in his criticisms because he was so bitter about not getting the recognition he thought he deserved. Although I did not think of Papa as less worthy than Dick at Maine or Bart at UCLA or any of my colleagues, I knew that his kind of work was indeed not wildly or widely appreciated by many of those (there are prominent exceptions now) whose work was "idea-driven"—as mine was. I thought I now had one of those "big ideas," and I was surprised and hurt by Papa's lack of interest in both the science of it and my success. It was especially painful to me (and maybe to him) when he would ask me to accompany him on collecting trips. I would almost always have "excuses" to stay home and watch the bees in our meadow. He would then lean on Dieter Radtke, who would come up from Florida.

The story of bumblebee thermoregulation and flower ecology was ultimately written in discrete little pieces for academic peers in specialty journals. These papers were necessarily esoteric and uninteresting to Papa (who as far as I know never read one, although for awhile I did send him reprints), and to any of my friends. It would have been hopeless to explain the excitement unless the theoretical concepts were first in place. I could never hope to provide that in a casual conversation. But I hoped that at least somebody might someday see. I thought it a shame that the whole picture was not available in one piece, since it would be very unlikely that somebody reading about the "rectal" temperature or the blood circulation would also read about why there are so many different kinds of flowers in a meadow and when they bloom, etc. I therefore decided to write a popular account for a general scientific audience, in *Scientific American* magazine. The magazine accepted it and even had an artist paint a cover picture of a bumblebee for the issue in which it appeared.

Herbert Mitgang, editor of *The New York Times* op-ed page, wrote to me in April 1973, "I read your fascinating article on the energetics of the bumblebee in *Scientific American*. There are so many things in it about energy, survival, ecology and balance of nature that I wonder whether you could do a less technical and more simplified essay for *The Times* Op-Ed page on the same subject." He wanted it in ten days.

Thirty-one weeks later I still hadn't lifted a pencil—I was still too busy chasing bees. During the summers in Maine I eventually captured and marked 940 individual bumblebees with little numbered and differently colored tags so that I could identify each bee individually. Then I followed as many of them as I could in the fields of our farm. There were several that I ended up seeing every day for up to thirty-two days, and I followed many individuals of various species for hours at a time, and then again the next day for hours. For example, one *Bombus fervidus* worker that I followed on one foraging trip for 122 minutes visited in that time 454 goldenrod inflorescences, 329 of aster, and 16 jewelweed flowers. It followed a regular trapline, but after I put some syrup in some aster flowers on its itinerary, it switched to aster, visiting 673 aster and only 99 goldenrod on its next foraging trip (of 110 minutes).

The real value of the study depended on making continuous observations to find out how the bees changed their behavior and when or why and to what they reacted. To keep track of a bee required not taking my eyes off it for even a second; I also took nectar samples of all the available populations of flowers, and learned which flowers bees would specialize in and why. To test my hypotheses that foraging paths and flower-handling skills are learned, I built a screened-in enclosure; marked all the individuals from a colony I put inside; then let one bee out of the nest at a time and followed it until it came back inside the nest; then let out another, etc.—all day long, day after day. I learned how bees sample a variety of flowers on their life's *first* trips out of the hive, and then they specialize in "majors" and "minors" much as students do in college, or as people do in choosing a profession. I was electrified by my results and felt they were so interesting that they could indeed be read by nonspecialists. So instead of writing just one op-ed piece for Mitgang, I eventually ended up writing three.

In one titled "What Bees and Flowers Know," I took the undoubtedly unpopular position that we were using too much energy. Did more make us happier? What is the limit when we occupy every square inch of soil? If we build up our population on the energy base of coal and oil or nuclear power, we would be likely to end up exceeding our ecological bounds and destroying our natural environment; and to think only in terms of decades or centuries is shortsighted. Paul Landon, editor of a journal I had never heard of called *Business and Society Review*, read the op-eds and then wrote to me—"I was very impressed by your article in the 21 January 1974 NYT Op-Ed 'What Bees and Flowers Know' "—and asked if I would write an article titled "The Limits of Adam Smith's Theory of Greed" for his journal.

Adam who? I was one of the last persons in the world to have any concern for or concept of economics and had never read anything about it. So I went to the library and was very surprised that this Adam Smith was none other than the "father of capitalism" who had published a very famous 1,097-page book (in 1776) titled *Inquiry into the Nature and Causes of the Wealth of Nations* (or *Wealth of Nations*, for short).

When I read this book I was surprised and pleased to discover that Smith's description of the labor of people and the economic organization of society has some uncanny resemblances to what I had been learning about bumblebees. To the bee, time is honey. It is their work distilled from flowers, and in the hive the honey is "exchanged" to allow different skill specializations of different individuals. Some of the analogies of humans specializing for a diversity of jobs and bumblebees specializing for a diversity of flowers that resulted in energy economy and the common good were apt, and striking. Yet what Smith was describing also included vast differences from what people now refer to as "capitalism." In Smith's model of capitalism, as in the bees' economy, *every* individual has an equal and fair chance to compete; everybody gets the same pay for the same amount of product resulting from his or her labor; and, most important, no costs of the competition can be foisted onto the shoulders of a competitor or a spectator. The last item seemed to me to be one of Smith's major points, because without it, capitalism would be like having a marathon where the competitors try to get ahead of each other not by their merits but by their demerits: strewing banana peels, withholding food and water, etc. To protect the common good would require restraints, and these could come only from government to channel the competition for equal opportunity. Competition is a great thing, but it has to be fair. Stealing natural resources that belong to everyone clearly cannot be allowed.

When my article came out, the editor had changed the title from "The Limits of Adam Smith's Theory of Greed" to "The Invisible Hand Loses Its Grip." In order to make it even more sensational, it also contained a full-page cartoon that turned my message on its head. This showed a caricature of Adam Smith with tattered coat standing in the rain selling pencils. A couple of months later the July 14, 1975, issue of *Time* magazine featured Adam Smith on its cover with the message, "Don't count me out, folks!" and the bold title, "Can Capitalism Survive?"

In my "capitalist bee" theory the individuals compete against each other in their foraging for nectar (and pollen). Specialization improves efficiency, which feeds into the colony's economy by providing it with more honey. No bee ever interferes with or takes anything from any other bee, except that to which all have equal access. No bee ever spreads poison or attacks any other. Since I liked the analogies with "real" capitalism, I wrote *Bumblebee Economics* to try to tie together a rather vast amount of dense technical detail into a simplified nontechnical story. This book, my first, was and is strictly about bumblebees. But if one wanted to read something into it, well, then it showed what capitalism is supposed to be: an environment improved, not degraded, by competition. My model remained the ecologist's model of community with interdependence of organisms and mutual limits on growth.

In 1977, I sent my book manuscript to several publishers. They returned it, saying that it was "too scientific." I would not have wanted to make it less "scientific" for any price or prize, because like Papa's work on the ichneumons, which was then still in progress, it was about the animals that I loved.

I had almost forgotten about publishing *Bumblebee Economics* when I happened to meet Harry Foster, then the science editor at Harvard University Press. Foster asked to see it. And he took it on.

The book didn't receive any prizes, but it was nominated for the American Book Award in 1979 and again in 1982, and it has had a long shelf life. Harvard University Press reissued it in 2005, and I included a new preface to summarize and update the research. This research happened to be primarily on the "rectal," i.e., abdominal, temperature of arctic bumblebees (*Bombus polaris*), who "incubate" their eggs even before laying them, thereby getting a jump on colony development. They are able to do this trick (of heating their abdomen) using the intricate physiological mechanisms of alternating current flow that I had worked so long to unravel.

twenty-two   Home to Natural History

It were as well to be educated in the shadow of a
mountain as in more classical shades.

—HENRY DAVID THOREAU

We are never tired, so long as we can see far
enough.

—RALPH WALDO EMERSON

AS BOROWKE HAD DONE FOR PAPA, OUR FARM IN
Maine anchored me, even during the fourteen years I spent in California.
I longed to go home at every opportunity, and imagined that someday I
would return for good. Yet Papa had lost Borowke, and I would end up
unable to return to the old Dennyson place.

Young animals face a difficult situation when they reach the threshold of adulthood and are about to leave the nest. They have grown up
amid suitable conditions (or else they would not have been successfully
reared); they were for a long time provided with adequate space and resources to live. But at the cusp of adulthood both food and other resources
usually become an issue because there is or will soon not be enough to go
around. If the grown young stay home after "fledging" they may come
into conflict with their parents, with each other or both. Dispersal is a
common option. But how is it achieved? Escalating intolerance is one

mechanism that helps spread apart, just as love is a physiological mechanism that helps bring together.

Neither party needs to know why it is evolutionarily programmed to love, hate, or desire freedom. Emotions are means that serve generally unconscious, adaptive goals, and they are fairly impervious to logic. Adolescent ravens, for example, experience wanderlust a few months after fledging, but if they don't voluntarily leave the home territory, their parents become less and less supportive, then become intolerant, and finally become aggressive in their attempts to drive the young out. The parents need the territory and its resources for themselves and to raise the next batch of young.

Many young birds wander for a few years before reproduction, yet they are eventually drawn back home where they grew up, to the world they were imprinted on and to which they are adapted. If they have been away long enough, they may find their parents' "homestead" empty. If not, they search for an opening nearby.

The other common "strategy" that many young animals (especially some crows and jays) use is for some individuals to stay with the parents and make themselves useful. These "helpers at the nest" are tolerated, and may later be rewarded by inheriting the territory. In the meantime, they must remain subservient.

Whenever I came back home from California, after the first hello I was off over the fields and into the woods. In the summers when Kitty and I came to visit, I could be out of everyone's hair by spending all day out in our fields with "my" bumblebees. I sometimes stayed with them all day, and Kitty came out at noon to bring me a sandwich. I could eat, but I had few opportunities to take my eyes off a bee I was following. There is no stopping in the middle of an experiment, or else the previous work is null and to be redone. Kitty was patient, and she had to do the hard part. She stayed in and around the house while caring for our baby daughter, Erica. Tensions with the "territory holders" developed, and she soon felt mercilessly attacked by Mamusha for what seemed like trifles but what to Mamusha's mind were undoubtedly large offenses. Mamusha brought Kitty to tears on several occasions. Papa wrote to me (on November 20, 1972):

If one person comes as a visitor and stands as a member of the household for a couple of weeks, this can be pleasant and practically bearable. But if an entire family with dog, child, and ravens moves in for a symbiosis of several weeks, or perhaps a couple of months, this results in a terrible stress.

To make matters worse, I often did not comply with Papa's wishes to accompany him on ichneumon-collecting expeditions. In short, I was not a good "helper at the nest," not even when it extended to the out of doors, to our meadow where I worked.

Our meadow was like a flower garden. It was where I first studied my bees' foraging behavior (which I later tied in with a nearby study site—a pristine bog), and Papa would go regularly to catch ichneumons. From spring until fall this unkempt field at the old Dennyson place supported a colorful and varied progression of plants that feed a vast assemblage of bees and other insects. These plants include red and yellow hawkweeds, blue self-heal, pale yellow potentilla, red and white clovers, bedstraw, purple New England asters and other asters, several goldenrods, milkweed, and many grasses. All formed a mat over the soil, and through it poked spirea, willow, and a sprinkling of aspiring bushes that would become trees. The return of the forest was inevitable. When a tree begins to shade the ground, it weakens the hold of the grasses and forbs. Tree seedlings can then gain a foothold, and even more shade and more trees can grow. Soon the meadow is gone and the forest is back.

We had lots of forest, but only one wild meadow; so, hoping to please Papa (and also to maintain my bumblebee foraging grounds), I hired a "bush hog" (a tractor-powered mechanical brush-cutter) to clear the rapidly encroaching trees. When Papa saw what I had done he was furious at me for "disturbing nature." My next, less intrusive attempt to help was also unsuccessful. In a long letter to Papa (January 8, 1975) I acknowledged solidarity, promised to catch him more ichneumon wasps, and offered to provide financial support:

> Dear Papa, I have to admit I'm glad to hear that you think I'm a comfort to you for carrying on the tradition in biological sciences. I am glad that I'm carrying on a tradition, rather than standing in the same place. Together we have come far. You were in the original tide of exploration of the new and distant lands, and then you made a worldwide reputation in taxonomy of a difficult group. For a while

we drifted on the same wave. The wave has gone on, and I guess I'm lucky to be on the crest of it, still. Traveling to "far and distant lands"—I've probably already passed *through* physiology, and am now at the borders of behavior and ecology and evolution. From such wide wanderings one gets, perhaps, an *intellectually* detached viewpoint—in the sense that it is free and not committed here or there to theory, idea, viewpoint, or approach.

Such free play, however, is in sharp contrast to my emotional, biased attachment. Perhaps it is necessary to have such "roots" in order to "sway." You said I was "born to carry a tradition—for a king-dom (you) called "home." I feel very strongly a part of that tradition. For me it begins in Wilton, Maine, where you and Mamusha started with a crosscut saw one winter in that beautiful hardwood forest back of the brook. There I shot my first deer, found the nests of broad-winged hawks, nuthatches, ovenbirds, and various warblers. I know "every" tree of large size, and where the big hawks have nested for many years back. This is my home. I could begin to purchase a similar lot at any time. But it would have no "meaning" to me.

You say your social security pays only half of your and Mamusha's expenses. I can well believe that. But for God's sake, don't *ever* sell a piece of the land—*never!* The farm has to be kept intact. If there is anything I can do to prevent that [breakup] it would preserve my piece of mind—and if I can give you some money for that end, I would consider it well spent. I think it is very important that you do *not* draw on your savings.

This offer was not accepted. I don't know why, exactly, but the cor-respondence suggests that there were conflicts between Papa and Mamu-sha. In any case it became clear to me that, like a young raven who has wandered widely and wants to come back home, I could not settle back in my first real home.

Kitty was not from Maine, but she knew that I felt strongly bound to Maine and to the farm, and she was more than tolerant when I suggested we purchase a piece of land at the foot of nearby Mount Blue. It had once been farmed and had been a sheep pasture, but the trees were growing back. The property was crisscrossed by long stone walls, and I found two rock-lined wells that had once served the house and the barn, respectively, of a homestead that had burned down in the 1930s. A little one-room tarpaper-covered shack (with a weathered sign over the door designating it as "Kamp Kaflunk") was the only structure on the 300 acres of land. The shack left much to be desired, even after we removed the porcupine feces. The trees surrounding it, though, were grand. There were stands of spruce, fir, beech, and maple; a few scattered great white pines; gnarled old sugar maples along the stone walls; and thickets of young pine, ash, and maple shooting up in the abandoned fields, pastures, and apple orchard. The hill overlooked a forest that extended down across a brook. And up toward the nearby spruce-topped Mount Blue and Bald Mountains is the home of kinglets and Swainson's thrushes.

These were the same woods, mountains, and swamps that was the magnet that drew me with my two friends when we ran away from the Good Will School. Phil Potter and I had hunted deer in this area. The Adamses' camp had been nearby, and our family had made memorable outings in this territory during the fall of our first year, picking blueberries. I had worked one summer at Camp Kawanhee at the edge of nearby Webb Lake, and I had many fond memories of square dancing in the nearby village of Weld. Papa had often taken me to collect ichneumon wasps here, and he described one new species I had discovered here as *Platylabus berndi*. If I could not be on my family farm, then I would make this the "home of my soul." I would plant connections—a sprout of black locust that came from a tree Phil Potter and I brought back from one of our fishing trips when we stopped at a farm where he had worked as a boy, specific oak trees for my kids, and some American chestnut trees of our ancestral American forests. Eventually also a colony of foxglove plants with pink flowers that the bumblebees love foraging on, from seeds that Ernst Mayr gave me; he had received them from of the garden of the Dutch behavioral biologist Niko Tinbergen.

Three years after purchasing "the Hill" (in 1977) I took the leap and resigned my position as professor of entomology at the University of California to take a new position in the zoology department at the University of Vermont, to be close to the north woods of home at last. I thought seriously about making a homestead and living there permanently. It didn't work out, though, and that was my fault. Kitty wanted to pursue a career in psychological counseling, and told me she would not live up on a hill in the Maine woods. I knew I'd have to go back there at some point. Kitty and I had walked down the aisle together to the haunting music of Pachelbel's Canon. We had camped in the Anza Borrego desert. Still, I wondered how she would handle walking through the deep snow to the outhouse in the winter and pulling weeds in the garden to the whine of the blackflies and mosquitos. I realized it would not work out, and said so. Telling Erica was the hardest part. She cried. In shock, Kitty almost immediately took Erica and Josh, our new dog, and they caught the train back to California. It felt like a double death.

When I remarried a year later, my new bride—Maggie—and I first lived in Burlington, Vermont, but we spent our first summer on the Hill. Maggie was a zoology student who had become disillusioned with academe, and she totally bought into my plan. She drew diagrams for building an ample (two-story) log cabin. Since I had already once started one in the woods at Good Will and had seen log buildings in the logging camps in northern Maine where I had worked as a student, her independently making these plans convinced me that we were soul mates. We started clearing a field and making a maple-sugaring grove. Maggie talked of buying a bulldozer, and this too made me think she was really serious. We never got the bulldozer, but we did rent a couple of oxen for hauling logs out of the woods, as I got busy with my ax, chopping down trees. We were well on our way to building a homestead.

Research has always been a big part of my life, but after fifteen years of deciphering delicate physiological mechanisms in the laboratory, I was starting to change directions. I was coming back to my roots in natural history. Maggie and I wanted to use our new home as our private "research station." It would become my new Hahnheide, and I would explore nature once again, but this time from a scientific perspective.

Our first summer started with dreams, hopes, optimism, and lots of activity, and one of my increasing joys was working with like-minded colleagues. In a previous summer I had helped teach a "field ecology" course with faculty friends from the University of Minnesota, at its field station at Lake Itasca. It was, as they say, "a blast," and a very productive one at that. I became fast friends with the graduate students in the course, especially with Dan Vogt, a fellow Mainer from the nearby town of Bethel. Dan was also a runner and a strong canoe paddler, and we collaborated on a study of the curious aggregation behavior of whirligig (gyrinid) water beetles on Lake Itasca. Thanks to expert and extensive canoe paddling, we captured many hundreds of these fast surface-skimming beetles and marked them with white paint spots, and then traced them all over the huge lake. We eventually determined that the beetles aggregated for safety (from fish predators) in the daytime, and often swam for miles on the water surface while foraging at night, to aggregate again by morning.

Maggie and I decided to take with us to the Hill about 100 ant lions (the predatory larvae of a primitive flying insect that dig pits to trap ants), to continue our then ongoing study of ant lions. Taking them with us to the Hill "field research station" was easy. But taking along sufficient habitat for them was more challenging. Their larvae live in loose sand, and there was no sand on the hill. We collected about fifty pails of sand at the beach of Lake Champlain, where the ant lions had come from, and we lugged it all up by hand along the half mile of steep trail through the woods to our cabin, so we could study the ant lions throughout the summer.

The ant lions were insurance. I had originally planned to work mostly with bumblebees. So, in addition to the sand for the ant lions, we also had to lug up fifty bulky cages for queen bumblebees to build their nests in. As is usually the case in the field, it was fortunate that we had started several projects at the same time, because even though Maggie spent about three hours each day lovingly feeding and fussing over our fifty queen bees that we had captured in the field, they refused to build colonies in our boxes. Big red ants destroyed the only two bee colonies that we did manage to get started, and I became fascinated by their slave-making activities on another species of ant.

At that time, for reasons that seem obscure to me now (except that I was forty-one years old and for the first time felt the clock ticking), I started to run once a day, as a diversion to clear my mind and make me ready or even content to sit down and relax at times. Papa was always ready to go to Burma, or America, with his insect net and his thirty-two rattraps; similarly, for me running is intense life stripped to bare essentials. All one needs is a pair of shorts and sneakers. Even less clothing will do, if you dare.

And so, once a day I dedicated two to three of the fourteen hours of daylight to a new project: getting prepared to try an ultramarathon. After the daily training jog (and sometimes run), drenched in sweat, I was usually enveloped in clouds of blackflies, mosquitoes, and sometimes deerflies, and I unwound by taking a dip in our brook. As if that were not enough exercise, I also chopped down trees for the log cabin we planned to build and to make our clearing. We needed to weed out the rapidly growing young pines, birches, ashes, and scrubby feral pear and apple trees in a maple-sugaring grove that we were nurturing.

On May 16 I wrote in my diary: "Before supper we worked on [clearing] the old field again. As usual I cut around seventy-five trees, and Maggie kept the bush cutters in full operation. We uncovered a new stone wall. We can now see a field coming into being. Perhaps part of this will also be a good site for a small orchard. Thinking about the future is now giving meaning to the present. A purpose brings anticipation with each day. I can't wait for tomorrow to begin, simply to create just a little bit more, in the research, and of the clearing."

Two days later I was devastated. I slipped and, owing to the momentum of my swinging ax, twisted and tore a cartilage in my left knee. Having years earlier experienced basically the same accident with the other knee, I recognized the symptoms, and that day Maggie drove me to the nearest hospital, in Farmington, Maine, to have the operation: I knew that if there was to be any chance at all of running in my planned 100-kilometer (62.2 miles) race, then every single day of healing was important. On May 19 I pleaded with the surgeon, Dr. Bitterauf: "Please do a good job—I want to run a hundred-kilometers footrace on October 4; it's the U.S. national championships." The nurse took my weight (156 pounds) and my vital signs. She had a strange look on her face.

"What's the matter?" I asked.

"Your heart rate is only thirty-eight; your blood pressure is ninety over fifty-five; and your body temperature is ninety-six-point-eight. I've never seen anyone in such low gear."

"Well, I'm relaxed and not doing anything right now—I'm on vacation."

"You look too skinny to be on vacation."

"Skinny?!—I want to lose at least six more pounds before running the hundred-kilometers championships in the fall."

"You won't be running for awhile. But if you do, you'll have to perk up some on your vital signs."

Three days after the operation I hobbled back up to the cabin on crutches.

Because of my injury, I was forced to remain indoors. I went back to the ant lions. I began by wondering how an animal with an almost microscopic brain could manage to build such a complicated tool as a trap that could catch insects. I filled most of the floor space of our small cabin with buckets of sand, and placed a couple of ant lion larvae in each one. I sat back to watch. Each larva backed up, flipping sand over its back with its closed pincers, which were used like a shovel. After each flip the larva made another jerky motion backward, again covering its head and pincers with sand, etc. As a result, it left a furrow in the sand while it traveled backward. In a few hours furrowed "tracks" of the larva curved and crossed over themselves all over sand in the bottom of the bucket.

It seemed to me that the tracks created in locomotion were, in effect, a shallow, long, trench-like pit. The larva's mandibles protruded from the sand at the apex of the pit. Could the tracks themselves aid in capturing prey? To find out I released several ants into the bucket with the larva. The small ants (three millimeters long) used the larva's tracks as walkways for several centimeters, and when an ant reached the larva, it was frequently caught. Ants that were larger (five to six millimeters long) and longer-legged, by contrast, seldom oriented to the track and were not captured. These observations seemed to suggest how the amazing pit trapping could have evolved from simple antecedents of the insect's hiding behavior.

I noticed that when a larva made a circular loop of about five centimeters in diameter, it got caught in its own track and then repeated another circular loop within that track. Now, however, instead of flinging sand to both sides, it flung the sand only to the outside of the loop. The groove got deeper, and the larva continued to make several more circuits in its own groove. After it made several circles in approximately the same groove, a small depression was formed. Small ants were detained for short periods in the "partial pit," and the ant lion caught them.

The size of the initial circle, to a considerable extent, defined the size of a finished pit. An ant lion could later enlarge its pit by leaving the apex, backing partway up the wall, and then digging in a ring around the wall. This maneuver undercut the walls above the ring, and gravity delivered sand from the edge of the pit, thus enlarging the top. The larva spiraled back down, flipping sand all the while, until it reached the apex of its new, larger pit.

The pits are built to catch food, but both the fed and the unfed larvae enlarged their pits within the first three days of building them and then they did not enlarge the pits further. Unfed larvae left their pits every few days to build another pit, but fed larvae did so just as frequently. Was there no difference in behavior between the two groups of larvae because the fed larvae also remained hungry? Although we kept larvae healthy for fifty-six days without food, we fed others as many as twenty-five large ants in four days. How does one define hunger in these animals? When we had watched 105 pits in the field for five one-hour periods back in Vermont, Maggie and I had observed forty-four captures of prey (and fifty-four escapes), corresponding to an average of one pit capturing one prey every twelve hours; but most of these prey were several times smaller than the ants we were using for our experiments. Maybe the larvae try to catch as much food as possible regardless of whether they have recently fed.

By June I was again chopping down spruce and balsam fir trees for the cabin, and by the end of the month I had finished thinning out what we hoped would decades later be a maple-sugaring grove. I looked forward to seeing these trees every year, watching them grow bit by bit. The portion of the old former field that we re-cleared now became a suitable habitat for chestnut-sided, yellow-throat, and Nashville warblers, indigo buntings, cedar waxwings, and white-throated sparrows. It became much like my favorite field at the home farm, and even a male woodcock would from then on claim the clearing every spring as his platform for sky dancing.

The school year had started, and I had to read the monthly research papers published in some fifteen different technical journals that I have to keep up with. My desk was stacked high with manuscripts sent by journal editors for me to review. There were several thick grant proposals to review, letters of recommendation to write for students and colleagues, and hundreds of reprint request cards to attend to. There were also potential students and postdocs who asked me to read drafts of their manuscripts, advise them on their research projects, or provide them with money and room in my laboratory. There were invitations to give talks and visiting professors to host. The temperature control room had to be fixed, research results had to be written up, and I had to work on a grant proposal with a deadline of November 1. Meanwhile, there were departmental and college meetings to attend and twenty freshmen advisees to see. I also continued my training, which now required me to spend at least two hours running up to twenty miles every day in order to have any chance of doing my best in the intended 100-kilometer race. Meanwhile, an administrator had me sign a form saying that I spent 100 percent of my time teaching.

Maggie and I had a son, Stuart; and when we returned to the Hill during the next two summers, we were also joined by an adoptee: Bubo, a great horned owl. Bubo was even more engaging than the ant lions had been. I returned to the Hill again in the fall and saw something preposterous: a group of ravens feeding together on a moose carcass near the camp. From what I knew of the natural history of these birds (who were then still rare in Maine) they were doing something for which there was no current scientific explanation. The ravens that I knew were aggressively territorial birds, yet here they were together and making loud calls I had never heard before. If I had been attracted to them, surely other ravens would be also. Were these birds recruiting strangers to share the feast? Couldn't be. Or? But what could the alternative be? I had stumbled onto a riveting question, one that might lead to a rich discovery. I wrote a very modest grant proposal to the National Science Foundation (NSF) about this new topic, and it was funded—after my last six tries to get funding to continue work in my specialty, insects, were rejected. It was a start in a new direction.

Senator Proxmire must have been following my career closely, because I was "honored" with a second Proxmire Award. Like the first, this one was also headlined in the *National Enquirer*. It said that I was seeking funding to find out "how ravens get dinner dates." The effort to answer my burning question about the mechanism of sharing behavior, presumably based on selfish motives, would be the toughest scientific challenge I had faced so far. It became all too obvious, and all too soon, that "the raven problem" was near impossibly difficult. As with the bumblebees, I needed to identify individuals and also keep track of them for long periods of time. Ravens were rare, and I couldn't even get within a quarter mile of one, if one had been located. But the lure was irresistible, and I decided to proceed as I had done with building the cabin and running 100 kilometers in 6 hours and 38 minutes and 20 seconds—one log and one step at a time—and asking myself every day if there was something else that could be done.

I now needed to be at camp with the ravens, even in the winter. Almost every week, I would leave Thursday after my class at the University of Vermont and drive the 200 miles to Maine; then I'd return to Burlington on Monday night. I needed to move cow carcasses miles into the woods, climb the tallest thick pines to band raven nestlings, and lie in snow-covered blinds for days. Such a massive effort, it turns out, is not conducive to married life.

The physical demands of the raven study gradually became far greater than anything I had ever done. So were the rewards. After completing each especially challenging part of the project, I experienced a tremendous feeling of accomplishment. It was not the same intense elation that I had while lying in my hospital bed after running 156 miles, 1,388 yards in a twenty-four-hour race at Bowdoin College (on a quarter-mile track where officials counted and timed every lap), setting the American record for the longest distance covered in that time. At the gunshot signaling the end of the race I fell like a shot deer, and in the hospital afterward I stayed awake for hours and said to myself 100 times, "I did it—I did it—I did it—." The next day, as I was recovering, a spectator left me a note saying, "I just want to express my utmost admiration and appreciation of your incredible twenty-four-hour run. I saw your mind, body, and spirit unite to accomplish the impossible." It was signed "Kimberly Beaulier (a fellow ultrarunner)."

It is sometimes necessary to try the (nearly) impossible, with the will and desire to pull through. I always had Papa as a *Vorbild* for that. As with the running, success with the raven project was also not a matter of talent. It was no sure thing, and I risked as much. The satisfaction was accordingly great, but it was no big burst; it was more like a slow long burn that never leaves. I have never experienced such feelings of satisfaction, without prior cost or deprivation. But this time with the ravens the cost may have been excessive.

One Monday night in midwinter, when I returned home after spending a week freezing in the snow with my birds, Maggie met me at the door with the ultimate cold surprise. She said, "I'm leaving you." I was shocked and angry, and even smashed some furniture. Fortunately, we sought counseling, and so we were able to part amicably and with understanding. It was very difficult for me, nevertheless. I did love Maggie, and I was sad that Stuart would grow up in a different house—because, like Papa, I was anxious to have a son who would relate to my world, a man's world. But things worked out. Maggie got a teaching position in the computer science department at the University of Vermont, and Stuart became a star straight-A student there with a double major in computer sciences and electrical engineering.

The raven study in Maine required about ten years of dogged work in the field. I recruited a brilliant and dedicated postdoctoral fellow, Dr. John Marzluff. He and his wife, Colleen, helped for three years. We wrote the research up in scientific journals, and I wrote two scientific books that were also meant to be understood by nonspecialists. This material was much more accessible to a broad audience than the physiology and behavior of bees, and it attracted a lot more attention. In particular, it led to contacts that provided most of the pictures and information resources that made a large part of the present book possible, and much more.

If ravens had been partially to blame for my losing my wife, they were also pivotal in bringing me a new family. After my book *Ravens in Winter* was published (in 1989) I gave a presentation on ravens at the "Darwin Festival" at Salem State College, in Salem, Massachusetts. It occurred to me that Salem is next to Boston, and I had just received a letter from an enterprising biology graduate student who had worked for years at Shark Bay in Australia studying wild dolphins. She was helping to teach an anthropology course at Harvard, and had a fellowship at the university's Bunting Institute to write a book about her experiences. She had read my *Ravens in Winter* and sought advice about writing about her animals. I was glad to oblige; and then on second thought I wrote to her, "On February 13 [1996], I am giving a talk titled 'The Sharing System of Ravens' at Salem State College. Why not stop by?" To my great surprise, she was in the audience and afterward introduced herself. We walked off campus to the Witche's Kettle or some such café, and after we had settled and I had had time to admire her pale green eyes she opened a large manila envelope and laid out for me her pictures of Holyfin, Puck, and her other dolphin friends. I think we both knew then that this was "it"—I fell in love just as easily as Papa had been famous for doing. Unbenownst to me at the time, Rachel Smolker had been engaged to be married, or I would not have invited her to visit me on the Hill to see my now wild ravens, Whitefeather and Goliath, who greeted us both at the campfire next to the cabin and who nest nearby in the pines to this day. But she did come, and we go there still with our two children, a girl and a boy who are now at the same ages Marianne and I were at the Hahnheide.

At our hill in Maine, I feel the present alongside the past, and our two kids remind me of a parallel existence a half century earlier. The good fortune we have experienced and the happiness we have felt are rooted in our contact and involvement with wild nature. Nature is now something people go to see mostly in parks or zoos where they may look, but it's either taboo or illegal to touch, smell, or taste. Here, I still feel a touch of wilderness, and of freedom.

Cutting-Edge Typology

Der Anfang aller Kunst ist die Liebe.
(The beginning of all art is love.)

—HERMANN HESSE

PAPA'S RELIGION CENTERED ON ICHNEUMON
wasps; and the road to their temple, the forests, passed through the science of biological nomenclature. He had few companions along this
road; it was paved with physical, financial, and intellectual obstacles; and
there were no laurels to be expected when he reached the end of it. Furthermore, like any long-distance run, this journey necessarily started and
continued at a slow pace. By the time it ended, almost everyone had probably forgotten why anyone would have bothered to enter it.

It began with popular enthusiasm. In the 1700s and 1800s the life
sciences were already animated by a branch of religion called "natural
theology." Its premise for studying nature and the relationships among
diverse living things was to reveal God's plan in the creation and eventual
order of the universe. At first that made the study of nature almost holy.
However, serious students eventually discovered that the creator's "plan"
included disreputable characters (such as ichneumon wasps and other
parasites), suggesting not a benevolent mind at work in the creation, but
instead a deviously sadistic if not malevolent one. Nature study lost some
of its shine, except when it involved curing diseases or devising other

means to gain power over nature that was now seen as "red in tooth and claw."

To contemplate even nature's pariahs, like ichneumons and tapeworms, objectively requires an unbiased perspective, and according to a definition by the novelist Gustave Flaubert (in the late 1800s), a truly scientific mind is characterized as *sans haine, sans peur, sans pitié, sans amour, et sans Dieu* ("without hate, fear, pity, love, and God"). Papa was driven—by his passion for ichneumons, as they are and for their own sake. However, Flaubert was wrong about love being absent from a truly scientific mind; even today few biologists who study animals are without a deep fascination for them, which I think is a prime phenotype of love.

Papa's main scientific fault was that he remained mostly isolated from biologists outside his specialty, so that he was not at the forefront of science, where the herd races to be because the support troops are deemed mere peons. I suspected that this "deficiency" was almost unavoidable outside the immediate environment of a university or research institution, and I sought the opinion of a professional, my friend and colleague Professor Ernst Mayr. On April 5, 1997 (long after Papa had died), he wrote to me:

> Yes, I probably know more about your father than most people in the United States, but mostly from hearsay. In Germany he had a reputation for being *schwierig* ["difficult"]. And he was called *Pascha* ["pasha," a Turkish chief]. He gave orders and everybody had to obey. I don't know what kind of college education your father had, probably very little biology; that is, modern biology. He probably had some natural history education. He must have felt very isolated intellectually in Maine and this aggravated certain tendencies in his character. I don't think your father had much *Meinungsaustausch* [an exchange of opinions] with other scientists, except about technical taxonomic questions. I doubt, for instance, that with his background, he could have had a real understanding of the Darwinian paradigm.

I could not have said it better. But given the circumstances, I could easily envision myself in Papa's position. As he said himself, "I grew up in the nineteenth century" (to explain why at age eighteen he felt he had to marry Elly after their tryst at the Baltic Sea while he was a soldier stationed in Russia). At that time—and even now—the Darwinian paradigm was glimpsed only superficially. Worse, it was misunderstood and also used cynically to promote nefarious political agendas. Its true meaning and implications had not yet become amalgamated into the zeitgeist. The zeitgeist of Papa's time, and that of centuries before him, was still "progress toward perfection." Perfection meant man, and not just any man; and ultimately the measure of all men was God, in the likeness of man. To Papa's credit, he never fell to that level. Still, he did subscribe to something similar. Given that he worked with ichneumons, he was forced to subscribe to the ideal of the "type" of a species, as though it were the best or model against which other individuals are measured. Others with different organisms and with different ways of classifying would derisively describe this process as old-fashioned "typology."

Even where my formal education had been deficient I was still privy to updates in a field rushing forward at lightning speed. I could always take professional contacts in a diversity of disciplines for granted. For example, there is the standard university practice of weekly afternoon seminars at which biology professors and graduate students gather to listen to, question, and debate one of their own or an expert in any of a diversity of subjects. Our seminar speaker at the biology department of the University of Vermont one day in the fall of 1992 was Dr. Alan Templeton from Washington University in Saint Louis, Missouri. Alan is tall and bearded, and his youthful appearance masks many years of experience as a molecular biologist working in animal taxonomy. His talk that day was titled "Genetics and Conservation Biology," as applied to the African elephant through DNA technology. He reviewed how it is now possible with new techniques in molecular biology to study the DNA of the tiniest tissue fragments, even if they are totally dried. Each cell of our bodies contains the entire genetic blueprint required to build the whole organism, and this blueprint specifies the organism's structure and much of its behavior.

Methods for accessing that blueprint are becoming ever more routine, as I learned near that time from my daughter Erica. As a junior—and an honors student—in our biology department, she had worked one summer at Genentech, a biotechnology company in California that specialized in making insulin from cultured cells that have been genetically engineered to contain insulin-making genes so as to make massive amounts of this hormone that is crucial for the treatment of diabetes.

Alan told us that day about one of his Kenyan graduate students who was finding that, using the new molecular techniques, he could trace the locality of where an elephant had been killed from the tiniest bit of tissue collected from one of its tusks. Determining the source of ivory would then enable one to distinguish legal from poached ivory, and this knowledge in turn was a tool in elephant conservation. Before this advance, merchants and poachers could merely claim that their ivory came from, say, Botswana (where some elephants could be killed legally), when it could have actually come from Kenyan and other populations where all the animals are endangered and protected.

Alan explained the problems and possible solutions of elephant conservation that the new molecular techniques allowed, and I thought of the immense problems of insect identification and taxonomy that similar molecular techniques could also help solve. In a recent paper in the *Canadian Journal of Zoology*, Felix A. H. Sperling from Cornell University reported that fifteen till-then presumed rare species of swallowtail butterflies were actually not species at all; they were hybrids of forty-two other species.

Papa sat in his desk and differentiated this or that ichneumon wasp from another species mainly by minute differences in body markings or shape, much as Vladimir Nabokov had focused on the distinctive hooks, teeth, spurs, etc., visible under the microscope on the male genitals of butterflies, the Plebeiinae (the blues). All of his and others' thousands of pages of taxonomic "keys," as they are called, were concerned with such external characteristics, and they were the accepted grounds for naming new species. Papa didn't use penile shapes in his species descriptions, but he used his keys to tentatively match the males and the females, which often varied enormously in size, color, pattern of body markings, and dimensions of every single body part that one might care to examine. Take color alone in *Eutanyacra saguenayensis*, a common species from Maine, as an example. In some species of ichneumons the sexes look similar. But in *saguenayensis* the females are almost entirely rust red in color, while the males are instead bright yellow with black rings around the legs and abdomen. It may seem hard to believe that such different forms belong to the same species, especially since in many species the males and females do look similar. There is no proof for many of the matches that Papa made, and they remain conjectural. However, the invisible genetic blueprint of a male and a female would match closely.

BIOLOGICAL CLASSIFICATION, WHICH HAD FOR hundreds of years relied mainly on appearance, received a huge boost in the late 1980s through new developments in molecular biology that permitted direct access to the genetic code on the DNA. The new spurt in the molecular revolution started innocently enough when Thomas Brock and Hudson Freeze of Indiana University described a new species of bacteria, *Thermus aquaticus*, that they found living at near the boiling point of water in a hot spring in Yellowstone Park, and they were not just content to name it.

Until then it was not thought possible for life to be able to grow and reproduce at such high temperatures, and they went on to examine this bug's unusual heat-resistant enzymes. Quite serendipitously, this set the stage for the isolation of one enzyme from this bacterium that literally changed the course of biology as we know it. The enzyme, a DNA-polymerase (by definition one that assembles the nucleotide bases to make complementary strands of DNA on existing strands), is one that essentially all organisms have and use routinely to replicate their DNA, but the variant from *Thermus aquaticus* (isolated in 1976) performs the DNA-making operation at temperatures near the boiling point of water, where the polymerases of other organisms are destroyed.

Researchers had long struggled and managed to replicate minute amounts of DNA "in the test tube." They had to heat the DNA to high temperatures in order to separate it into its two complementary strands so that the bases could then be assembled onto one of them, to create the new strand. But the polymerase then available (and required to facilitate this process) became inactivated. As a result, each round of DNA replication required that new enzyme be added. This was an impossibly lengthy and tedious process for making usable amounts of DNA. It was then that Kary Mullis, a California research technician, came up with the idea of using the heat-resistant "Taq" (for *Thermus aquaticus*) polymerase for *continuous* replication of DNA (of any other organism and of any chosen sample) at high temperature. It worked. And by 1988, when this, the "polymerase chain reaction" (PCR), came into use, the smallest piece of DNA, such as that recovered from even a fingerprint, could be copied billions of times in a few hours!

The "Taq" DNA polymerase opened new frontiers of science, medicine, and technology. It ushered in "DNA fingerprinting," which was used not only in identifying elephant populations, but in human individuals for solving crimes. My colleagues and I also used it in our raven studies in trying to determine who shared with whom. Later it became possible with "restriction enzymes" to excise specific genes or sections of DNA, and the PCR could then be used to amplify this DNA and generate sufficient copies of the gene to use for insertion into a "plasmid" (bacterial origin with circular DNA) that could be used to "infect" mammalian cells. Clones of these infected cells could then be selected and grown in vast amounts in culture, to then make commercial quantities of a specific protein product, such as insulin at Genentech, where Papa's granddaughter (my daughter) Erica worked. She is now a heme pathologist, and is relying increasingly on the PCR in her detective work of diagnosing genetic diseases. "Primers" (sections of DNA used for PCR replication) that code for specific mutations "recognize" and anneal (bind) to DNA of tissue samples with the same mutation. As Erica told me in 2006: "PCR diagnosis is exploding, and there is no end in sight."

Before the PCR, these possibilities were undreamed of; in the mid-1960s, Gunther Stent, from Cal Tech, one of the most prominent molecular biologists, wrote an article in *Science* predicting the demise of molecular biology. At that time I was inclined to believe him. I had been raising vats of *Euglena* for a year, trying to get even a trace of mitochondrial and chloroplast DNA (and not succeeding) before my research on the historic origins of these organelles could even begin. By the 1990s, however, it became possible using only a few cells to get as much DNA as one wished to work with.

One could get enough DNA to isolate and amplify also non-coding sections of DNA (termed "junk" or "fossil" DNA) for use in studies of evolution. In particular, the PCR technique has been used to amplify "microsatellites" (many times repeated sequences of nucleotides on *noncoding* DNA). This DNA has no known function to the organism, and therefore the mutations that show up on it are neither selected for nor weeded out, and hence they accumulate at a steady rate. They act as markers, and the amount of differences that accumulate trace the passage of generation times. They can be used like a clock to examine the time elapsed since a split from a common ancestor, who started from the same base sequence. The PCR techniques, applied to this DNA, have become a very useful tool in taxonomy. They now show us, for example, that skin color, which we have previously used as a marker to group humans into "races," is not really as valid a marker as it might appear. There is more genetic variation among people with black skin color, or white, than that between blacks and whites. The microsatellite data of our DNA shows that we diverged only very recently (about 20,000 to 140,000 years ago) from a tiny population originally living in northern Africa.

I wondered what Papa would have thought if a graduate student, using a now commercially available kit, would try to rely on species identifications from a fragment of DNA from cells in a piece of a wasp's ground-up leg? What would he think of recent proposals for sidestepping descriptive biological detective work to try to define species using DNA "bar codes"? Not much, I am sure. The new DNA technology would have been very helpful for Papa in ironing out some but not all taxonomic wrinkles of the Ichneumoninae. However, I doubt that his classification scheme would have been significantly altered. Rather, I suspect it would have been confirmed, and it can now serve as a basis for interesting problems that DNA technology can solve. But DNA alone is never enough. We each have our own individual DNA "fingerprint," yet we share 70 percent of our genome with the fruit fly (of which there are hundreds of species), and presumably about as much with ichneumons as well. The DNA shows us that we are all kin, to varying degrees, but what matters in differentiating species is not so much how much DNA is shared or not shared, but more important, what that DNA does. At least theoretically, there is no lower limit to the number of base sequences that need to differ to differentiate species according to the *ecological* concept that is the ultimate basis that defines a species.

LOUIS AGASSIZ (THE SWISS-BORN, GERMAN-educated Harvard professor who has been credited as one of the founding fathers of the American scientific tradition, and who was famous for his work on fossil fishes and his formulation of theories of glaciation) wrote in 1867 what most of my professors probably thought almost a century later: "The discovery of new species is now almost the lowest kind of scientific work. [It] does not change the feature of the science of natural history, any more than the discovery of a new asteroid changes the character of the problems investigated by astronomers. It is merely adding to the enumeration of objects." Papa's work may not have been considered sufficiently serious even by other taxonomists, because he didn't concern himself directly with evolution and new taxonomic techniques. I remember him visiting a taxonomist (though not of ichneumons), and he came away *tödlich beleidigt* ("insulted to death") and resolved never to speak with him again, and I'm sure he didn't. He didn't elaborate to me what this insult was about. He didn't need to. It could only have related to his treatment of his beloved ichneumons, because he would never have been insulted by how he treated anything or anyone else. It would have been about not being taken seriously. A cutting-edge taxonomist had implied that by being a typologist he was someone who contented himself with looking through a microscope and counting the tiny hairs on wasps' legs.

In *The Intelligent Man's Guide to Science*, the popular science writer Isaac Asimov further discouraged any actual or would-be ichneumon taxonomists when he proclaimed, "Biology is a system that proceeds from biochemistry to associated subjects of neurophysiology and genetics. All else is stamp collecting."

There is some truth in these assertions. But beyond its intricate physiological mechanisms, biology is history. No amount of chemistry can explain how or why the different ichneumon wasps came into being, or what roles they play in different ecosystems. At each level of complexity above the atom—to molecules, tissues, organs, organisms, communities, ecosystems—totally new and unpredictable properties emerge that are not presaged by the previous level. To some extent this is true even in the physical sciences. A chemist who is knowledgeable about the properties of oxygen and hydrogen could not predict the properties of water, much less those of ice, or the shape of individual snowflakes. Also, collecting ichneumons has scarcely any resemblance to collecting stamps. True, Papa savored the moment behind every ichneumon that he captured, and he enjoyed keeping the animal as a tangible memento of a fine sunny day when his senses were open and he smelled the flowers, heard the buzz of bees, and saw flashy butterflies pass. But this does not make his work "stamp collecting," unless the means are confused with the ends. The "ends" of Papa's ichneumon collecting were not just to stick insects into an "album" with preexisting slots.

Science rests on empiricism, and the "enumeration of objects" (or phenomena) is the first and necessary step in any science. Only afterward can one proceed to try to compare and search for patterns and regularities, and only then can one try to make sense of it all by formulating hypotheses, testing them, and ultimately searching for the causes and effects of a given phenomenon. One can't do one before the other, any more than one can cross the finish line of a race before taking all the steps leading up to it. And in any organismal biology, before any enumeration can take place, the observer must first name. *Naming* before knowing.

Modern biology can fairly be said to have originated with the invention of a practical and internationally recognized system of naming the earth's bewildering diversity of living forms. The system we now use already embeds considerable information within the names themselves. This system is called the binomial (two-name, referring to genus and species) nomenclature system in which the genus identifies the "address" of a species on the evolutionary tree of life. The system has been in place since the mid-1700s, when the Swedish naturalist Carolus Linnaeus tried to name all of the earth's plants and animals in his magnum opus, *Systema Naturae*. By its tenth edition, in 1758, he had named 4,400 species of animals, including the first ichneumon.

Now we know that there are tens of thousands of ichneumon wasps alone. When Papa began his work, the majority of them had no names. For all practical purposes of biology, they didn't exist. In his lifetime, however, he described 1,479 new species, in six major works covering 1,200 pages. Even now there is still not even a "field guide"—or anything even remotely approaching one—that would make ichneumons accessible to the nonspecialist. Those ichneumons that are named are described in a handful of "keys" such as those created by Papa, which I found coated with guano in our barn in Maine. The keys are published in a scattering of obscure journals, such as *Les Ichneumons de Madagascar*, of which I could locate only one copy. How does a nonspecialist like myself then identify a specimen he or she may find somewhere in the world? The answer is simple: it's practically impossible. The next question is: How would a *knowledgeable* lifetime ichneumon specialist (perhaps half a dozen exist in the world) identify the species status of a specimen that seems unfamiliar to him or her?

To answer that question it is helpful to keep in mind the current definition of "species." The generally accepted definition that Ernst Mayr and others have championed is the "biological species concept" (BSC), in which a species is defined as an interbreeding population that cannot breed with any other group. If the two defining words are "population" and "interbreeding," then obviously this definition is of no practical use to you if you hold in your hand a single specimen that you may never expect to see again and that nobody has ever seen before.

In practice you are forced to infer from your lifelong knowledge and familiarity whether or not your specimen is likely to satisfy the definition of a species. The most reliable and practical (and usually the only available) criterion is probably physical appearance (i.e., phenotype). Suppose you find Papa's key to North American ichneumons and, using it, you deduce that your specimen is "probably" *Barichneumon californicus*. (In this case you would have had to use Supplement 2 of *Canadian Naturalist*, Volume 98, pages 959–1026, 1971.) In the species description (on page 1013) you learn that the type specimen (the individual that was used to write the original description) was captured in Los Angeles on October 5, 1969 (by B. Heinrich). Only one specimen is now (still) in existence. It is in the second (the postwar) collection of G. Heinrich (which is in the Bavarian state collection in Munich), where it would be identified as the type specimen with a little red label, among 23,000 other ichneumons in that collection. Chances are that even if you think you have a representative of this species in your hand, you may still find a small discrepancy between the written (and on very rare occasion illustrated) description and your visual assessment. And you will wonder if that discrepancy is due to individual variation or whether it might signify a different subspecies—or an altogether new species. To evaluate the various alternatives, you will need to compare your specimen directly with the original type specimen, the one with the red label, on your next trip to Munich. If you then decide that your specimen is sufficiently mismatched from the type, you may wish to describe it as a new species, and you would then assign your specific specimen as the type of the new species, by publishing its description, putting a red label on your specimen, and storing it in a public collection.

ABOVE: Age twelve, with Jacob and a young hawk at the Old Dennyson Place in Maine in 1952.

BELOW: With Jacob, and wearing an "I Like Ike" button, on an old bicycle I had fixed up, in 1952.

Next to our house, catching ichneumon
wasps, 1951.

Papa and Mamusha cutting pulp wood,
during our first Maine winter.

Mamusha with her pet monkey and her
pet raccoon, around 1955, in the driveway at
our farm.

RIGHT: Papa and Mamusha, on their way to Mexico.

BELOW: Papa pinning gophers at his campsite on an expedition to Sierra de Guadalupe, Mexico, 1953.

BOTTOM: Mamusha taking a break at the campsite on the expedition.

Papa with helpers, on an expedition
in Angola, around 1955.

ABOVE, LEFT: Marianne and me leaving to go to Good Will Home, School, and Farm, 1952.

ABOVE RIGHT: My Good Will School high school graduation picture, 1959.

RIGHT: The end of a weeklong canoe trip with friends down the Allagash River in northern Maine, to celebrate graduation from the University of Maine, June 1964.

BELOW: Home for a visit, clowning with Papa, around 1966.

LEFT: With Papa and an owl, at our campsite in the Uluguru Mountains of Tanganyika, 1964.

BELOW: Skinning birds under the tent flap at our camp in the Uluguru Mountains.

LEFT: Taking notes on bumblebees during observations in a large bee enclosure built on our meadow in Maine, around 1972.

With my nephew Charles H. Sewall, lining bees at the Dennyson Place, around 1970.

Me *(left)*, Kitty, and friends at UCLA, and baby Erica, around 1972.

Returning to Borowke for the first time, in 1992, with Mamusha and her nephew Roman Bury.

Mistakes happen sometimes. For example, for a long time two entirely different-looking grasshoppers were thought to be two species, until someone discovered that they were instead examples of two separate phases (with entirely different appearance, physiology, and behavior) of the same migratory "locust" (*Schistocerca gregaria*). One phase changes to the other by growing up in crowded conditions. Similarly, although amateur and professional entomologists alike can generally identify butterflies by their striking color patterns, there are some butterfly species whose spring and summer broods have different coloration.

Butterflies can be easily reared from caterpillars and identified by wing colors and/or penis shape. But mistakes in identifying ichneumons would be difficult to notice and demonstrate. Papa admitted one mistake in print, and as far as I can recall, he never admitted any other. The one he did admit concerned the misclassification of *Platylabus berndi*. He described the species from one male specimen I caught near our farm in Maine. Later, he decided that the specimen was really *Platylabus albidorsus*, a new species he had already described earlier using a female specimen. He said that this species regularly occurs in two phases: a rare dark-colored one and "a more frequent red one." As a rule, he said, the platylabini display no or very little color differences between the sexes, but in this case "the rule was broken" (page 1018 in the same publication quoted above), so he assumed that my reddish male was a mate to the sometimes dark-colored phase of *P. albidorsus*. Does this sound confusing and contradictory? I think it should. The above is perhaps a typical example showing that mistakes are inevitable in taxonomic work on a group where one knows next to nothing of the populations, so that the "species concept" as defined by Mayr as the only valid criterion is as yet impossible to apply, and one is forced to resort to typology.

Type specimens can be misunderstood, yet they are a start and a means. There is little possibility of achieving a solid, useful foundation of work without them. Papa keenly recognized their imperfections. As early as 1934, he had devoted five pages of print in his *Ichneumoninae of Celebes* to describing the "chaotic" nature of the literature in previous attempts to classify tropical ichneumons. He wrote then:

Whoever advances to the systematic work of the Ichneumoninae must truly spend unlimited time and spare no money to be able to arrive at some sort of useful result. A revised and systematic treatment of the tropical Ichneumoninae would be a work for which one life would scarcely suffice, but it could mean a valuable foundation for zoological science, because without the foundation of a useful system of classification it would be impossible to come close to the interesting zoogeographical and developmental problems for which this group of parasites are especially interesting.

A quarter century later, in his 1960 *Synopsis of Nearctic Ichneumoninae Part I* he wrote: "All I can do is open the gates. Beyond them lies a wide-open field still waiting for further research."

At the end of his publication on the Celebes ichneumons, Papa mentioned that he had, in working up his Celebes catches, examined type specimens from collections in Oxford, London, Brussels, Calcutta, Bombay, and Colombo, and he concluded: "I have done all that is humanly possible, but the general lack of material and the confused state of the systematics do not allow a clear or perfect result." True, but his was a first step. In any project, as in any race, the first step is always the most important, and often the hardest.

Papa never got to work on any of the "zoographical and developmental problems," and for that reason he was marginalized and may have been justified in complaining about the "academic stuffed shirts." I think that he chafed at the reality that he was one of those who did hard work, often at his own cost, while others—such as those who sat in museums with vast collections at their disposal—profited from it.

There was his former friend, Ernst Mayr, who had had an early background almost identical to Papa's—Mayr remarked that for him "the key to paradise" came from Stresemann, as Papa himself might have said. Yet one was called "the Darwin of the twentieth century," while the other remains unknown in scientific circles.

I believe both got an equal measure out of life, and at an equal cost. I sensed this after I came to know Ernst as a personal friend, shortly after he and Papa parted company. Ernst sent me copies of all his books and autographed most of them: "To my friend, Bernd," and "with admiration." I do not recall what I wrote in mine that I sent him, but my sentiments were similar, although I felt Ernst was almost as inflexible and self-righteous as my father.

The young Ernst came from a family of physicians and was a bird watcher. He had seen a rare bird and stopped in at Berlin to tell Stresemann about it. Stresemann urged the young man to write a short paper, which was published, and the publication led to an offer to lead an expedition to New Guinea under the sponsorship and support of Stresemann's good friend Leonard Sanford, then at Yale University. The success of the New Guinea expedition led to Mayr's becoming curator of birds at the American Museum of Natural History, where he escaped the horrors of World War II. Perhaps as significant for Mayr was Lord Walter Rothschild's having to sell his collection of 280,000 bird specimens to Sanford, then at the American Museum. The birds, a treasure trove of scientific insight, fell into Mayr's lap.

There was immense scientific wealth in this collection: so many new species to describe, and the breadth and diversity of the specimens, plus the time to spend with them. With this collection at his fingertips for a large part of his entire career, Ernst had the empirical evidence to frame the concepts of the biological species and biogeography, for which he is famous.

Compare this with Papa's study of ichneumons, which are so rare, so little known, and so little collected that there is virtually no leeway for theoretical work relating to these topics. Furthermore, whatever insights might have resided in their study depended on their expert, Papa, doing all the grunt work (i.e., the collecting). It is relatively easy to assess the resident birdlife in any one area in a few months, but to do so for the thousands of resident ichneumons in the same area would be almost impossible in one man's lifetime. Even after decades, one would still continue to find individuals never before seen.

But Papa largely got the life he wanted—he experienced the great nature of Europe, Persia, Burma, Sulawesi, Halmahera, North America, Mexico, Angola, and Tanzania. Meanwhile Mayr was tied to the collections in the bowels of a museum. Of course Papa's freedom came at a cost, too; he had to give up scientific renown. But if he had been consciously faced with a choice to live one life or another, I wonder if he would have chosen differently.

Papa was a naturalist—that is, he would be defined by Doug Futuyma, a biologist at the State University of New York at Stony Brook, as "a person who is inexhaustibly fascinated by biological diversity, and also who does not view organisms merely as models, or vehicles of theory, but as the [thing] that excites our admiration and our desire for knowledge, understanding, and preservation." But it is in theory that fame resides—and in the articulating and especially the untiring persistence and energy of promoting simple models until they penetrate the general consciousness and then become associated in people's minds with the names of their sponsors. A hint of a misappropriation of credit can be gleaned from Papa's "Snoring Bird" expedition to Sulawesi. This expedition was pivotal in the careers of Stresemann and Mayr as well as Papa.

One of the key elements of the "species concept" that Mayr championed is geographical isolation. For land animals like birds, such isolation is most pronounced on oceanic islands, of which there are thousands in Indonesia. Smaller islands have fewer species than larger ones, and Mayr's thesis was that there is a balance between colonization and extinction. The ornithologist François Vuilleumier, who recently summarized Mayr's career and influence, claims that Mayr's ideas came from Stresemann's monograph (1939) of the Celebes (Sulawesi) birds. In this monograph, Stresemann wrote: "The circumstances of geographical distribution [of Celebes birds] seem to me to speak such clear language that we have to mold our theories to the facts; should we do it the other way round, we would base our views on theory!" With the Celebes birds there was so much fact that Stresemann theorized, and Vuilleumier wrote, "There is no doubt that Stresemann's Celebes monograph is the most significant of his many publications in biogeography"; and "The intellectual relationship between Mayr and Stresemann was so strong that the connection can and should be made. Mayr cited Stresemann's Celebes paper on a number of occasions." Stresemann had written that for a variety of reasons, Celebes is a key in biogeography because it represents a mixing area of different faunas (Asiatic and Australian). Vuilleumier continues: "Mayr made extensive use of Stresemann's ideas" in the Celebes monograph (which was not summarized in English); but I never saw anyone mention the fact that Stresemann's work was based on the bird specimens that Papa, Anneliese, and Marlis collected. Every one of the thousands of specimens bears Papa's personal label. Furthermore, throughout the two years of collecting, Papa was in almost constant correspondence with Stresemann, and much of this correspondence concerned both the birds themselves and the publication of what they revealed. Indeed, Papa wrote some of the paper himself, and Stresemann acknowledges this in the publication. The issue of authorship came up in the correspondence between Stresemann and my father. Papa wrote that he assumed he might be a coauthor of the paper because he and his helpers had taken on the burden of financing the expedition (on the basis of selling what they caught), and because they had taken all of the considerable physical risks, done the planning, and produced all the birds and raw data. However, Stresemann wrote back to suggest that he would do the gentlemanly thing of "assuming responsibility" of authorship, as though sparing Papa from potential embarrassment. I think Papa knew full well that there would be nothing onerous in being a coauthor of a work that had consumed him and his loyal crew for several years. He simply went along with no fuss, conforming to a German tradition of the time that the professor was the boss when it came to publishing

THROUGHOUT HIS COLLECTING CAREER, STARTing with the woodpeckers in the Dobruja in 1925 and especially with the gophers in Mexico in 1952, Papa repeatedly showed his sensitivity and awareness of the "biological species concept." That Papa was not able to apply it to ichneumons is not a testimony to ignorance. Even in birds—the best-studied organisms on earth in terms of their diversity, habits, distribution, physiology, and ecology—where the classification and enumeration of species is for all practical purposes complete, the naming of species does not and did not in practical terms always satisfy the biological species concept. In those species whose habitat is on islands, such as the different mountain populations of birds we collected in Tanzania, the species designations of birds of similar appearance are still judgment calls. The answer to whether differences among these populations of similar-appearing groups are sufficiently different to indicate species status is subjective, because one cannot know if the animals *would* interbreed if they had the opportunity to do so. Nor is there an *exact* point when populations can be definitely stated to be sufficiently isolated in space and time so as to be separated into full species, since evolution occurs along a continuum.

The species issue is not resolved to this day, if that means having one exclusive definition. Mayr's BSC allows for multiple forms, but stipulates that if those forms *can* interbreed, then they are one species. The problem is how would one know if they can or cannot interbreed if they exist geographically separated? There is no way to make a test.

Under a new "phylogenetic species concept" (PSC), the distinct populations with different evolutionary histories (because of their isolation) are considered species rather than subspecies *regardless* of whether they do or could interbreed. Recently some of these multiple forms that had been lumped into one species—such as a number of the birds we collected in Tanzania on different isolated mountain ranges—have indeed been split into separate species designations. I recently went to the museum in Berlin—where Stresemann spent his life and where Papa got his start—to examine birds he had brought back from Celebes. I was especially interested to see one of his favorites, a new ground thrush that he had discovered and collected on three separate mountain chains. At least to my eye, the females from different mountains show differences in both size and coloration, which is indicative of long reproductive isolation. Stresemann had named the thrush *Heinrichia calligyna*, splitting it into three subspecies. I believe it will eventually become two or three species, provided someone takes the initiative to look closely and declare the results in a peer-reviewed journal.

When Papa complained to me in 1971 about the "stuffed shirts" in academe (at the time, he probably feared I was in danger of becoming one of them) who from their ivory tower looked down on his activity as "typology" or worse, as "stamp collecting," I tried to cheer him up. I sent him a quotation from William Morton Wheeler, an illustrious professor from Harvard who in 1923 said that the world of nature provides an "inexhaustible source of spiritual and esthetic delight," and who wrote:

> We should be happier if we were less completely obsessed by problems and somewhat more accessible to the esthetic and emotional appeal of our materials, and it is doubtful whether, in the end, the growth of biological science would be appreciably retarded. It quite saddens me to think that when I cross the Styx I may find myself among so many professional biologists, condemned to sit forever trying to solve problems, and that Pluto, or whoever is in charge down there now, may condemn me to sit forever trying to identify specimens *from my own specific and generic diagnosis* [my italics]

while the amateur entomologists, who have not been damned professors, are permitted to roam at will among the fragrant asphodels
of the Elysian meadows, netting gorgeous, ghostly butterflies until
the end of time.

Papa replied, "Thank you for Wheeler's quotation. Quite cute! I
would like to hang it on my wall, nicely art-printed and painted. I am now
still *before* the Styx, already quite saddened by keeping trying to 'identify
specimens from my own specific and generic diagnosis,' feeling in the
depth of my soul the senselessness of it all. I had rather liked to be a physician."

It was probably a disappointment to Papa that circumstances had
not permitted him to become what he wanted to be professionally (and
that I would not fulfill his wish to carry on his work). However, when he
wrote this he still had more than a decade left to live, productive years
when he would race the Styx and keep on with his task of almost half a
century—the generic descriptions of the "Oriental" ichneumons that he
had buried in a sealed metal box in the woods of his beloved Borowke.

twenty-four The Oriental and Russian Manuscripts

And you, my father, there on the sad height,
Curse, bless, me now with your fierce tears, I pray.
Do not go gentle into the good night.
Rage, rage against the dying of the light.

DYLAN THOMAS

AS I TRACKED PAPA INTO HIS EIGHTH DECADE
and wondered about the craziness of the world and our attempts to navigate through it all and come out intact, the scent of the trail became faint. I had been increasingly distant from him while I was busily ensconced at the opposite end of the continent, at Berkeley. When I did on occasion come home to Maine, I spent most of my time outside with the bumblebees. However, as before, I sometimes had occasion to sit opposite him at his desk to talk and perhaps sketch an insect that he pointed out to me as being of special interest.

I recently found one of these watercolor sketches. It was of a yellow-orange ichneumon wasp with a black-tipped abdomen. I did not remember ever seeing such an ichneumon, and it struck me as an odd-looking beast. Nor did I recall painting it. The sketch had no label to identify it, and I wondered what the story behind it could be. In frustration I got up from my desk at our home in Vermont to go for a jog down our country road, hoping to oxygenate my brain to try to retrieve the circumstances of that particular sketch.

It's mid-January 2005, a clear, cold day after a recent snowfall. After my footsteps fall into their automatic rhythmic pattern, my mind clears a bit and a scene unfolds: I myself did not pick this particular wasp to draw. Papa must have had some reason for pointing it out to me. I don't think this looks like a European or an American wasp. It must be from either Africa or New Guinea. In my mind I leaned toward New Guinea, because the African ichneumons have a preponderance of dark blue and purple coloration, and almost no yellow spotting or banding typical of the "Orientals."

Several days later, while sorting through piles of Papa's letters, I happened to find one (dated February 9, 1971) that made me think again of the orange-yellow ichneumon with the black-tipped abdomen.

He had written: "Your ears should have been ringing during the last week! I described *Listrodromus berndi*, from Madang"—an ichneumon I had caught for Papa while on a bat-collecting expedition with George Bartholomew in New Guinea—"a rather oddly colored new species of a European-Oriental genus (occurring neither in Africa nor in America). An interesting new record. I intend to 'smuggle' this description into the next issue of 'The Ichneumoninae of Burma,' where the genus will be included anyway."

"The Ichneuminae of Burma," a work that he more often referred to as the "Oriental manuscript," had already consumed more than forty years of Papa's life, and it seemed appropriate that I was now steered back to this manuscript because of its central position in his story. A month later, on March 9, 1971, he had written to me:

> The last weeks were devoted to relentless work on the Burma ichneumons. To drive the Burma publication ahead as far as possible is the last large task, which is really "close to my heart." It is for me a kind of compulsion or obsession or testing—however you may call it. Too many years of happy adventures, of hard work, of changing times and history are involved to let it *all* get lost and vanish with me, as if it had never existed. Once, when we should have ample time together—if this should ever happen—I will tell you the long story of my exploration of the ichneumons of the Oriental region—

it is indeed almost as eventful and interesting as an Ellery Queen novel.

I never did seem to have the time to hear more than a few fragments of his story. I became interested in it only two years ago, when I retrieved the bound volumes of the Burmese ichneumons (i.e., those of the Oriental region) under the trash and debris in the loft of Mamusha's barn. They are in two parts, and in gold lettering on their green covers they are titled: "Burmesische Ichneumoninae." Beneath the titles in similar lettering is printed "Gerd H. Heinrich, Dryden, Maine, U.S.A."

When I first found these volumes under the rubble and guano, I never suspected that I could have contributed a specimen—or even a new species—to Papa's opus. Now, after reading his letters, I decided to take a closer look at the history of the manuscript. Most of our family history is bound up with it, and it was a deciding factor in our emigration to America when I was ten years old.

A large part of this work on "Oriental" ichneumons originated in 1930–1932 from Papa's collection in Celebes. That was the beginning. Shortly thereafter, in 1934, Dr. René Malaise collected ichneumons in northeastern Burma for Stockholm's Naturhistoriska Riksmuseet, and he brought Papa an abundance of interesting and even spectacular specimens. Papa then studied what was known of the fauna of the so-called "Oriental region" (especially eastern Asia), and Burma's rich fauna beckoned to him. So at his own cost, he undertook the Burma expedition in 1937–1938. When war clouded the horizon, his dream of a comprehensive survey of the entire Oriental region was dashed, but he wanted to publish at least his work on the Burmese ichneumons.

Back in Poland, Papa focused his attention on the manuscript. He worked with "ever-increasing intensity," as he himself put it, trying to get it done while publication was still possible. War began in 1939, before he had finished writing; and, incredibly, he always kept his "Oriental manuscript" by his side and continued working on it throughout the war. It was just about ready for publication in 1943, when he wrote the introduction to it, as follows:

> . . . The proclamation of "total war" was (just) given and with it the prohibition of printing anything not contributing to the war ef-

forts—this makes the publication of this monograph impossible. While I am writing these words, the war has already entered a new phase. Under the continuous bombing raids, one city after another, with all its irreplaceable scientific works and art treasures, is falling into rubble and ashes. I cannot dare hope that this, my own small contribution and the results of a lifetime endeavor, will survive the general destruction and be saved for publication.

As the Red Army was closing in, Papa sent the manuscript to his physicist friend Max Vollmer, who lived in a villa on the outskirts of Berlin. Papa then picked a secret place at Borowke, and there he buried a copy of the manuscript and "about 500" of the most valuable type specimens that supported the work. He described details of that burial in his introduction as well:

> I did not want any "treasure hunter" to be tempted to dig out the boxes expecting gold! [They were metal boxes of the type he had used on expeditions, and they had been made and were soldered shut by Josef Kowalewski, the blacksmith at Borowke whose life Papa had saved by rescuing him from a Nazi internment camp.] I had chosen a small spruce forest, still young enough so that the trees would not be cut for at least ten years, growing on a dry, sandy soil, and slightly elevated ground. The sand that I had excavated I carried to a nearby water-channel where I sank it. How to camouflage the place I had learned from the digger-wasp *Ammophila*: the filled ground was covered with moss, small twigs, and a few pebbles until it looked exactly like the environment and could not possibly be discovered. I had prepared a sketch of the location, and two copies, and sent the sketches to three friends asking them to rescue the "treasure" later on, should I not be able to do it myself.

Papa was prescient. As he had feared, he was indeed never able to return to reclaim his treasure, even though he miraculously survived the war. After our dramatic escape to northern Germany, when we had been living in the Hahnheide for about a year, Papa wrote to Professor Vollmer, but the letter never reached him. In what had turned out to be a race among the Allies in the last days of the war, the Russians beat the Americans to Berlin, and Vollmer was recruited by the Russians to work on their nuclear program. (He later died horribly in Russia, from radiation sickness.) Papa learned "by a stroke of good luck" that before Vollmer willingly went (or was taken) to Russia he had forwarded Papa's manuscript "to an address in Thuringia." After the war, that part of Germany was sealed behind the iron curtain in the Russian-occupied sector of Germany. The manuscript seemed permanently out of reach.

On one of Papa and Mamusha's collecting trips to southern Germany during the second year of our residence in the Hahnheide, Papa met an adventurous zoology student named Dieter Kramer, to whom he revealed the story of the lost manuscript. Kramer was intrigued and became sufficiently inspired to try to cross the dangerous border into East Germany and attempt to retrieve it. Choosing literally "a dark and foggy night," as he recounted it to me, he slipped through the border, hiked to the village in Thuringia and to the address Papa had learned from Vollmer, and miraculously returned the next night carrying the 2,000-page manuscript under his arm. Soon, it was residing with us in our hut in the Hahnheide.

Papa immediately tried to get it published, but Europe was too poor and uninterested in anything "not concerned with food and immediate needs." Not that anybody in Europe at that time had the means to print it, anyway. But in 1948, three years after the end of the war, when mail connections to the rest of the world were finally reestablished, Papa wrote Professor Henry Townes, the ichneumon specialist at the University of Michigan. Townes replied that he believed the publication of the "Oriental manuscript" would be possible in the United States, and he therefore suggested that Papa immigrate to America together with his manuscript (and family). So I have a 2,000-page manuscript about parasitic Asian wasps to thank for my life in America.

But publication faced as many difficulties here as there. When we arrived in Maine in 1951, Papa was informed that in order to be published here the manuscript would have to be translated into English. Papa's English was not adequate, and he did not have the means to pay a translator. However, the real obstacle was much greater: the 500 type specimens from Indochina, Burma, and Java, the basis of the work, were—if they still existed at all—buried behind the iron curtain in Poland. In 1969, nearly a quarter century after the war, tensions had eased and Papa took a gamble. He sent a copy of the map showing the location of the burial site to the Polish Academy of Sciences in the hope its members would locate the specimens and permit him access to them.

Unless the types were available to the world, publication of his work would be pointless. The project had reached a dead end. I recently found out some details about how it was revived so that the work came to be published after all. I fast-forward to thirty-five years later.

IN THE SUMMER OF 2004 I WAS INVITED TO GIVE A talk about my work on ravens at a conference in Metelen, Germany. There I met numerous biologists from all over the world, who were also studying the behavior of corvid birds—crows, magpies, jays, and ravens. One biologist was Nuria Selva from Poland, who presented her studies of ravens' scavenging behavior in the Bialowieza forest in Poland.

Almost a year later, Selva contacted me in Vermont by e-mail to seek my advice about the release of two tame ravens at her institute, and also to tell me about some interesting raven behavior (wild ravens picking spruce twigs and using them to cover a pile of carrion). I had always dreamed of visiting the Bialowieza reserve, mainly because it is a wild area that contains the largest population of bison in Europe. I replied that if I came I would want to make a stop in Warsaw to see my father's ichneumon collection.

Selva immediately responded that her partner, Adam Wajrak, a journalist covering nature and the environment who works for *Gazeta Wyborcza* (one of Poland's leading dailies), "immediately connected [your] surname with the wasps and an article in the *Gazeta*" about the entomological collections at the Warsaw Institute of Zoology at the Polish Academy of Sciences. Adam had learned that Papa had had an extensive correspondence with the director of the institute, Bohdan Pisarski, an expert on ichneumons. In a letter dated January 3, 1959, Papa had written (in Polish) to Pisarski, thanking him for sending specimens from Papa's collection (which the Poles had taken, or rescued, from Papa's main collection at the Borowke house after we left). They had arrived in Maine "in perfect condition" and Papa sent American species in exchange. The correspondence discussed continued exchanges of ichneumon specimens. Another letter, dated September 16, 1959, contained something more significant. It was a long letter in perfect Polish by Papa. Most of it concerns fine details of taxonomy of specific species that Papa was working on for the Burma manuscript. But what interested me most was written in a postscript at the end:

> I attach a drawing showing the place (more or less) where during the war I buried about 500 [specimen] types from my collection. It is very probable that they did not survive these 15 years, but it would be worthwhile to go check if maybe some parts can be rescued/saved. The types are in a welded tin box, that was itself once more packed inside a bigger metallic box. The box is located about 20-30 cm under the ground. With the help of "mine detectors," it will not be difficult to find.

Pisarski, along with Maciej Rejmanowski, the "cultural conservator" from Bydgoszcz (formerly Bromberg) and two sappers with a mine detector, went at once. With Papa's map, they easily found the spot, dug

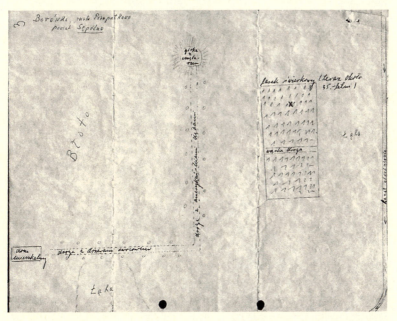

Papa's map of his buried ichneumon-type specimens that were needed for his Oriental manuscript.

down, and retrieved the collection, which contained 517 type specimens (along with 193 reprints of scientific publications and six boxes of valuable books). According to Pisarski, all were "in a very good state."

I had previously heard from Papa that the type specimens had been retrieved, but this was all I knew. I now told Selva how amazed I was that "our ravens" had been instrumental in giving me these precious details, and I ended my note with a wish: "I sure would love to see that map, but I suppose it is long gone."

It wasn't. A week later, on April 13, 2005, I received an e-mail from Selva telling me that she had contacted Wieslaw Bogdanowicz, the director of the Institute of Zoology (Pisarski had died seven years earlier). He told her that he had the map and would send it to me as an e-mail attachment. I could not believe it—an image of Papa's map and his letter to Pisarski. The type specimens are now berthed, with his prewar collection, at the Polish Academy of Sciences in Warsaw; at the end of the war Papa's main collection had been taken from the house at Borowke to become the property of the Polish government. The cold war notwithstanding, since the late 1950s the Poles lent Papa his own collection, mailing it to the house in Dryden, Maine. Now, with the type specimens on hand, he could update and finally try to publish the manuscript. And yet, despite all the miracles that had occurred to reunite Papa with his ichneumons, to his great disappointment he found that he could still not get the manuscript published in the United States.

Once again, Europe smiled benevolently on Papa. The Swedish entomological journal *Entomologisk Tidskrift* agreed to publish his work (in the German language, as written) in a series of papers, at the rate of one or two installments per year. About half of the manuscript (issues 1 to 7) appeared between 1966 and 1970. But when Papa sent in the eighth issue for publication, he received a letter from the editor saying that further publication would have to be discontinued. The Research Council of Sweden had decided that the content of the *Tidskrift* would have to be restricted to articles dealing with the Fennoscandian (Nordic) fauna, or else it would have to be of special interest to Swedish readers.

This news was a blow to Papa, and it was followed by more bad news. Officials at the University of California (with whom I had put him in contact) informed him curtly that there was "no money for taxonomy of the ichneumons" of the western states, which he wanted to work on. When he failed his eye test as he tried to renew his driver's license in 1971 at the age of seventy-five and consequently lost the license, he wrote, "I'm now sentenced to house arrest. With lack of a license all is lost. I am thrown on the dump . . . at least two to three years too early." He had finished writing four additional manuscripts and could not get them printed. He declared angrily: "Things are clearly deteriorating. . . . All I am still longing to do is to be in the woods somewhere during the summer, and hunt my fascinating insects."

He then wrote to me about yet another distraction from his ongoing work on Oriental ichneumons: Dr. Alexander Rasnitzyn, a professor at the Paleontological Institute of the Russian Academy of Sciences in Moscow, had become interested in ichneumons as a sideline and had sent him very interesting material collected in Siberia:

> Rasnitzyn wrote me that he was going to send me the types from the Amur area [the Russian far east and Manchuria] and that he would add 250 specimens, also from that area, and furthermore 750 specimens from European Russia. He asked me to do him the favor of identifying these specimens for a series of books on "The Fauna of European Russia." Believe me, when I read this, and when the four huge boxes arrived by air a few days later—I was sick, stunned, beaten, kaput! This is a work, even worse than writing a book, a big task of perhaps four months. Ichneumons are not beetles or butterflies. . . . I do not know what to do. For one thing: I hate the job. For another: I need something which grants me at least a modest living. I hate to live off my savings. I rather would like to do something pleasant: chop wood or see bumblebees pollinating flowers. But maybe I can get a grant for doing the Russians this favor, under the aspect of "cultural exchange"—then for an honest to god systematic-monographic exploration of the fauna of a state like Florida. Most likely! Perhaps I should follow up that line?

Papa said that he hoped, however, "to get things [the Oriental manuscript] printed in Russia." Also: "I intend to try also Poland. Never give up!"

In January 1973 he was still working on "that damned book" [the Oriental manuscript] and said, "I suppose in hell I shall have to write taxonomic works from eternity to eternity." He wanted "to be free at last to see some woods again."

Instead, he applied to the National Science Foundation (NSF), explaining his qualifications and saying that he had been sent over 1,000 specimens of Ichneumoninae by Rasnitzyn, as well as additional material from museums in Moscow and Leningrad, and had been asked to identify them: "The Russian request honors me and the work would certainly be of scientific value, and also at the same time contribute to international scientific friendly relations and cooperation." He was asking for $1,000 for an anticipated one year of work.

The request was not granted. Later Papa asked me to "help" him write another proposal for NSF. But I knew that if the first proposal had not been funded, then there was no hope at all, since his original request had been so modest and it had the added attraction of helping to promote international goodwill. So I told him I could not help; I felt that when he was inevitably rejected again by NSF, he would blame me. It was a lose-lose situation. I think he felt that I was making a big break from "the tradition of home" and family, which to him revolved around "the Ring."

The family ring had belonged to Oma, his mother, who wore it when she was young. She later gave it to her son, Papa, when he went off to fight in World War I. Afterward, his daughter, Ulla, became "Oma's girl," and she got to wear it, and in his old age Papa tried to get it back from her—Ulla had worn it for thirty-seven years already because to her it was "a tangible memory of my connection to Oma, who I loved above all others." The ring has a black onyx inset which shows (under a magnifying glass) a lion holding up a six-sided star in one paw, and a human figure holding up a five-sided star in its right hand and a large bird with a straight long bill unlike the short curved bill of a raptor (is it a raven?) in the left hand. Ulla declined to give, or return, it to him, because she said she wanted to give it to me; and he said, "Not to Bernd!" (Was this because I had not shown enough interest in "tradition"?) After Ulla refused to give it to him, even when he offered what he considered the astronomical sum of $1,000, he instigated a lawsuit against her in order to get it. But he lost his case. I got the ring (of course not the old Dennyson place).

By then, Papa had spent almost a full year working on the Russian ichneumons (with Mamusha's help, as always), and he had covered all the expenses connected with the work by drawing on his savings.

Over a period of four years, Rasnitzyn and his Russian colleagues sent Papa over 2,000 specimens (with thirty-nine new species out of sixty-seven species, as per the ultimate publication) rather than the several hundred originally discussed. Papa was later, apparently, pleased with Rasnitzyn's "surprise packages" of material he and several others had collected on paleontological expeditions. "It is quite exciting," he said, "to see so many 'new faces,' and the imagination flies away to their homeland, the mountains of the Caucasus, to the Altai Mountains, Transbackaleci, to the Primore Provinces, along the Ussuri River and the Kirgiz-Steppe." Was he remembering, through the insect fauna, his years in Russia as a soldier during two world wars?

Papa persisted with the work on the Russian ichneumons because Rasnitzyn had promised him that he could keep Rasnitzyn's privately collected ichneumons. Rasnitzyn had found many species that were unknown, and Papa wrote to him that he had named several species after him. Instead of a quick, enthusiastic reply, there was an unusually long pause in the correspondence. Eventually "R," as Papa called him, did reply, but the content and style of his letter was "not at all the old cheerful R—I had the impression that somebody else had leaned over his shoulder, while he was writing." "R" had written, "I am sorry to tell you that if you describe any new species from the museum material [sic] all the types have to be delivered to the museum." Papa answered, "Naturally all of the types of the museum material which were sent to me will be returned to the museum. However, if you refer to the unprepared material which you have sent me as a gift for my work, then I disagree." Papa asked for an immediate answer, but never got one. He said, "I am afraid R is in trouble" in communist Russia for having offered the specimens as a "gift" for my "favor." Apparently the specimens, through Papa's work, had assumed value in the eyes of the authorities, who wanted them back.

"I have been robbed to the bones for the sake of communist ideology," Papa said. "They have stolen my huge first collection, my house, my home—my life, so to say. I see no reason to let me be expropriated of the type specimens." He found the situation so disappointing and disturbing that he stopped work on the Russian wasps. "But at last I came to the decision to finish the manuscript anyways."

· · ·

1973, he wrote to me:

> I had to stop my taxonomic work, simply because my brain was too
> clouded. Was very sorry to lose time, but, believe it or not, one can
> just not come along in taxonomy without the brain! I have other
> sorrows, too. The printing of taxonomic publications apparently has
> come to a complete halt, because the flow of money so far available
> for the entomological journals has apparently been averted from
> taxonomy to "more important objects." Again I can see the trend
> and its origins. Do not tell me that there is no such thing as a
> "trend," as you did with reference to collecting [the stopping of].
> Such trends are as real and tangible as the warm and cold Gulf
> Stream. And you are part of it . . . . Whether these "trends" are
> "right" or "wrong" is a second question. Discrimination against tax-
> onomy and taxonomists helps to salvage money for the "more im-
> portant and more interesting" (one's own) projects. All this goes
> probably subconsciously or "instinctively." The only real blunder is
> that taxonomy is a far cry from being concluded. This is a fact appar-
> ently not known to many of the leading biologists. Or is taxonomy
> altogether dispensable? I would like to see how you would write a
> publication on warmth-control in bumblebees and Sphingidae [the
> moth family to which my sphinx moths belong to] for example with-
> out having names for the insect species and plant species you are
> talking about. "Bug" and "weed" would probably not be sufficient
> determination of the items involved. [The above] may sound bit-
> ter—and it is. The main reason that I am so disgusted is the discon-
> tinuation of my monograph on the Ichneumoninae of Burma ["The
> Oriental Ichneumoninae"], a work which has taken a good part of
> my life.

The Russian work, on the other hand, was published in a Russian
journal and in the Russian language; this almost guaranteed that not one
ichneumonologist (or anyone else) would read it. Even Papa could only
assume it said what he'd written.

In 1975, at the age of seventy-nine, he was still at work on several
ichneumon manuscripts; and on February 14 of that year he wrote to me:

I am dragging my feet and my mind is clouded on some days. One day you too will, rather suddenly, realize that the rest, which might still be allotted to you, has shrunken to a small number of years. This was particularly stressed in my case by the steady, though slowly, and inexorably advancing blindness.—I made an inventory of what I wished to finish still before the curtain falls, and there was quite a lot! So I went to work at once, systematically, and am now on my way, more exactly said: on my race with time. I explained it all in detail, because I want you to understand.

On Easter Sunday morning of that year Papa wrote to tell me about a meeting with a museum representative from Ottawa whom he had consulted about selling his collection "on my terms." (He had also tried, unsuccessfully, to sell the collection on his terms to the Smithsonian Institution, where he had spent considerable time in the national collections; and to the Museum of Comparative Zoology at Harvard, where he had also worked on the insect collection.) After a digression in which he talked nostalgically about his walk "over the bumblebee meadows into the wood—this walk is traditional," he continued the letter:

Thanks for your letter, which I enjoyed very much. We were in agreement, this time, and this is a comfort and a blessing among friends. I particularly liked what you wrote about your feelings when the first crows return to Maine and let you hear their calls. I too love these first announces of the coming spring very much. They have another advantage: to be so smart, that the trigger-happy so-called "hunters" and the .22 [rifle]-boys will not easily kill them off, as they did with most other animals around. The crows will stay with us for a long time to come. . . . But enough of this sad story. I am now reading the book on Africa you gave me [Peter Matthiessen's *The Tree Where Man Was Born*, 1972]. There, too, the days of the marvelous large beasts are counted. *Homo sapiens* is certainly the greatest pest the earth has ever borne.

There was then a bright spot. A very bright spot. Finally a publisher had been found for the rest of the "Oriental manuscript." Ironically the Polish Academy of Sciences had accepted it (still in German, as written).

The final installments (numbers 8 through 11) then appeared in the academy's journal, *Annales Zoologici*. The eighth was published in 1974 and the last in 1980. This part of the work was later collated into Part 2 of *Burmesische Ichneumoninae*.

Presumably my *Listrodromus berndi* from New Guinea (which he had written to me about in 1971, and which I thought I had sketched; see page 408), would be found in this second half. I was glad that he had written to tell me of his plan to "sneak it into" the work, or I might not have been inspired to pursue the story of the "Oriental manuscript." Now I faced the difficult task of trying to find the description of this wasp within the endless pages of Part 2.

The two volumes consist of eleven separate identification keys of ichneumons. They are taxonomically arranged, but there is no index. The page numbers are not consecutive, because they refer to journal pages, not to placement within the two collated volumes. I started flipping the pages, hundreds of them, and eventually (in Volume 2, on page 458) came to the genus *Listrodromus*. As Papa had written to me in the letter, this genus was here described as being of Palearctic and Oriental origin, and absent from America and Africa. Papa wrote that although the genus had not been demonstrated in New Guinea, its distribution there can be assumed "with certainty." Apparently because of that rationale he had included my new species from New Guinea in this work on the Burmese Ichneumoninae. I thought his "rationale" might be a stretch, but for once I did not raise objections.

Listrodromus berndi (pages 460–461) is described by dozens of details. Most of them I don't understand, but I saw that the first abdominal tergite (dorsal plate) is given as "light yellow"; the second and third are "light orange"; and the fourth to the seventh are black. I now examined once more my watercolor sketch, which so piqued my interest, to see if it matches the written description. Yes, this unusual pattern of color *seemed* to match, although I wasn't clear if the color I painted with was orange or brownish yellow. Had the orange faded to yellow? At the end of the species account (there is no illustration) Papa had included this line: *Benannt zu Ehren des Sammlers, Dr. Bernd Heinrich* ("Named in honor of the collector . . . "). So this species was definitely a candidate for me to have sketched. Papa did not need to name this ichneumon after me, and was it necessary to include this accolade in a scientific publication? It would have no meaning to anyone keying out a species. It could only have been meant for me to read.

That day, January 30, 2005, marked the first time that I had finally seen "my species" in print—the ichneumon I had so casually netted in New Guinea. It was also the first time that I had taken a more personal interest in the content of Papa's—and all of our family's—life-changing manuscript. Twenty-four years earlier, when he had written about naming the ichneumon after me and putting its description into the Burmese manuscript, my ears were not ringing with the announcement. Now, I felt honored and excited by the possibility that my mystery watercolor sketch could indeed have been this wasp.

But I'm a scientist. I need proof. And just to be sure that it really was the wasp, I sent a copy of my sketch to Erich Diller, the ichneumon curator at the Munich Museum where Papa's postwar collection is housed. To my great surprise, Diller informed me that my picture was not of L. berndi from New Guinea, but of Ctenochares bicolor, from Africa. Thinking that I had made a very poor drawing or Diller had made a mistake, I then asked him to send me a picture of L. berndi. He did, and there was then no doubt: I had been fooled. Papa would have laughed. Although the two ichneumons have a similar but (for ichneumons) unique color pattern of yellow body with black-tipped abdomen, they are different species—a remarkable coincidence that I don't think is a coincidence at all. Could Papa possibly have made a deliberate attempt to humor me from the grave? I wondered. The notion seemed too bizarre and too "sophisticated" at first glance. On the other hand, putting myself in his mind-set, I began to realize that his life had been totally dedicated to his ichneumons, and he knew that I was the only family member with a scientific link to this, his world. He had told me that he had conflicts with his own father, but that they had mellowed when he got older. Did he expect the same with me? If so, maybe he realized that I would indeed become interested eventually, especially as I was a practicing entomologist working with the Hymenoptera. Who else in the family would ever read any of his reprints in the barn attic? All of his correspondence and notes were left in the house in his files, available for Mamusha's perusal—even his love letters. Why were these, arguably the most important letters for family history, incongruously left hidden among his "discarded" scientific reprints? He must have known I'd pore over them someday, especially if I was trying to find the identity of an ichneumon I had sketched while I sat beside him at his desk.

Papa's highly questionable inclusion of my one specimen from New Guinea in this large work devoted to the Oriental fauna now added a fresh twist to his declaration that the manuscript involved "a history like an Ellery Queen mystery." On the day in the barn loft in Maine when I first picked this manuscript up out of the debris and dust, I had an inkling it might be important, but I never dreamed that it would put such a smile on my face.

AFTER THE CHATTY EASTER LETTER IN 1975 I heard from Papa only infrequently. He wrote of visiting his colleague Dr. Robert Carlson at the Smithsonian, where he viewed slides of Korea and "felt a little sad" because "the land is mountainous, but the mountains have been stripped bare of the woods, and my imagination saw no ichneumons! Now further woodcutting has been forbidden—in the right moment—because there is nothing left for cutting. And this way things usually go, and also in other countries."

I'll quote from three more letters.

May 21, 1977. I have not cashed the check you kindly gave me for the printing of the last Burma [installment], which Warsaw promised to print. Would you agree that I cash it now for helping to cover the other pending expenses? Money is a strange matter with rules of its own and strange qualities. It is modern *Homo sapiens*'s primary and single source of energy. This source must be tapped by work. But since the invention of "inflation," its value is without duration; in reality, it is imaginary. A gigantic fraud, which keeps the wheels turning—for some time—mainly on expense of the disabled and oldsters.

Papa's meager savings were dwindling. He had no pension and was suffering from a colon blockage due to a large tumor. I don't know what he was afraid of, but he slept with a loaded revolver under his pillow and his loaded shotgun next to his bed, behind a locked door of his brick ichneumon house. He replaced the old bumper sticker on the fender of his Chevrolet, which said, "This Car Climbed Mount Washington," with a new bright red one that read, "Stamp Out Pornography." Once a week, always at the same time, he went out for a chat and an ice cream at Goodwin's Restaurant in Farmington with a professor at the college there who was a fundamentalist Baptist and who (successfully) agitated for Papa to be awarded an honorary PhD.

> August 5, 1979. I have just finished the pinning and spreading of the Ichneumoninae you had kindly collected for the old man in Colorado and Lake Itasca, Minnesota. . . . But: I have just made a decision: not to continue to work and collect the American fauna. . . .
>
> All the best to you! Do not work too much! Take time enough to enjoy the beauty of this world. I am afraid that the hereafter will be rather boring.
>
> > Best wishes and regards,
> >
> > Yours,
> >
> > Papa

> June 23, 1980. "Dear Bernd: The collection [this is the collection he made in America and Africa since coming to Maine] has been packed carefully in seven large transportation boxes for the journey to its last and "eternal" rest in München [Munich]. It is raining here—A truck will come today to fetch it all. Marianne was here yesterday to fetch the Dillers and bring them to the plane for Europe today. My eyes went wet when the car with them went down the driveway and away around the corner. *Es ist eine Zeit zu kommen und eine Zeit zu gehen* ["There is a time to come, and a time to go"]—a Dutch proverb.
>
> —The old man

Dr. Carlson, of the Smithsonian Entomology Division, where Papa had worked periodically, confided to me (in 1999): "I hated to see it [the collection] go overseas." Indeed, Papa had wanted to leave it at Harvard or the Smithsonian but had been offered only $20,000, which insulted him. For one thing, as Carlson said, "He was so emotional about his ichneumons—he really loved his animals." He had invested his life in them—they were not something extraneous to it, as with most collectors who could lean on academic or government positions for financial support throughout their working life and even into retirement.

From 1975 on, Papa was increasingly unwell. He was in and out of the hospital and there were several alarms, all false, about his impending death. So when, in the winter of 1984, Mamusha wrote to me that Papa was very sick and was again in the hospital in Farmington, and that it was "serious," I did not take her warning to heart. I was not nearby on the Hill doing research; I was in Burlington, Vermont, where I was teaching at the university. I did not jump into the car and drive to Maine. I should have. A short time after Mamusha wrote to me, Marianne took Papa to the hospital, and the next morning, December 16, 1984, he was gone. He had died alone, with no wife, no daughter, no son at his side.

My good-bye, as it turned out, had been made the previous summer at the hospital. I had seen a quiet, apparently confused old man who hardly seemed to recognize or acknowledge me. And truth be told, I could not recognize in him the gallant knight of the skies who had fought for Germany in the losing war against the world, in what he considered the ultimate battle for honor and duty.

There were so many currents running through my mind that I didn't know what to think. I was sure of nothing. My emotions were still raw from Papa's slights and our increasingly divergent views and perceptions of nature. I had, in the last decades, tried consciously, and perhaps also unconsciously, to distance myself emotionally from his orbit of influence. Yet I am what I am partly because of him. He was an absent father in my early childhood, but this was a "normal" western cultural pattern, especially at that time.

What was not so "normal" in western culture was that he took me on his excursions into the woods, with his thirty-two rattraps and cheese in his backpack, or with his insect net in hand. We were companions in the Hahnheide and later in the woods at the farm. He had showed me and made me experience nature, as few fathers ever would or could. Whether intentionally or not (probably not) he taught me about the wonders of nature that today still bring me such joy. Admittedly, he was never fair. He neglected his daughters. He saw me, as his only son, not as a unique individual but as someone to be valued in terms of whether I would become an extension of him. I received advantages despite his intentions. I think

I even benefited from our arguments—which focused my thoughts and helped me clarify and articulate them. He saved me from being condemned to fifty years of boredom. Despite my resistance to him, he instilled in me the mind and the values of a naturalist: to be open to all possibilities, to be a close and careful observer, to discipline my interpretations with facts, and to work hard at my passions so that they might bear fruit.

twenty-five   Echoes

All we have, it seems to me, is the beauty of art
and nature and life, and the love which that
beauty inspires.

—*The Journey Home*, Edward Abbey

THE ROOTSTOCK OF THE SCIENCE OF BIOLOGY IS a love of life and its basis in nature. Papa loved above all else to be out in the woods and fields, provided he had an "excuse" to be there. His main excuse was collecting ichneumon wasps. For me, it started with the collection of beetles, eggs, caterpillars, and sphinx moths. These activities were the foundation of what we now call "natural history," a tradition that has led to understanding organisms in their natural environment. For Papa natural history observations led to his profession as an ichneumonologist, because the naming of animal diversity was then still at the forefront of biology, and it had almost bypassed the wasps. For me, it led to physiological ecology, and then to studies of animal cognition and intelligence.

Over the centuries, there have been vast changes in the perception of nature and in the type of research problems that biologists think about and do experiments on. Those changes range not only within the fields of nomenclature, but also in the way we see ourselves and our place in the world. My family history is entwined with these changes. Much as students of biology might visit their shrines, such as Wallace's original journals or Darwin's homestead in England, so in retracing Papa's life and my family's history, I wanted to revisit our own emblematic sites.

Around 1990, when I was starting to pursue my father's roots, I received a letter from Jobst von Nordheim of Trittau, our destination after the war. Jobst had just read my book *Ravens in Winter*, and he wondered if I might possibly be Gerd's son because "all tracks in the raven book I just read lead to Trittau and the Hahnheide and the years of the *Flüchtlinge*." Jobst was the son of Papa's friend Hans von Nordheim, who had made the experience of the Hahnheide possible by saying that if we had to flee Borowke we were welcome at the Sonnenhof, the family property in Trittau. Jobst and his wife Condy had recently renovated and moved into the Sonnenhof, and they invited me to visit them. I remembered it as a slightly run-down place with lots of fruit trees and a path nearby through the woods. I was eager to see it again and to meet Jobst and his wife.

I came to Trittau by bus from Hamburg. Then I took a walk down "memory lane" from the place where the schoolhouse once stood, past the linden trees I still recognized, down to the *Mühlenteich* (millpond), where the mill still stood that had once ground the grain grown on the surrounding farms. I recalled the beechnuts that we had gathered in the forest and that we brought here to be pressed into oil by the power from the water overflow. I heard a chaffinch singing; and seeing a wood pigeon nest on the linden tree next to the mill, I climbed up for the familiar sight of two white eggs. I looked into the reeds along the pond's edge, hoping to spot a mudhen or a rail. There were coots in the millpond, and mallards, as always. Continuing on the gravel path from the pond to the Sonnenhof, I first smelled roses, then linden blossoms. The linden trees were in full bloom with their pale aromatic yellow flowers. The trees seemed alive with the hum of bumblebees. I knocked on the heavy wooden door of the large brick house with green-painted beams and white window trimmings draped with ivy. A small gray-haired woman with smooth pale skin and a broad smile, wearing tinted glasses, answered the door. It was Frau Condy Nordheim. She asked me if I was the one who had written the book on ravens. Hearing my "Ja," she welcomed me in.

I was dazzled by the high ceilings and the spacious living room, whose windows faced a garden edged by the path where Marianne and I had walked hundreds of times, carrying our schoolbooks and my sling-shot. There were portraits on the walls, and they caught my eye. One was of a blond man with a mustache. He was shown wearing a military uniform with the left sleeve of his jacket empty. An Iron Cross was pinned on his chest—a military decoration exactly like the one Papa had framed next to his desk in Maine. This was Jobst's grandfather, *der Oberst* (the boss, the patriarch). He had penned a poem (published in a history of Trittau) about his ancestors, who "sank into the grave as knights, swords in hand, to fight for king and country," and about how he had not bowed "like the crowd" to "an empty hat"—how he "did not raise his hand and shout 'Heil!' to worthless trash." He was "not an SA [storm trooper] of Jesus Christ," because "politics spoils when it is in the church." The *Oberst* would have been seventy-seven years old at the end of the war in which Hitler had brought disgrace to Germany. Hitler had achieved popular support, and then seized power, by draping himself in the flag of the country the people loved, and the veil of the church they worshipped.

The other portrait, of Papa's friend Hans, the judge, showed a thin, pale, gray-haired man with even teeth and a thin smile. Condy served tea, and we sat on upholstered chairs under the portraits around a low table. Soon, our conversation turned to our fathers, Jobst's and mine, who had been good friends in Berlin. Jobst had heard about Papa from his own father, who had told him that Papa "traveled with his women deep into the jungles where no white man had ever been to catch rats and rails. Nothing stopped him—not hordes of jungle leeches, nor disease—all to catch the Snoring Bird, *Aramidopsis plateni*." Jobst continued: "In the jungles your father looked like a savage—I remember seeing his picture—broad and athletic, and hairy from top to bottom with a great beard."

Jobst recounted how once when he was about twelve years old he had gone with his father and Papa to the Berlin Zoo: "We walked up to a bear cage, and he [Papa] called '*Dickie—Dickie*.' A big bear got up out of his lair and came up to the bars. He reached through and scratched the bear's head, and the bear made soft gentle whining sounds. Then we went to another cage and he called '*Peter—Peter*,' and a giant panther got up and slowly ambled to the bars of the cage where we were standing and began to purr. He reached in and scratched it behind the ears. There was nobody quite like your father," Jobst concluded. "He was a *Ständiges Gesprächsobject*" (an inexhaustible topic of conversation).

Apparently not only women were in Papa's thrall. After a while I excused myself and put on my Nike running shoes to go for a jog. The pasture was close, and I found the approximate spot where the cattle shed had stood when we arrived here safely after the flight. The field was planted in rye, and bright blue flowers and red poppies grew mixed in among the grain.

I jogged along at the edge of the field, past a dense hedge of oak saplings, hazel bushes laden with green nuts, mountain ash with green berries, blackberry vines with white blossoms, and raspberry vines drooping with ripe red berries. Green finches, chaffinches, blackbirds, and thrushes were cashing in on the feast, and I thought how we had foraged here among them, many years earlier. Leaving the field, I took a path onto a stone-cobbled road that leads into the Hahnheide forest of tall straight beech trees stained with green algae. Shafts of sunlight filtered through the tree crowns, and hoverflies hummed while holding steady positions. Here and there were giant spruces and oaks. A raven called from one of the spruces, and I found its nest high in the fork of a beech tree.

The forest floor on each side of me was matted with the hulls of last year's crop of beechnuts, dried leaves, fallen spruce cones, and fresh new pale-green leaves of sorrel. Dark green moss covered the exposed tree roots. The forest exudes the mushroom smells that trigger nostalgia. The scent of eagle fern flooded my nostrils, and I remembered the ichneumons that we hunted as they were sunning themselves on its long fronds. I could imagine Mamusha, Ulla, or Papa holding the reins of the horses when we came here along this road and how the horses pulling our wagon clip-clopped on the rocks.

After about a mile I started looking for the sandy side road off the stone cobbled road. I thought I'd be able to find it, but I couldn't be sure. The trees were bigger, but after a quarter mile I recognized the turnoff into the deep spruce woods. I paused, looked down, and saw a *carabus* ambling along. I was overcome with emotion, and dropped to my knees and wept. Rationally, statistically, it wasn't possible that we had ever gotten here; and I had never expected to be here again; the past flashed before me; it all seemed like a dream. Now I remembered every turn of this path as though I had taken it just yesterday, and not almost half a century ago, and after less than half a mile I even found the remnants of a narrow footpath leading off into the woods along our tiny black-pebbled brook. I entered a mixed forest of beeches and spruce—and, up ahead, a tiny low cabin was dwarfed by the trees. The clearing that used to be in front was gone. Now all was in deep shade, and I was immersed in memories.

The cabin looked unchanged. I almost expected to see Anneliese walk out as I approached it. Cautiously I opened the door, which was propped closed with a stick. There was the familiar tiny brick fireplace, with the metal cooktop and its one hole on which to set a pot. It could have been used yesterday but hadn't been used for half a century. How did Anneliese, Mamusha, Ulla, and Erna, cook here for almost six years? What a change it must have been from their days at Borowke! Yet for me this place was paradise, where my memories—and my life—really began.

Papa's equivalent of Hahnheide was Borowke, and he never left it in spirit. He left it after he buried his type specimens and closed his mother's grave, returning only for a short time in January 1945, when he was home on leave from the Luftwaffe. That was the last time he saw it, and so it remained in his memories exactly as he had known it for half a century—an unblemished paradise. Following the communist takeover after the war, Borowke became a collective farm. It didn't work; the people who were settled there couldn't even manage to grow their own potatoes. Then it was made into a state farm. People got paid for working there, and the state installed them in the family house. Later, with privatization, the state tried to sell the apartments to the tenants already living there, but they were too poor to pay. In the meantime, the beautiful estate, carefully tended for so many years, had gone to ruin.

I WENT BACK TO SEE BOROWKE ONE BLUSTERY day amid snow flurries. It was the end of March 1992, and I was traveling through Europe on my way home to Vermont after a couple of months spent at the Hebrew University in Jerusalem, where I was studying flowers in the Judean Desert and visiting my friend and fellow evolutionary ecologist Avishai Shmida. Mamusha, who had flown to Warsaw from Maine, was to link up with me. She had gone to Poland to visit her relatives and to try to get Borowke back, now that the communists, who had claimed it for so long, were no longer in power.

Her nephew, Roman Bury, drove us through a countryside of endless fields and pine and birch forests, and through scattered towns and villages. On the edges of towns and cities we saw acres of gray rectangular multistory tenements. These depressingly uniform, formless blocks housed thousands of humans like bees in big hives. Linking the cities were narrow lanes lined with fruit and linden trees that ran over rolling hills stretching to the horizon. The soil was brown and bare and showed a dusting of snow, but occasionally we saw the flushes of faint green from the newly sprouting grain.

We stopped near noon for tea at a restaurant-bar in Mrocza. Groups of men enveloped in clouds of cigarette smoke sat drinking beer. Just outside the door above the sidewalk numerous noisy rooks were refurbishing their large stick nests in the linden trees. These social crows have settled to build their colonies in almost every Polish city we had driven through; Chojnice, Keynia, Naclo, Kosciercyna, Lobez, Chostna, Barwice, Swidwin, Stonowice. Each town also had magpies and hooded crows. Jackdaws, small crows with bright white eyes and an ash-gray nape, the same kind that I saw at Beit Gubrin in Israel, perched on red roof tiles, chimneys, and television aerials. They nested in attics and broken-down buildings.

Finally we arrived in Sepolno and turned off the paved road onto a muddy dirt track. Mamusha showed me where she had often come on horseback to pick up the mail. Then we passed a little roadside shrine with a praying Jesus and a cross. "We used to put flowers on this and make a wish. It couldn't hurt." (True, but it didn't help.)

We followed tractor tracks veering around one mud puddle after another. After about a mile or two we passed through a brick gateway. Two rows of bare chestnut trees with green moss-tinged trunks continued beyond it. They were not nearly as large as I had imagined they would be.

A low barn near the entrance of the chestnut avenue looked abandoned, although we heard the grunts of a pig coming from it and saw a flock of sparrows fly out. The barn looked dwarfed by the pole next to it, which had a stork's nest perched on top.

We stopped in the mud in the middle of the Borowke farmyard, the spot where I imagined the well with pole and swivel for lowering and raising a bucket must have been. But I saw no well; nor did I see the pond where Margarethe had reputedly swum naked in the moonlight. There was, however, a large weeping willow tree and a depression containing a rusted car moldering in the weeds. A big white rooster and several brown hens were scratching among chestnut seeds, which had started to sprout. Snowdrops were starting to bloom, nodding their delicate white heads among scraps of broken bricks and the empty white shells of dead snails. Muscovy ducks wandered amid rusted machinery.

Only a stone's throw farther along the same muddy track stood the two-story house. It looked barren. Plaster had fallen off in patches, revealing naked brick underneath. As we walked closer a small blue enamel sign became visible by the door. In white lettering it reads: "Borowke."

Mamusha's emotions must have been raw. She said nothing about what she was feeling, but I saw hints of pain. Perhaps she was facing the reality of the mismatch between what she remembered and what she saw now. If she had come "to get Borowke back," it was surely not for herself; there was nothing she could do with it. Perhaps she was confronting demons of the past. Was she thinking of how she felt then when she gave me to Anneliese, who brought me here, where she and her sister Liselotte both doted on me as if I were their own child?

We wandered around the yard, quiet and perhaps a little subdued. A man dressed in shabby work clothes walked slowly by, acknowledging our "hello" with only a nod. Then we found an old woman—a babushka—hobbling around in back of the house, and Mamusha approached her to strike up a conversation. I saw a smile—two silver teeth in the front of the woman's mouth. The two spoke in Polish, and I did not know what was being said. We were invited in for coffee.

The babushka introduced herself as Agnes Moczadko, and she introduced us to her friend Rosalina Sas, who was also living in the Borowke house. We sat around a low table in a tiny wallpapered apartment. The old communal kitchen that they had showed us downstairs looked like a dungeon. Agnes and Rosa slept near the kitchen stove to keep warm. To save precious fuel they burned peat briquettes, which are still mined in the bog.

Agnes's family (she was born Krueger) had lived near Sepolno for over 250 years. At first the Germans had taken everything from the Poles. Later the Poles took everything from the Germans and put Agnes, along with her parents, into a detention camp. The property that was owned by her family for generations was confiscated. Then one day while her husband was out shoveling snow, the Russians grabbed him off the road and shipped him to slave labor in Siberia. She has been a widow for twenty-four years. Her greatest fear was that with privatization, her apartment at Borowke would now be taken as well, and she would have no place to go. As I learned this story, I hoped Mamusha would not pursue what she had come here to do. We had gained everything by leaving. Agnes had lost everything by staying. I wrote her a letter (April 9, 1992) to ask her questions, since I had not been able to follow the conversation, and I gave her a modest gift of a few dollars to help treat her illness. She wrote back and thanked me profusely and said that she had never in her life received "so big a present."

Mamusha and Agnes had talked of old times in Polish. Agnes had heard that an old woman here long ago "used to paint pictures" and she had also heard how that woman's son could walk on his hands. She knew that Papa had saved Josef Kowalewski, the blacksmith, from the Nazis, and she had heard a rumor that Papa had "buried a treasure" before we all left at the end of the war. There was also a painting of two children by Margarethe's friend the Spanish painter Castelucio, but none of her paintings of nature and flowers at Borowke were visible. All had been stolen from the house, or, as in the case of the ichneumon collection, confiscated (and rescued) by the Polish government.

We did not meet any other of the eight families who now lived in Borowke, which was now the property of the town of Sepolno. Few were left except single old women whose husbands had died and whose children had left. I noticed a photograph of four children on the wall of Agnes and Rosa's apartment, and wondered about the likely sad turns of fate their lives had taken without their fathers. Mine may not have been a model of the current ideal that we associate with being a good father. Instead, he had been a powerful and influential one, and to me, in those times, that had made a lot of difference.

I was not comfortable inside the house, and I walked out to the back, past a rickety pen with two geese, past old cherry trees that would have borne fruit fifty years ago but were now dying and being replaced by rapid-growing poplars. Two magpies were busily building a tall domed nest of sticks in one of them, and the sight of their activity cheered me. I brushed past hazelnut bushes festooned with recently unfurled, yellowish-brown catkins. Underneath them I saw the dark green leaves and unfurling fresh white blossoms of more snowdrops over a rich, dark loam. Papa had mentioned these flowers, and they now gave me pleasure.

Scarcely 200 yards farther along, the hazelnut bushes became intermingled with birches and spruce. And here, on a small rise, I found a lilac bush in a clearing, and next to it a holly and a sprig of English ivy. These three exotic plants caught my eye, because they were out of place amid the natural flora. Indeed, I later learned that they had been planted by my Oma.

At first I could scarcely recognize that I had entered a tiny cemetery. There were rough outlines of a crumbling wall of fieldstones, but looking closer into the rank growth of weeds, I saw smashed and shattered gravestones scattered over the ground. Papa's mother, Margarethe, and his sister Charlotte would be buried here somewhere. I had earlier asked Mamusha about Oma's burial, and she had told me, "There was a great funeral. The British prisoners who worked on the farm carried the casket of the old lady from the house, through the garden, and into the family cemetery."

"Were the prisoners forced to carry the casket?" I asked.

"They were probably asked if they would, but they were not forced. They wanted to. They knew the old lady and respected her. They had it good at Borowke. They got sent raisins and sugar from England, and they made whiskey right there in the kitchen. The prisoners had girlfriends in the village. They had to report to British officers at their headquarters. This was the military. There were international conventions. This was not political—like the Nazis."

Absorbed in thought, I barely noticed a partially open grave by a pile of rocks, with fragments of carved stone. Poking around the weed-choked debris I found, lying on its side and half buried in the earth, a female figure chiseled out of stone. Her head and arms were missing. The statue was carved to look as though her body was enveloped in a loosely flowing robe. A granite block, probably the base of the statue, was overgrown with patches of bright green moss. Clearly, this figure was the Nike of Samothrace, the Greek goddess of victory, which Oma had put at her daughter's grave and which her friend Erna Starke had written about in her remembrances (see page 22). I saw an inscription in the base, and with a pointed stick I scraped away the moss and read the following words, which I could just barely decipher from the weathered stone: *Sieg, Wahr-*

heit, Güte, Liebe, Schönheit ("Victory, truth, goodness, love, beauty"). The base of the Nike had, in smaller letters, an almost illegible and eroded inscription that I eventually deciphered as "The Wanderer's Night Song," by Johann Wolfgang von Goethe. It depicts a sleeping quiet landscape where even the birds are still, and says to wait because soon you too will rest:

> *Über allen Gipfeln*
> *Ist Ruh,*
> *In allen Wipfeln*
> *Spürst du*
> *Kaum einen Hauch;*
> *Die Vögelein schweigen im Walde.*

Papa had told me about the Nike, and seeing the grave and the lofty sentiments so crudely desecrated by war and wrong convictions made me uneasy. I did not care to look inside the dark hollow of either Margarethe's or Lotte's final rest, where robbers had apparently thought that they might find the rumored buried treasure.

While trying to decipher the stones, I heard the continuous, joyful-sounding music of a skylark in the distance, and from nearby came the cheery staccato chattering song of a chaffinch. Sprinkles of light shone on late-spring snow that had just fallen. But the gray clouds parted and the sun shone briefly to leave droplets of sparkling water visible on shoots of fresh grass starting to sprout from the earth.

As if on cue for the occasion of sunshine, I heard trumpeting calls from the direction of the forest beyond and below the graveyard on which I stood. Cranes! The sound was exotic and unexpected, but I had heard (sandhill) cranes before in the western United States, so I knew what I was hearing. My surprise would hardly have been greater if I had heard the call of an ivory-billed woodpecker. Cranes are of the order Gruiformes, which includes the rails. Many of the world's cranes are rare and endangered. Curiously, while rails are the epitome of secretive birds, their relatives the cranes are a shining example of demonstrative showiness, with their large size, their plumage, and their fantastic courtship dances. In courting cranes, the individuals of both sexes jump and bow to the accompaniment of loud bugling.

It occurred to me that cranes can live to be over fifty years old. Could

The cranes at Borowke.

the birds I was now hearing be the very same individuals that Papa knew? It was unlikely. But so much of what I had already found to be true about our family had been equally unlikely. Papa had talked to me so often about the pair of cranes at Borowke that they were part of my mental picture of the place, and I believe they were a significant part of what made this his "home of the soul." I had fervently wished to see them, but I was so sure there would be no trace of them that I had blotted the possibility from my mind. To hear their loud trumpeting at this very moment, at the desecrated grave, made me feel as though I were hearing phantom spirits.

Lightness entered my step, as if I had experienced a transfusion of energy, and I walked down the sandy ridge into the thicket of hazelnuts, spruces, and birch. Six or eight wild boars erupted as one out of the underbrush, and I saw their black backs disappearing into a low-lying alder swamp. Again a raven croaked in the distance. I walked on trampled paths where the boars' fresh tracks mingled with those of roe deer and elk.

Seeing spruces on the sandy ridge, I immediately thought they were probably the "small spruce forest" that was "still young enough so that the trees would not be cut for at least ten years, growing on a dry, sandy soil and slightly elevated ground"—and "the sand I had excavated I carried to a nearby water-channel where I sank it" because "I did not want any 'treasure hunter' to be tempted to dig the boxes out, expecting gold! . . . A few days later—the exodus of millions of people toward the west began." That's when Mamusha and Ulla had left in the night with Marianne and me; and our adventurous flight to the Hahnheide had started. It was also the site where one night in January 1945 he had buried a copy of his Burma manuscript and the precious type specimens from his ichneumon collection.

Near the foot of this rise is the swamp—the swamp where Papa had made a hiding place for himself at the beginning of the war before submerging himself in the Luftwaffe. It was also the swamp where he hid his treasure. He wrote (in his ichneumon manuscript),

I crossed the swamp and entered a sedge meadow that would within a month be aglow with bright yellow buttercups, white and pink daisies, delicate yellow primulas. I looked up—two cranes, standing still, were watching me. I admired their tall, erect posture, the soft gray plumage of their backs and bushy tails, and the white sides of their heads, which resemble those of our whooping crane. I noted the bold black stripes on their heads and the crimson red patches too. In this magic moment of eye-to-eye contact, they both leaped up and with slow wing beats lifted off into the air. They flew side by side and vanished around a copse of woods, trumpeting all the way in wild jubilation.

Epilogue

O Father! Grant thy love divine
To make these mystic temples thine!
When wasting age and wearying strife,
Have sapped the leaning walls of life,
When darkness gathers over all,
And the last tottering pillars fall,
Take the poor dust thy mercy warms,
And mould it into heavenly forms!

—"THE LIVING TEMPLE," OLIVER WENDELL HOLMES

I believe in God, only I call it Nature.

FRANK LLOYD WRIGHT

WHEN I THINK OF OUR JOURNEY FROM BOROWKE
when I was only four years old, and my return nearly fifty years later, and
everything that happened in between, I am humbled. How was it all possible? The last time I saw Papa, in the summer of 1984 when I visited him
at the hospital in Farmington where he had gone for a checkup because
of spells of dizziness, I felt awkward. What could I do to cheer him up,
other than show him I was still around, despite our disagreements? I
found it impossible to imagine what he had experienced in his lifetime,
and I had no idea what I would have done in his shoes. I could only

The portrait of his mother, Margarethe, that
Papa kept on his desk. It was sketched by his
friend Willie Prüssing.

acknowledge a debt to him, for many things he may not have known he
gave. I think he thought he failed, because I was "disobedient." But going
against him so often deepened my thoughts, steeled my resolve, and kept
me awake at night thinking about the imaginary "rules" and "laws" of
nature that he assumed to be as solid as those of Frederick the Great. He
was on the right track, and had found enchantment, but he was not yet at
the destination. This is a new, ecological age, and its universal religion
will probably become like his, that of nature on a global scale. Our moral

choices will be informed by that vision of the whole, which is greater than all of us humans combined. Individually, we are like cells of a giant organism, the earth's biosphere. There is a physiology of the whole, with disease and health, and thanks to Wallace's and Darwin's insights and the work of thousands of researchers from all over the world, we now have an inclusive creation story for all of life, and it implies universal spiritual and moral imperatives for living. Now, however, I remembered us living the immediate, the prosaic. I remember him wearing a baseball cap and white shorts as we knelt down together among rotting tree stumps and gradually peeled off the moss and bark looking for hibernating ichneumon wasps. I remembered being terrified when he was at the wheel of his green Chevrolet, and how he never trusted anyone else to drive. I remembered him driving at forty miles an hour straight through an intersection, looking neither right nor left, because we "had the right of way" (as we got hit by another car). I remember him congratulating the pilot of a commercial plane for landing smoothly. I could not visualize him flying his "Blechesel" over the trenches of Verdun, or his biplane in the battle of Mount Grappa, or being asked by the Red Baron to join his squadron; nor could I see him as the intrepid explorer hunting the Snoring Rail in the thorn thickets of Sulawesi. I saw an old man like any other. A dear and unique one, with whom all warring, and so much more, was over.

Thirteen years earlier he had written to me about his visions of mortality. He complained even then of "spells" and general weakness and depression. He added in the same letter:

> During the night, when I am laying sleepless, threatened by various symptoms, I am downright scared, not about dying as such—I had my share of life!—but about having cancer or something alike, and then being *tortured* slowly to death in a hospital by senseless application of all kind of methods to prolong breathing another day, another hour. This is my nightmare!

He was spared that nightmare, but he did unnecessarily prolong the life of his adenoma. His friend Harry Brinkman, the local surgeon, had made a diagnosis six years before and recommended removal of his bowel growth. Papa refused. He made some lame excuses. But I think he

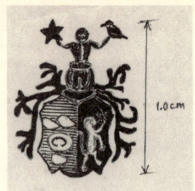

1.0 cm

The insignia on Papa's (and now my) ring.

had some notion of having lived already and being at peace. Most important, he had finally published the Burma manuscript. It had been the
focus of his life for over half a century, and now it was finally done.
As Hermann Hesse wrote: "We demand that our life have meaning
and purpose, but it has only exactly as much purpose as we ourselves
give it."

A year before his death Papa's friend, the Romanian ichneumonologist Mihai Constantineanu (then eighty-seven) had written to Papa: "You
had a very nice life on this earth, and published many famous scientific
ichneumonological works—You have lived a true life, and therefore nature has helped you to accomplish so many scientific works. I congratu-
late you heartily."

I'm not sure what a "true life" is, but I think Papa's core value was to
live according to what he saw as "truth." Since there are many religions,
adherence to one as such cannot serve truth. I suspect, therefore, that he
tried to find it in nature. He was of the mind of British biologist Thomas
H. Huxley, who wrote in *Evolution and Ethics* that science

> sees the order which pervades the seeming disorder of the world;
> the great drama of evolution, with its full share of pity and terror,
> but also with abundant goodness and beauty, unravels itself before

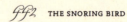

her eyes; and she learns, in her heart of hearts, the lesson, that the foundation of morality is to have done, once and for all, with lying; to give up pretending to believe that for which there is no evidence, and repeating unintelligible propositions about things beyond the possibilities of knowledge.

The problem is that the possibilities of knowledge for an individual are extremely finite. We are forced to make the best of it by having trust in at least some authority. I believe in baseball scores (because, as someone once pointed out, "there are thousands of witnesses"). And I believe in science, because it is the only other major human preoccupation besides sports where there are universal objective and consensually derived standards enforced by a community of peers. I believe Papa chose one man above most others to help guide his life, and in practical terms it was Stresemann, whom he called *Meister*, and who came from a system—science—that is derived from a method of consensus gained from the world community, and hence has a grounding in general reality rather than personal revelation.

A life is ultimately not philosophy. It is specifics; and when I look back, I can refer to specifics of Papa's life. Some of the things he did, could or would do in his prime:

1) Eat lots of ice cream
2) Choose oatmeal as his lifelong breakfast meal and stay with it
3) Stand on his hands
4) Laugh at a good joke
5) Roll naked in stinging nettles to "master discomfort"
6) Show anyone who cared his ichneumon collection
7) Be good company
8) Fall in love at the drop of a hat
9) Hunt ichneumons from dawn till dusk, day after day, year after year, whenever opportunity presented itself and even if it didn't
10) Never give an inch in a disagreement.

Perhaps even more characteristic in my mind are the things he did not and would not have done:

1) Show up with 80,000 fans at a ball game
2) Go to a political rally or demonstration
3) Dangle a worm on a hook to try to catch a fish
4) Have a beer
5) Read a newspaper or a work of fiction made to sound convincingly real
6) Go "hiking" without looking for something
7) Wash the dishes
9) Buy or wear any kind of jewelry
10) Eat a peanut butter and jelly sandwich.

His favorite joke was about a taxi driver in New York who has never had an accident. A news reporter asks this man how he has managed to have such a perfect driving record. The answer: "I drive assuming everyone is crazy." Papa thought this hilariously funny. I do not know why.

PAPA'S ASHES ARE SCATTERED ON THE OVER-grown field behind the barn where he loved to hunt ichneumon wasps as they were sipping the sugary anal secretions from the aphid colonies on the young white pine trees. He first frequented this field in 1951, when we moved here, and he continued to do so until he was barely able to see. The field now has a scattering of apple trees seeding in where deer feed. Blue and purple New England asters and yellow goldenrod bloom here and are visited by honeybees and bumblebees until the maples turn red in the fall. It is the dancing ground of the woodcock in the spring, and meadowlarks hide their nests here in the previous year's matted grass, which is crossed by the tunnels of voles. The field is bordered by a crude New England stone wall, and in a granite fieldstone within that wall a small brass plaque was implanted in 1985. Already it is encrusted with moss and lichens, and in summer Virginia creeper and a tangle of wild asters and goldenrod practically obscure it; as with most things in nature, you have to look closely and study it awhile to understand it. Underneath the foliage and the lichens is the inscription: "Dr. Gerd Heinrich, Nov. 7 1896–Dec. 17 1984. New Borowke. A Dedicated Researcher, Who Lived a Full Adventurous Life—His Way." His remains have by now become part of the ecosystem. He has returned to what he loved.

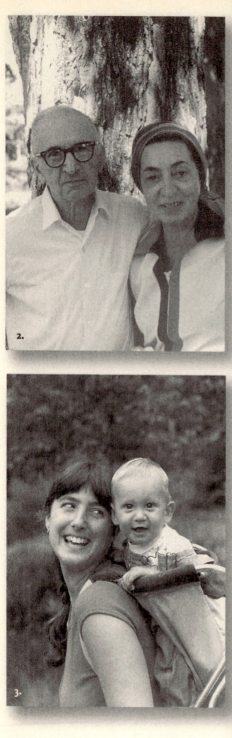

3.

2.

1.

4.

5.

1. Marianne Sewall (now Perry) and her sons, Charlie and Chris. 2. Papa and Mamusha, 1978. 3. Maggie Eppstein and baby Stuart, 1984. 4. A party at my cabin on my hill. *Left:* Me and UMO running mate Bruce Wentworth. *Center:* Raven colleague Duane Callahan and Good Will Pike cottage mate Bob Nagy. *Right:* Half-sister Christel Lehmann and Marianne's husband John Perry. 5. Rachel and baby Eliot, 1999. 6. Young Eliot at the cabin, 1999. 7. Stuart's high school graduation, 2002. 8. Lena and Eliot, 2003. 9. Lena and raven, 2006. 10. *(Left to right)* Gabe, Erica, and Liam.

Author's Note

When we have passed a certain age, the soul of the
child we were and the souls of the dead from
whom we have sprung come to lavish on us their
riches and their spells.

—MARCEL PROUST *IN SEARCH OF LOST TIME*

I RETURN NOW TO THE CHARACTERS OF THIS
book, to give a brief account of what happened to them. All of our friends
and neighbors of 1951 in Maine who were Papa and Mamusha's contemporaries are now deceased, except "Dr. C.," who became a multimil-
lionaire. Mamusha continues to lives at the farm alone (with a flock of
chickens, two cats, and a dog). Erwin Stresemann, who had been important in shaping Papa's life, died twelve years before him, behind the iron
curtain in East Germany. Liselotte, Anneliese's sister, married in Poland
and stayed there all of her life. Deceased also are Anneliese; her daughter
and my half-sister, Ursula (Ulla); and Ulla's husband, Leon. All three immigrated to and settled in Chicago around the time that we came to
Maine. Anneliese did not remarry, and she became close to a family, the
Brombergs, for whom she worked as a domestic. Ulla worked as chief
accountant for the W. E. Long Company, a bakery cooperative in Chicago.
Her passion was the daisies that she grew and often photographed in her
backyard, now that horses were out of reach for her in the city. Leon
worked as a shoe store manager in Chicago; and their son, Thomas, was

for decades a real estate agent and home builder, before studying for the bar and becoming a prosecutor for the city of Rockford, Illinois. His son, David, teaches high school math.

Dieter Radke, the boy from Germany whom Papa had invited to join him on an expedition because he felt Dieter was surrounded by good luck, and who on a chance of one in a million ended up coming through Borowke on the first day of World War II, was later sent to the Russian front. Being in the infantry, he felt he would never have survived if it weren't for Papa, who had advised him to study languages. He had done so, and thus was enabled to swing an assignment relatively out of harm's way as an interpreter and then a code cracker. After the war Dieter was an interpreter for the Allies at the Nuremberg trials, where he personally saw some of the top Nazis tried for war crimes. There he met Americans, and through them he immigrated to Oklahoma and eventually settled in Fort Myers, Florida, where he still lives at age ninety, vigorous, healthy, and happy, raising chickens and pineapples. His two daughters, Margaret and Rosemary, are teachers and live near him. His first wife is deceased; however, as of this writing he is planning to remarry. Mamusha is to be "maid of honor" at the wedding. Dieter maintained his close relationship with Papa all his life, and they went on many ichneumon-collecting trips all over eastern North America. On one trip he found entangled in a spiderweb an ichneumon that turned out to be a new species. Papa named it *Protichneumon radtkeorum*.

The Good Will Hinckley School and I are now on good terms; the school even asked me to serve on its board of directors. The L. C. Bates Museum there has now reopened and is a going concern. Instead of my hidden caterpillars in the basement of fifty-four years ago, the museum put on an exhibition of my pencil sketches and watercolors, and the kids have made a new nature trail that they named after me.

My full sister, Marianne, retired recently after a long career teaching grammar school in western Maine, and she now does volunteer work. Her two sons, Chris and Charles, both graduated from Bowdoin College. Chris has three kids—Fiona, Gwenyvere, and Shea—and he works as an organic chemist for a colloid firm in Rockland, Maine. Charlie has a PhD in toxicology from the University of North Carolina and works as a researcher for Merck Research Laboratories in Pennsylvania. He has a son named Baxter, after the governor of Maine and the state park by that name. Both Charlie and Chris are avid fishermen, and Charlie has become my traditional deer-hunting partner in Maine every fall.

When I started to assemble the material to write this book, my son Stuart was a second-grader interested in drawing Ninja turtles and dinosaurs, and said that when he grew up he would be either an artist or a digger of dinosaur bones. He has just—in June 2006—graduated from the University of Vermont with an A average as a double major in electrical engineering and computer sciences and worked at the Argonne research labs in Chicago before entering graduate school in Raleigh at North Carolina State University in the fall to major in artificial intelligence. His mother, Maggie Eppstein, still teaches computer science at the University of Vermont. My older daughter, Erica, graduated with honors from the University of Vermont after majoring in biology, and then earned an MD from the medical school there (where I had applied to be a student and had been rejected but was later accepted as a full professor). She is now finishing her hematopathology fellowship at the University of Michigan, and is about to begin working as a pathologist at the Maine Medical Center in Portland. She has two sons, Gabe and Liam, who are almost the same age as Rachel's and my two young kids, Eliot and Lena. Eliot wants "to become a conservation biologist and save endangered species." Lena writes an illustrated journal.

Erica's mother, Kitty DeRusso (Panzarella) became a family counselor and is happily remarried in California.

My half-sister Christel, whom Papa had denied the chance to participate in our African expedition, now travels to Africa regularly from her home in Germany. She has made many friends there, especially among the Masai in Kenya, and she says, "Africa is now my second home." She has three sons, Ingo, Volker, and Eckard, who are respectively a hotel cook, a computer specialist, and an architect and musician (and all of them are pacifists).

Long after my years at Good Will ended, I returned to visit Lefty Gould. On May 10, 1993, Lefty and I stayed up half the night, and he joked in the morning that I had kept him awake. When I left after a hearty breakfast of bacon and eggs, his wife Harriet said, "Next time bring your tape recorder!" I promised I would. But when I came back the next time it was already too late, because the old soldier had died. The last time I saw him was also the only time that I saw him in his Army uniform. He was lying in a casket in the front of Moody Chapel at Good Will.

My other inspirational mentor in Maine, Professor Dick Cook at the University of Maine in Orono, had advised me to always be careful when I sucked perchloric acid (PCA), a strong carcinogen, up into a pipette (by mouth). Perhaps he had slipped up or not followed his own advice—he died tragically of throat cancer. My PhD mentor at UCLA, George Bartholomew, who got me on a life track as a physiological ecologist, has since long retired; he lived in California and traveled to see the world. He, too, passed recently (November 2006).

Rolf Grantsau, who illustrated ichneumons for Papa in the Hahnheide and helped him trap water shrews, who put the first paintbrush into my hand and taught me the art of slingshot use and maintenance, answered an ad to work as a lumberjack in British Columbia. Rejected for the job (because of his weight—too light) he instead studied biology at Kiel University under Professor Karl Wilhelm Koepke. When Koepke immigrated to South America, Rolf followed him. For many years Rolf worked for an auto manufacturer in São Paulo and also pursued his naturalist's passion for South American hummingbirds, bats, fish, and snakes.

Erich Diller, who illustrated ichneumons for Papa in Maine and often collected with him, recently retired as the curator of entomology at the museum in Munich where Papa's postwar collection is now housed. He is married to Juliane, the daughter of Karl Wilhelm Koepke. She has a PhD, specializing in the study of bats, and she also works at the museum where Papa's new (American and African) ichneumon collection is located. On Christmas Eve 1971, when she was seventeen and was on a flight over the Andes en route to visit her father at a biological station in Peru, the Lockheed Electra carrying her, her mother, and ninety other passengers encountered a freak storm, The plane exploded and Juliane's seat detached from it. She fell 10,000 feet (twirling while still belted in) and landed in the rain forest canopy of the Peruvian jungle. Ten days later, as the flight's sole survivor, she found her way out by following a stream, as she had been taught to do by her father in case she should ever be lost in a jungle. She found the stream, she told me, by recognizing the call of a hoatzin, a bird that inhabits stream banks.

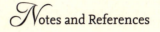

Notes and References

CHAPTERS 1, 2, AND 3

B. Heinrich. *In a Patch of Fireweed*. Harvard University Press, Cambridge, Mass. (1984). An abbreviated memoir with my pencil sketches, illustrated.

Papa's experiences in World War I are from his unpublished memoir and in: G. Heinrich, *Von den Fronten des Krieges und der Wissenschaft* (Dietrich Reimer, Berlin, 1937).

CHAPTER 4

The bird observations of the Dobruja expedition that Stresemann accepted were published in *Journal für Ornithologie*, vol. 75:6–37 (1927). Papa's bird work from the Elburs expedition of 1927, *Journal für Ornithologie*, vol. 76:237–313 (1928).

A recent summary of the issue of species versus subspecies question: J. Mallet, "Subspecies, Semispecies." In S. Levin et al. (eds.), *Encyclopedia of Biodiversity*, vol. 5 (Academic, 2001) pp. 523–526.

History of egg collecting: see L. Kiff, "Oology: From Hobby to Science," *Living Bird Quarterly*, 8–16 (Winter 1989).

For the adaptive significance of the color-coding of birds' eggs: B. Heinrich, "Why Is a Robin's Egg Blue?" *Audubon* 88(4):64–71 (1986). Also: N. B. Davies and D. Quinn, *Cuckoos, Cowbirds, and Other Cheats* (Poyser, London, 2000); and T. J. Underwood and S. G. Sealy, pp. 280–298 in D. C. Deming (ed.), *Avian Incubation: Behaviour, Environment, and Evolution* (Oxford University Press, Oxford, 2002).

The values of making bird collections: Kevin Winker, "Bird Collections: Development and Use of a Scientific Resource," *Auk* 122: 966–971 (2005).

Papa's first publication, of the ichneumons of the Dobruja, the first of about ninety publication on ichneumoninae spanning over fifty-four years: "Beitraege zur Kenntnis der Ichneumonfauna der Dobrudscha," *Deutsche Entomologische Zeitschrift* 397–400 (1926).

Persian expedition: G. Heinrich, *Auf Panthersuche Durch Persian* (Dietrich Reimer, Berlin, 1933).

Burmese expedition: *In Burmas Bergwäldern* (Dietrich Reimer, Berlin, 1940).

CHAPTER 5

General remarks, characteristics of ichneumons, and state of knowledge: G. Heinrich, "Synopsis of Nearctic Ichneumoninae Stenopnusticae in Particular Reference to the Northeastern Region (Hymenoptera)," Part 1 (1960).

Cues of ichneumons during host hunting: Felix L. Waeckers, "Multisensory Foraging by Hymenopterous Parasitoids," PhD thesis, University of Wageningen, Netherlands (1994).

A general popular account of wasps: H. E. Evans and M. J. W. Eberhard, *The Wasps* (University of Michigan Press, Ann Arbor, 1970).

Life cycles of ichneumon wasps: H. C. J. Godfray, *Parasitoids: Behavioral and Evolutionary Ecology* (Princeton University Press, Princeton, N. J., 1994). An unusual parasitic life cycle: 789 adults of the polyembryonic wasp *Pentalitomastrix plethroides* have been counted emerging from one caterpillar of a naval orangeworm (Pyralidae). See p. 229 in H. V. Daly, J. T. Doyen, and P. R. Ehrlich, *Introduction to Insect Biology and Diversity* (McGraw-Hill, New York).

G. Heinrich, "Les Ichneumonides de Madagascar, III, Ichneumonidae Ichneumoninae, Mémoires de l'Académie Malgache." Fascicule 25, 139 pp. (1938).

CHAPTER 6

A. B. Meyer and L. W. Wiglesworth, *The Birds of Celebes and the Neighboring Islands*, 2 vols. (Berlin, 1898).

About Wallace and the early Victorian naturalist explorers: Peter Raby, *Bright Paradise: Victorian Scientific Travellers* (Princeton University Press, Princeton, N. J., 1996).

The Wallace line: H. C. Raven, "Wallace's Line and the Distribution of the Indo-Australian Mammals," *Bulletin of the American Museum of Natural History* 68:179–283 (1935).

Publications of birds from the Celebes expedition: E. Stresemann, "Die Vögel von Celebes, Tail I und II, Biologische Beitrage von Gerd Heinrich," *Journal für Ornithologie* 87:299–425 (1939). E. Stresemann, "Teil III, Systematik und Biologie," *Journal für Ornithologie* 88:1–135, 389–487; 89:1–102 (1940–1941).

Review of the Celebes expedition in America: M. M. Nice, "Siewert's 'Storche' and Heinrich's 'Der Vogel Schnarch,' " *Condor* 35:82 (1933).

K. D. Bishop and J. M. Diamond, "Rediscovery of Heinrich's Nightjar," *Kukila* 9:71–73 (1997).

Recent books with information about and illustrating Heinrich's nightjar: D. T. Holyoak, *Nightjars and Their Allies* (Oxford University Press, Oxford, 2001). Nigel Cleere, *Nightjars* (Yale University Press, New Haven, Conn., and London, 1998).

Current book about the Sulawesi birds: B. J. Coates, K. D. Bishop, and D. Gardner, *A Guide to the Birds of Wallacea: Sulawesi, the Moluccas, and Lesser Sunda Islands, Indonesia* (Dove, Australia, 1997).

K. D. Bishop, "The Decimation of the Once Glorious Sundaic and New Guinea

Lowland Forests (manuscript, 2003). But *Aramidopsis* still lives: F. Lambert, "Some Field Observations of the Endemic Sulawesi Rails," *Kukila* 4(1–2):34–36 (1989).

G. Heinrich, "Die Ichneumoninae von Celebes," *Mitteilungen Zoologischen Museum Berlin* 20(1):1–263 (1934).

CHAPTERS 7 AND 8

Abbreviated history of World War II from: "V-E Day," *Time Magazine Sixtieth Anniversary Issue.* (Time Books, New York, 2005).

Hans Sluga, *Heidegger's Crisis: Philosophy and Politics in Nazi Germany* (Harvard University Press, Cambridge, Mass., 1993).

Kenneth C. Davis, *Don't Know Much about History* (Crown, New York, 1990).

Hajo Holborn, *A History of Modern Germany 1940–1945* (Princeton University Press, Princeton, N.J., 1969).

Göring quotation. See Gustave Gilbert, *Nuremberg Diary* (Farrar, Straus, and Co., New York, 1947). Gilbert was an Allied-appointed psychologist who visited with Göring and other prisoners at the Nuremberg Trials, and took notes of his conversations, ultimately writing the book about them.

CHAPTERS 9 AND 10

Paul Shepard in his book *Coming Home to the Pleistocene* (Island, Washington, D.C., 1998) on the potential educational experience (p. 160): "Toys in modern society may be a burden to children in ways we do not yet understand. Toys are precursors to material possessions—they objectify the world as passive and subordinate to ourselves and, despite childhood pretending, are nonliving. Toys may be symptomatic of social deprivation, solitude, and isolation." Shepard also notes (p. 168) that going out and foraging in one's home area is a better learning experience than "teaching modules about the rain forest" and that "most great naturalists were also hunters."

History of Trittau, including World War II: O. Mesch and H. J. Perry (eds.), *Geschichte und Geschichten: Beitrage zur Trittauer und Stormaner Region* (privately printed, 2003, ISBN 3-936091-05-6).

On what I experienced with regard to the Allies as occupiers see: H. Mitgang, "Chocolate Grenades," *Newsweek* (February 26, 1996), p. 15.

Excellent photographs and descriptions of carabids can be seen in E. Wachmann, R. Platen, and D. Barndt, "Laufkaefer: Beobachtung-Lebensweise" (Naturbuch Verlag, Augsburg, 1995).

CHAPTERS 12 AND 13

Quotation, "treated as reasonable people": see Charles E. Clark, *Maine: A History* (New England University Press, Hanover and London, 1990), p. 160.

One of the earliest references to bee hunting in America is found in James Fenimore Cooper, "The Oak Openings, or, The Bee-Hunter, *Works*" (Putnam, 1848). Cooper's story takes place in July 1812, in the "unpeopled forest of Michigan" where, because the Indians light periodic fires to clear the ground, grass and white clover

grow among scattered oaks. Ben Buzz (his real name is Benjamin Bodin), the bee
hunter, practices his art in almost the same way that Floyd and I did, 140 years
later, except that Ben placed a tumbler over a bee on a blossom, and holding his
hand underneath to prevent the bee's escape, he then placed the tumbler over a
piece of honeycomb on a stump, and his hat over the tumbler. Instead of a tumbler
we used a "bee box." The intricacies of constructing and using bee boxes are described by George Harold Edgell, in his little booklet *The Bee Hunter* (Harvard University Press, Cambridge, Mass., 1949). Edgell makes no mention of Cooper. He
describes his fifty years of experience lining bees in New Hampshire, starting
when he was age ten. He had learned his art from "an Adirondacker" who "smoked
and chewed at the same time and could spit without removing his pipe from his
mouth." Edgell maintained that the bee box is "the most important" equipment,
and his design is considerably more fancy than the ones we used in Maine. He
concluded that lining bees is "one of the most difficult, complicated, and fascinating games in the world." I certainly agree with his last point.

Robert E. Donavan, in *Hunting Wild Bees* (Winchester Press, Tulsa, Okla., 1980), writes:
"I've taught my kids how to hunt bees. It's not a skill kids need to know to get by in
life. . . . And they don't need to know how to make maple syrup, or how to cut
firewood, or how to tell a white oak from a black oak. But I think their lives are a
little less rich if they don't know these things. . . . Every now and then we need to
reestablish contact with things natural. Bee hunting is a good way to reestablish
that contact."

Our methods of bee hunting in Maine implied that the bees communicate; they
brought others back to share the feast—often hundreds came to our bait. In the
English translation of Maurice Maeterlinck's *The Life of the Bee* (Dodd, Mead, Cambridge, Mass., 1901), the author describes his experiments with paint-marked
bees. He notes, "The possession of this faculty (to communicate food location to
hivemates) is so well known to American bee hunters that they trade upon it when
engaged in searching for nests." Curiously, Karl von Frisch mentions neither
American bee hunters nor Maeterlink, in his *Bees: Their Vision, Chemical Senses, and
Language* (Cornell University Press, Ithaca, N.Y., 1950) or in his magnum opus *The
Dance Language and Orientation of Bees* (Harvard University Press, Cambridge, Mass.,
1967).

CHAPTERS 14, 15, 16

The "Good Will Idea" and its history are given by Good Will's founder, the Reverend
George Walter Hinckley, in *The Man of Whom I Write* (Galahad, Fairfield, Maine,
1954).

CHAPTERS 17 AND 18

Information on climate change that could relate to speciation in Africa: D. A. Living-
stone, "Postglacial Vegetation of the Ruwenzori Mountains in Equatorial Africa,"
Ecological Monographs 37:25–52 (1967). R. L. Kendall, "An Ecological History of the
Lake Victoria Basin," *Ecological Monographs* 39:121–176 (1969). K. W. Butzer, G. L.

Isaaz, J. L. Richardson, and C. Washbourn-Kamau, "Radiocarbon Dating of East African Lake Levels," *Science* 175:1069–1076 (1972). J. C. Stager, P. A. Mayewski, and L. D. Meaker, "Cooling Cycles, Heinrich Events, and the Desiccation of Lake Victoria," *Papaeogeography, Palaeoclimatology, Palaeoecology* 183:169–178 (2002).

Papa's letter to Stresemann from the Uluguru mountains was published in *Journal für Ornithologie* 103:122–123 (1962).

S. D. Ripley, and G. Heinrich, "Comments on the Avifauna of Tanzania," *Postilla* 95:1–29; 96:1–45; 134:1–29 (1966–1969).

Recent taxonomic work on birds including specimens from our Tanganyika expedition: P. Beresford, J. Fjeldsa, and J. Kiure, "A New Species of Akalat (*Sheppardia*) Narrowly Endemic in the Eastern Arc of Tanzania," *Auk* 121(2):23–34 (2004). R. C. K. Bowie, G. Voelker, J. Fjeldsa, L. Lens, S. J. Hackett, and T. M. Crowe, "Systematics of the Olive Thrush *Turdus olivaceus* Species Complex with Reference to the Taxonomic Status of the Endangered Taita Thrush *T. helleri*," *Journal of Avian Biology* 36(5):391–404 (2005). R. C. K. Bowie, J. Fjeldsa, S. J. Hackett, and T. M. Crowe, "Systematics and Biogeography of Double-Collared Sunbirds from Eastern Arc Mountains, Tanzania," *Auk* 121(3):660–681 (2004).

G. Heinrich, "Synopsis and Reclassification of the Ichneumoninae Stenopnusticae of Africa South of the Sahara," monograph, Farmington State College Press, 1–5:1–1258 (1967–1968).

CHAPTER 19

Gerhard Ritter, *Frederick the Great* (University of California Press, Berkeley and Los Angeles, 1954).

CHAPTER 20

My first publication, from trapping small mammals as an undergraduate, in: *Maine Field Naturalist* 17:24–25 (1961).

E. Schrödinger, *What Is Life?* (Cambridge University Press, Cambridge, 1944) sparked excitement in molecular biology; but later G. S. Stent, a dominant practitioner from Caltech when I was a graduate student at UCLA, would proclaim a dim future for the field: "That Was the Molecular Biology That Was," *Science* 160:390–395 (1968).

The idea of the origin of chloroplasts and other symbionts inside cells: Lynn Margulis, *Symbiosis in Cell Evolution* (Freeman, San Francisco, Calif., 1981).

The three papers that Dick Cook and I published on my master's thesis project on Euglena physiology: *Journal of Protozoology* 12:581–584 (1965); *Journal of Protozoology* 14:548–553, 1967; *Journal of General Microbiology* 53:237–251, 1968. This work discussed in D. E. Buetow (ed.), *The Biology of Euglena*, 3 vols. (Academic, New York and London, 1968).

Why publishing in peer-reviewed journals is important: Lynn Margulis, "Science, the Rebel Educator," *American Scientist* 93:482 (2005).

My study of the effect of leaf geometry on the feeding behavior of the caterpillar of *Manduca sexta* in *Animal Behaviour* 19:119–124 (1971) was later followed up by discoveries of adaptive feeding behavior to hide feeding damage from predators, *Oecologia* 40:325–337 (1979); *Ecology* 64:592–602 (1983).

The experiences with honeybees were later followed up by experiments showing the mechanisms whereby a swarm regulated its temperature, *Science* 212:565–566 (1981); *Journal of Experimental Biology* 91:25–55, (1981); *Scientific American* 244:146–160 (1981). Other follow-up experiments showed the mechanisms of temperature regulation of individual honeybees, *Science* 205:269–271 (1979); *Journal of Experimental Biology* 80:217–229 (1979); *Journal of Experimental Biology* 85:61–67, 73–87 (1980); *Journal of Experimental Biology* 208:1161–1173 (2005).

My friend and colleague Kenneth Morgan followed up on the tiger beetles with a study on my hill in Maine, *Physiological Zoology* 58:29–37 (1985). I followed up with African dung beetles in collaboration with my former professor, George A. Bartholomew, *Journal of Experimental Biology* 73:65–83 (1978); *Physiological Zoology* 52:484–494 (1978); *Scientific American* 241:146–156 (1979).

CHAPTER 21

The key lab studies showing control of body temperature by blood circulation in sphinx moths, *Science* 168:580–582 (1970); *Science* 169:606–607 (1970); *Journal of Experimental Biology* 54:141–166 (1971). Concepts from this work were subsequently applied also to large dragonflies, *Journal of Experimental Biology* 74:17–36 (1978); and to bumblebees, *Journal of Experimental Biology* 64:561–585 (1976); and to winter-flying moths, *Science* 228:177–179 (1985) and *Journal of Experimental Biology* 127:313–332 (1986).

Field studies of body temperature control and energetics of bumblebees, *Science* 175:185–187 (1972) and *Journal of Comparative Physiology* 77:49–79 (1972).

Flower evolution hypothesis, *Science* 176:597–602 (1972); *Scientific American* 228:96–102 (1973); *Evolution* 29:325–334 (1975); *Ecology* 57:890–899 (1976); *Annual Review of Ecology and Systematics* 6:139–170 (1975).

Mechanism of body temperature control in bumblebees involving exercise, *Journal of Comparative Physiology* 78:337–245 (1972); *Journal of Experimental Biology* 58:677–688 (1973); *Journal of Experimental Biology* 61:219–227 (1974); *Journal of Comparative Physiology* 96:155–166 (1975); *Naturwissenschaften* 78:325–328 (1991); *American Naturalist* 111:623–640 (1977); *Physiological Zoology* 52:484–494 (1978); *Journal of Experimental Biology* 58:123–135 (1973) and 73:65–83 (1978) and 127:313–332 (1987); *Physiological Zoology* 52:484–494 (1978) and 56:552–562 (1983) and 59:273–282 (1986); *Science* 228:177–179 (1985).

Mechanisms of body temperature control in bumblebees involving blood circulation, *Nature* 239:223–225 (1972); *Journal of Comparative Physiology* 88:129–140 (1974); *Journal of Experimental Biology* 64:561–585 (1976).

Foraging specializations of bumblebees, *Ecological Monographs* 46:105–133 (1976); *Behavioral Ecology and Sociobiology* 2:247–266 (1977); *Ecology* 60:245–255 (1979); *Oecologia* 140:235–245 (1979); *Journal of Comparative Physiology* 134:113–117 (1979); *Animal Behaviour* 29:779–784 (1981).

New York Times op-ed articles on bees, flowers, and energy: Sunday, November 25, 1973; Thursday, February 21, 1974; Friday, May 12, 1979. Relation to Adam Smith in *Business and Society Review* No. 12:30–34 (1975) and No. 17:86–87 (1976).

Milton Freedman in "The Real Basis of the Capitalist System," *San Francisco Chronicle* (April 10, 1979).

Recent updates on role of abdominal temperature in colony development of bumblebees, *Physiological Zoology* 66:257–269 (1993); *Canadian Journal of Zoology* 72:1551–1556 (1994) and 76:2026–2030 (1998).

CHAPTER 22

A history of the region near "the Hill": Vincent York, "The Sandy River and Its Valley" (Knowlton and McLeary, Farmington, Maine, 1976.)

D. S. Wilcove and T. Eisner, "The Impending Extinction of Natural History," *Chronicle of Higher Education* (September 15, 2000).

Our pit-trapping study was published in *Behavioural Ecology and Sociobiology* 14:151–160 (1984); and our bumblebee temperature regulation field studies in *Physiological Zoology* 56:552–567 (1983). The work on body temperature variation in dragonflies with Dan Vogt appeared in *Physiological Zoology* 56:236–241 (1983). The study with chickadees with Scott Collins was "Caterpillar Leaf Damage and the Game of Hide-and-Seek with Birds," *Ecology* 64:592–602 (1983). It was followed up with a multiauthor study using blue jays, *Animal Learning and Behavior* 12:202–208 (1984).

The running from that summer sparked a continuing interest in exercise physiology and resulted in a comparison of published performances of men versus women runners at various distances, in *Ultrarunning* (January-February 1985). An article, "Endurance Predator," was reprinted from *Outside Magazine* in B. Bilger and E. O. Wilson (eds.), *Best of American Scientific and Nature Writing* (Houghton Mifflin, Boston, Mass., 2001). My book comparing our running with other animals' and positing running as a human adaptation is *Racing the Antelope* (HarperCollins, New York, 2001).

The main research on and relating to the "left wing" sharing behavior of ravens done on the Hill, *Behavioral Biology and Sociobiology* 23:141–156 (1988) and 28:13–21 (1991); *Animal Behaviour.* 42:755–770 (1981), 48:1085–1093 (1994), 50:695–704 (1995), and 51:89–103 (1996); *Auk* 110:247–254 (1993) and 111:764–769 (1994); *Condor* 96:545–551 (1994). This research led to controlled studies of the mental aspects of the raven's adaptation to the environment, including intelligence, *Condor* 90:950–952 (1988); *Auk* 112:499–503 (1996) and 112:994–1003 (1995); *Animal Behaviour* 56:1083–1090 (1998) and 64:283–290 (2002); *Wilson Bulletin* 111:276–278 (1999); *Proceedings of the Royal Society of London* B:271:1331–1336 (2004); *Ethology* 111:962–976 (2005).

CHAPTER 23

Most often referred to for Biological Species Concept (BSC): Ernst Mayr in: *Animal Species and Evolution.* Belknap Press of Harvard University Press, Cambridge, Mass. (1963).

The question of enzymes and temperature regulation in insects: B. Heinrich, "Why Have Some Animals Evolved to Regulate a High Body Temperature?" *American Naturalist* 111:623–640 (1977). T. P. Mommsen and B. Heinrich. "Flight of Winter Moths Near 0°C," *Science* 228:177–179 (1985).

Selected references on the new molecular techniques:

M. A. Innis, K. B. Myambo, D. H. Gelfand, and M. A. Brow, "DNA Sequencing with *Thermus aquaticus* DNA Polymerase and Direct Sequencing of Polymerase Chain Reaction–Amplified DNA." *Proceedings National Academy of Sciences* 85:9436–9440 (1988).

K. Mullis, F. Falcone, F. Scharf, S. Snikel, R. Horn, G. Horn, and H. Ehrlich. "Specific amplification of DNA *in vitro*: the polymerase chain reaction," *Cold Spring Harbor Symposium of Quantitative Biology* 51:260 (1986).

R. K. Saiki, D. H. Gelfand, S. Stoffel, S. J. Scharf, R. Higuchi, G. T. Horn, K. B. Mullis, and H. A. Erlich, "Primer-Directed Enzymatic Amplification of DNA with a Thermostable DNA Polymerase," *Science* 239:487–491 (1988).

D. C. Queller, J. E. Strassman, and C. R. Hughes, *Microsatellites and Kinship. Trends in Ecology and Evolution* 8:285–288 (1993).

M. S. Blouin, M. Parsons, V. Lacaille, and S. Lotz, "Use of Microsatellite Loci to Classify Individuals by Relatedness," *Molecular Ecology* 5:393–401 (1996).

DNA technology used for the raven study: P. G. Parker, F. A. Waite, B. Heinrich, and J. M. Marzluff, "Do Common Ravens Share Food Bonanzas with Kin?" *Animal Behaviour* 48:1085–1093 (1994).

S. Beardsley, "Bending Bar Codes," *Scientific American* 26–27 (May 2005). Felix A. H. Sperling, "Natural Hybrids of Papilio (Insecta: Lepidoptera): Poor Taxonomy or Interesting Evolutionary Problem?" *Canadian Journal of Zoology* 68:1790–1799 (1990).

Virendra Gupta, "Contributions of Gerd Heinrich to the Study of the Subfamily Ichneumoninae (Hymenoptera: Ichneumonina) Together with a Bibliography of His Publications," *Oriental Insects* 23:337–348 (1989).

D. J. Futuyma, "Wherefore and Whither the Naturalist?" *American Naturalist* 151:1–6 (1998).

Mary LeCroy, "Ernst Mayr at the American Museum of Natural History," *Ornithological Monographs* No. 58:30–49 (2005).

François Vuilleumier, "Ernst Mayr's Biogeography: A Lifetime of Study," *Ornithological Monographs* No. 58:58–72 (2005).

M. R. Lein, "Ernst Mayr as a Lifelong Naturalist," *Ornithological Monographs* No. 58:17–29 (2005).

W. M. Wheeler, "The Dry-Rot of Our Academic Biology," *Science* 52:61–67 (1923).

CHAPTERS 24 AND 25

Hans Sluga, *Heidegger's Crisis: Philosophy and Politics in Nazi Germany* (Harvard University Press, Cambridge, Mass. 1993).

Gerhard Ritter, *Frederick the Great* (University of California Press, Berkeley and Los Angeles, 1968).

Regional convergence of Ichneumoninae in G. Heinrich, "Synopsis and Reclassification of the Ichneumoninae Stenopneusticae of Africa South of the Sahara," Vol. 1 (1967).

Ernst Mayr, in *This Is Biology: The Science of the Living World* (Harvard University Press, Cambridge, Mass., 1997), argues that new scientific knowledge about life "inevitably leads to ethical considerations" in order that we "never make life more difficult for future generations." Similarly, the sociobiologist Edward O. Wilson, in *The Future of Life* (Vintage, New York, 2002), cites numerous examples, from Saint Francis of Assisi to the poet Janise Ray, to suggest that our relationship to nature and the ecological crisis are moral issues. Science and religion converge on common ground. Father Thomas Berry, in *The Dream of the Earth* (Sierra Club, San Francisco, Calif., 1988), comes to a similar understanding from his perspective as a priest. He talks of "rights" of existence of each member of the life community—the rivers, the ocean, the forests, the fish—and says that international rivalries have received more concern than earth's life systems, and we are only now becoming aware of what is at stake.

As Paul R. Ehrlich and Richard L. Harriman have pointed out in *How to Be a Survivor: A Plan to Save Spaceship Earth* (Ballantine, New York, 1971), in order to maintain our standard of living and our freedoms we must increase the size of our "spacecraft" or limit the number of occupants: "Those who claim that the government could never intrude into such private matters as the number of children a couple produces may be due for an unpleasant surprise. There is no sacred 'right' to have children. The argument that family size is God's affair and not the business of government would undoubtedly be raised—just as it was against outlawing polygamy. But the government tells you precisely how many husbands and wives you can have and claps you in jail if you exceed that number"(p. 33). Of course voluntary means are preferable. But why should those who are moral and responsible be asked to bear the burden created by those who are not? I'm hypocritical. It just goes to show that mere good intentions are not good enough.